2002

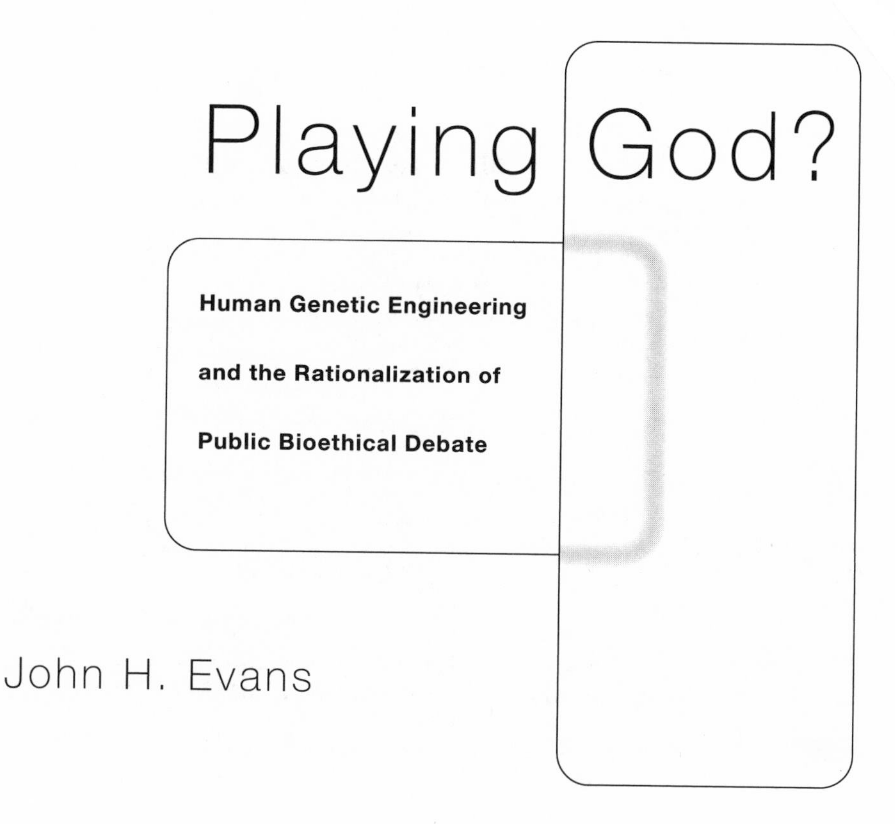

Playing God?

Human Genetic Engineering and the Rationalization of Public Bioethical Debate

John H. Evans

The University of Chicago Press
Chicago and London

JOHN H. EVANS is assistant professor of sociology at the
University of California, San Diego.

The University of Chicago Press, Chicago 60637
The University of Chicago Press, Ltd., London

Printed in the United States of America

11 10 09 08 07 06 05 04 03 02 1 2 3 4 5

ISBN: 0-226-22261-6 (cloth)
ISBN: 0-226-22262-4 (paper)

Library of Congress Cataloging-in-Publication Data

Evans, John Hyde, 1965–
Playing God? : human genetic engineering and the rationalization of public bioethical debate / John H. Evans.
p. cm. — (Morality and society)
Includes bibliographical references and index.
ISBN 0-226-22261-6 (alk. paper) — ISBN 0-226-22262-4 (pbk. : alk. paper)
1. Genetic engineering—Moral and ethical aspects. 2. Bioethics. I. Title. II. Series.

QH438.7 .E935 2002
174′.25—dc21 2001026688

⊗The paper used in this publication meets the minimum requirements of the American National Standard for Information Sciences—Permanence of Paper for Printed Library Materials, ANSI Z39.48-1992.

Contents

Acknowledgments

A very large number of people helped me produce this book. Tom Gieryn, Steve Hart, Matt Lawson, Jonathan Moreno, and Erik Parens all read chapter 4 in a very early form and provided critical comments. A larger number of scholars helped shape this book in informal discussions: Jeff Alexander, Robert Baker, Daniel Callahan, Ron Cole-Turner, Peter Conrad, Diana Crane, Wendy Espeland, Rene Fox, Bruce Jennings, Paul Lichterman, C. Ben Mitchell, Kelly Moore, Tom Murray, Henry Small, and Bob Zussman. I presented rough drafts at the Hastings Center in 1997 and 1999, and at the 1998 Louisville Institute Winter Seminar. Wayne Baker, Rene Fox, Erik Parens, Jim Walter, and David Yamane graciously gave comments on an entire early draft. Kieran Healy, Alan Wolfe, and the anonymous reviewers for the University of Chicago Press provided insightful suggestions for improvement of the last draft.

The librarians at the National Bioethics Reference Library at Georgetown University also deserve thanks for their persistent helpfulness and for allowing me to gather one of my data sets in their library by photocopying only a few pages each from hundreds of books and articles—making quite a mess in the process.

The early years of research and writing were made possible with the support of colleagues: Courtney Bender, Bethany Bryson, Julian Dierkes, Jackie Gordon, Jason Kaufman, Erin Kelly, Kim Kracman, Matt Lawson, Mike Moody, Abbie Saguy, Brian Steensland, and Maureen Waller. Gene Burns, Frank Dobbin, Michèle Lamont, Bruce Western, and Viviana Zelizer also gave advice at critical moments. The intellectual seeds of this book go back more than a decade to my thesis at Macalester College, supervised by Chuck Green.

The first of the many drafts of this book was a dissertation at Princeton University. Anyone who is familiar with Miguel Centeno, Paul DiMaggio, and Bob Wuthnow—or who has read the acknowledgment pages of previous Princeton dissertations—will know that it goes without saying that my committee was absolutely generous with their time and brilliant in their general mentoring. Is there any limit to the helpfulness and support of these three? I have yet to find it.

I have been the recipient of generous financial support, without which I would have been unable to complete this project. Funds for research expenses from the Association of Princeton Graduate Alumni and the Center of Domestic and Comparative Policy Studies of the Woodrow Wilson School of Public and International Affairs at Princeton University, as well as a National Science Foundation Doctoral Dissertation Research Improvement Grant, were gratefully received.

More extensive support from Princeton University—in particular the Center for the Study of American Religion and the Princeton Society of Fellows of the Woodrow Wilson Foundation—kept me in food and shelter through the research period. A writing fellowship from the Louisville Institute at Louisville Theological Seminary allowed for a year of uninterrupted dissertation writing. The Robert Wood Johnson Health Policy Scholars Program at Yale University offered critical support and pushed my mind in a more policy-relevant direction while I was rewriting. Thanks to Rogan Kersh, Taeku Lee, Ted Marmor, Gary McKissick, Eric Oliver, Robin Rogers-Dillon, Mark Schlessinger, and Mark Suchman for intellectual community at Yale. Finally, I am grateful to the Department of Sociology at UCLA for giving me the leave at Yale and for providing a collegial and intellectually stimulating environment.

Alan Wolfe, as series editor, was supportive of the project from first hearing, as was Doug Mitchell, who graciously led me through the publishing process. None of this would be possible without the love and support of my spouse, Ronnee Schreiber. The arrival of baby Danielle the day after chapter 4 was first drafted puts everything into perspective. This book is dedicated to these two people in my daily life who make it all worthwhile.

Introduction

> Bioethics has badly neglected the aims and aspirations of research as an area worthy of moral exploration. It has not taken on, for careful examination, the implicit models of human life and welfare and the human future that lie behind the biomedical research enterprise. If it has not been utterly captured by that enterprise, it has mainly stood on the sidelines, wagging its finger now and then. That is no longer good enough.
>
> **—Daniel Callahan, "Cloning: The Work Not Done" (1997)**

On 23 February 1997 the news broke that Scottish researchers had taken a cell from an adult sheep, placed it into an egg, and implanted the egg in another sheep. Thus was produced Dolly, the now infamous cloned sheep.[1] The public was immediately concerned. Could this technique be used to clone humans? Could a woman give birth to her own genetic twin? Could a family decide that they wanted a baby who would be a genetic version of a grandfather? Could Michael Jordan be cloned, or, more ominously, Adolf Hitler?

As astonishingly new as cloning might appear, it is actually just one technique in a class of techniques developed over the past four decades. I will call these techniques human genetic engineering (HGE), which I define as the intentional transformation of genes in the body (somatic engineering) or the descendants of a person (germline engineering) through chemical manipulation.[2] Cloning was a sensation because it appeared to be such a leap in our HGE abilities: instead of merely changing some of the genes in a person's descendants, cloning appeared to allow people to select an entire genome for their descendants. We would no longer need to find the genes that regulate intelligence. We could just clone Einstein.

Human cloning, as well as the HGE of a person's descendants, is not yet possible. However, in 1990, a mere seven years before Dolly, the first successful act of *somatic* HGE took place. In the culmination of years of scientific research and ethical debate, a four-year-old girl with a rare genetic condition called ADA deficiency had a gene added to some of the cells in her body to correct a problem with an incorrectly functioning gene.[3] More recently, while at least one participant in a clinical trial has died from somatic HGE, there seems to be more proof that somatic engineering has benefited patients, suggesting that the technique will continue to be applied.[4]

Despite the apparent dangers, the motivation to apply this technique is strong. Consider the story of a parent of children with sickle-cell anemia, a genetic disease that many researchers have proposed treating through HGE:

> Our sickle-cell-anemic children are now young adults. Only rarely can we all be home together as a family. Usually one or two, sometimes all three, are in hospitals being treated for acute disease crises or for the debilitating effects of the disease. This is now a way of life, or I should say, a way of existence. . . . I have watched my children, young and frightened, face major surgery for the removal of their gallbladders. At one point my daughter experienced a blockage of blood vessels causing a thrombosis in the brain, which in turn produced a stroke. . . . I hear her cry for a normal life, and if this cannot be, I have heard her ask, "Why can't I die?"[5]

Stories such as this remind us that there are strong humanitarian reasons for a continuing effort to implement these techniques. While it seems clear that society should do whatever it can to help this family, should we genetically change these children to make them better, or to make their *children* better?

As I show in later chapters, the primary debate in recent decades has involved the technique called germline HGE. Although it has not yet been attempted on humans, scientists have conducted germline HGE on other animals. In 1999 it was announced that jellyfish genes had been inserted into monkey embryos. In the *New York Times* article reporting on the research it was noted that "if scientists can add genes to monkey embryos, it should be possible to add genes to human embryos."[6] If the rapid pace of change in the field of genetics in the past carries through to the future, germline HGE may well have happened between the time I am drafting these words and the time you are reading them.

The concern with germline engineering is the same now as it has been for decades: if we learn how to engineer genes that cause conditions that are universally considered "diseases," such as sickle-cell anemia, we also will learn how to engineer genes that seem to influence other components of human functioning, in both the soma and the germline. A prominent bioethicist, who was also chair of the National Institutes of Health committee that approves genetic engineering experiments, recently coauthored a book that discussed the possibilities of genetically engineering people or their offspring to make them bigger or stronger, to make them require less sleep, age more slowly, have greater memory or intelligence, and perhaps even be less aggressive.[7] This may seem like science fiction, but it is not. As we will see in the pages that follow, bioethicists are not inclined to hyperbole or futuristic speculations, and neither are government commissions. That a government commission convened a meeting in the fall of 1997 to discuss these genetic "enhancements" should give us pause.[8]

This is all taking place in the context of our rapidly growing knowledge about the functioning of genes. Our knowledge of the function of various human genes should become even more extensive following completion of the first map of the human genome in mid-2000.[9] It seems that every day newspaper headlines trumpet a newly located gene; genes for breast cancer and Alzheimer-type conditions have already been identified. Some researchers have even claimed to have identified a "gay gene" that "causes" homosexuality.

Despite the promise of this new knowledge, many people seem to have strong visceral reactions to the idea of human beings planning to change the human species genetically.[10] When the chair of the government commission writes that he is "dissatisfied" with the "violently aggressive characteristics of human beings," and that it would be "an improvement in human nature" to change any genetic influences on aggression in our species, I imagine that many people become uncomfortable.[11] What, after all, are the criteria by which we should "improve human nature"?

I believe that visceral reactions are not irrational tendencies to be suppressed, but unarticulated wisdom from which we can learn. We can reflect on why we have such a reaction to planning changes in the genetic makeup of the human species, by looking at the in-vitro fertilization industry. Why, we should ask, do people spend tens of thousands of dollars of their own money, undergo numerous painful procedures, and face numerous health risks using in-vitro fertilization when they could, for probably less money, adopt a child? The answer is that many people seem to have an almost existential desire for the intergenerational transmission of themselves into the future. For many people—perhaps most—it is not enough to transmit themselves into the future through teaching children their values; instead, they want their bodies to be somehow represented in the future through the continuation of their genome.

Even more broadly, we should reflect on why we care about the future at all. From a purely egocentric perspective, the present generation might as well totally despoil the planet. Why should we care what happens after we die? However, at least some people want to leave a habitable environment for future generations. If we have some deep impulse to provide a good environment for future generations whom we will never meet, we seem to have an even stronger impulse to protect the makeup of the individuals who will someday inhabit that environment.

Human beings care deeply about their genetic progeny, and that is why this issue of HGE is so important. Moreover, HGE has the potential to relieve the suffering caused by genetic diseases. Yet how do we as a society decide whom or what to engineer, or whether to engineer at all? There might be a consensus about removing a gene for Tay-Sachs disease, but there would be much less of a consensus about making human nature less aggressive or changing the "gay gene" should it actually exist.

Reviewing the eugenics movement that culminated in the genocidal policies of German National Socialism, an early participant in the HGE debate noted that what "has saved man from his limited visions in the past has been the difficulty of devising suitable means for reaching them."[12] In the past, society has had the luxury of saying, "We do not need to decide until it is possible." Now we *must* decide what to do, and we can no longer be reassured that the inhumane possibilities will be frustrated by our limited abilities. With the "suitable means" nearly within reach, we need to have strong guiding "visions" to reach conclusions about what our goals should be in HGE technology.

The general public does not generate opinion about a topic such as HGE on its own. Rather, the general public's views of the choices in a debate are structured by various professionals who get paid to think about HGE and similar topics, who both speak to and claim to represent the views of this public. In an earlier era such people would have been called "public intellectuals," but now they are better described using the language of professionalization.[13]

I shall show in this book that at precisely the present time, when the public needs to make crucial decisions about whether our society should engage in various types of HGE, the structure of the debate among the professionals has been eviscerated. I will discuss this at length in the pages that follow, but for now it is accurate enough to say that a bigger, deeper, more fundamental, or "thicker" debate has been replaced by a smaller, shallower, more superficial, or "thinner" one.[14]

At a time when society should be debating which goals or ends HGE would forward, from among all of the possibilities, the debate has been restricted to only a few institutionalized ends. At a time when we should be attempting to derive *societal* ends, we have shifted decision making to the autonomous *individual*. At a time when all options should be kept available, the nature of the discourse used in the debate makes one choice impossible—the decision *not to engage* in any HGE.

Do not be misled by the *shoulds* in the paragraph above. This book does not take a normative position on the act of HGE itself. It does, however, take the normative position, following the lead of Jürgen Habermas, that the public *debate* about HGE should not be artificially limited when such limitation favors the interest of one group over others. Close readers will also conclude that I agree with Michael Walzer on the importance of retaining a "thick" debate, at least somewhere, and with theologian James Gustafson, who believes that exclusive focus on the "thin" without consideration of the "thick" means that "significant issues of concern to morally sensitive persons and communities are left unattended."[15] I simply note my motivations here, and move on to explaining why debates become "thin," returning to these concerns in the concluding chapter.

I am certainly not the first to note that debates in our public sphere are becoming thinner.[16] There is a rich literature on this subject, which I will mine for insights later. In the following chapters I build on this intellectual tradition to explain why the *form* of the debate about HGE has changed over the past forty years to favor the thin debates over the thick. Moreover, I will demonstrate that the thickness or thinness of debate has led to particular conclusions about the ethics of HGE technologies—conclusions that will eventually lead to certain technologies being developed and used.

There are two dominant explanations for the rise of a thin debate over HGE, which I shall critique in the pages that follow. The dominant view among bioethicists is that the thin debate is the result of Americans' recognition, beginning in the 1960s or 1970s, that we live in a pluralistic society with many competing conceptions of the "good life." To accommodate these varying conceptions, the argument goes, Americans created a neutral, "thin" third language that was to be used in debates over policies that would affect all Americans, using values and forms of argument shared by all citizens. Each particular community—be it Roman Catholics, feminists, or African Americans—would still use the thick debate among themselves, but would translate their thick debate to the thin shared language for use in public. Thus, the debate has justifiably thinned as the United States has become a more democratic society.

It is correct that the underlying political philosophy of the HGE debate changed in the 1960s or 1970s toward using this "overlapping consensus" approach.[17] However, the debate changed not because of public demand, or for some inherent reason, but rather because the change forwarded the interests of some groups over others. The locus of the HGE debate was purposefully shifted away from the public to the bureaucratic state, resulting in a change in the operative political philosophy as well.

Another common explanation, both in the humanities and in the social sciences, is that thinner debates are the result of some inevitable process outside human control, such as "modernization" or the progressive enlightenment of human thought. Contradicting claims of inevitability is difficult, yet by revealing agents and interests behind supposedly natural processes, I can criticize this view. The multiple contingencies and moments at which the debate could have changed—supported by evidence that not all debates become thin—suggest that the inevitability narrative cannot be correct either.

In contrast to these dominant explanations, my perspective begins with the recognition that the debaters are not unencumbered individuals, but are embedded in a complex of social relations that influence their thinking and determine their interests. To the standard explanations I add a very simple interest for each individual: that each wants to promote his or her vision of what is ethically correct. Once we consider the actual debaters, we see that to be able to promote their vision, they need resources. At minimum they need the legitimacy to continue to be published. A more critical resource is access to those who actually make decisions about HGE. This raises the importance of the interests of those who supply those resources, who then may select the type of argument that they prefer. I also complicate the two dominant explanations by considering one set of social relations for each debater—the profession in which they are embedded. Therefore, it is authors, embedded in their profession, who have different abilities to raise the resources necessary to continue to make their arguments. Change in the form of the debate, and its conclusions, is then a story of competition between professions for the resources necessary for that profession to produce arguments about the ethics of HGE.

Using this perspective, the pages below will show that the debate over the ethics of HGE was originally the province of eugenicist scientists, whom a group of theologians quickly challenged. The debate during these early years was thick, with arguments about HGE ranging from the notion that it is the ultimate embodiment of human values to control our genetic inheritance, to the idea that to do so is to usurp God's purposes for humanity. The theologians and other challengers to the scientists called for the general public to become more involved—essentially to become engaged as the resource provider for arguments about HGE. Ironically, this call would be turned back against the theologians, as scientists pushed for a shift from public control over the ethics of HGE through legislative activity to bureaucratic control through government advisory commissions. This would be a crucial turning point, as the advisory commissions became the ultimate arbiter of whether HGE was ethical and thus whether it would occur.

State advisory commissions, for reasons we will see below, tend to select thin arguments. This tendency resulted in a shift in the entire debate toward thinness; it also led to the rise of a new profession, bioethics, which embraced this new form of argumentation. As the bioethics profession strengthened its jurisdiction over the ethics of HGE, thick arguments were first crowded out, then actively delegitimated. By the 1990s society was left with a thin debate among these professionals, a debate that seemed unable to ask the deeper questions that these technologies demand.

If the rate of technological invention in this area continues at its furious pace, society will soon be facing a decision about which features we would like to engineer into future humans. The health of our public sphere will be put to the test. Will we be able to have a debate that considers the full richness of the moral question before us? At this point I would say no. Still, through understanding why the debate has become as thin as it has, we may be able to question the need for this thinness and reconstruct a thick debate, such as existed in the past.

This book opens with an explanation of the theoretical framework that will be used to understand the thinning of the debate over HGE from the 1950s to the mid-1990s. To anticipate some strands of this longer discussion, what I have been calling thinning will be described as a shift from a "substantively rational" to a "formally rational" debate. To continue with this Weberian framework, I will also explain how I analyze the changing debate by looking at the ends pursued in arguments, the means proposed, and the relationship between means and ends.

Chapters 2 and 3 chronicle the early years of the debate, before the advent of the advisory commissions. Chapter 2 considers the debate from the late 1950s to the mid-1970s, beginning with a description of the reform eugenicist scientists who were arguing about ends. One of these ends was to reintroduce to the public a sense of meaning and purpose, which they thought had been lost after Darwin had destroyed traditional religion. This debate among scientists brought them perilously close to the jurisdiction of the theologians of the time, who saw the threat and responded, calling for a public debate over the ends that society should pursue through science. A largely substantively rational debate ensued, primarily between scientists and theologians, setting the stage for the more striking and rapid changes in the debate that followed.

Chapter 3, covering the debate from the mid-1970s to the mid-1980s, describes how scientists deftly responded to attempts by theologians and others to allow the public to decide the ethics of HGE. The response of scientists was to advocate the creation of government advisory commissions that would ease calls for setting the ends of HGE research through congressional action. These government advisory commissions needed a formally rational type of argumentation, and the professions competing for jurisdiction at the time were not capable of providing it. Thus was born the new profession of bioethics, which would meet this need. Chapter 3 also discusses the first change, resulting from professional competition, in how the means of HGE is described. Fearing that the expansive ends pursued in HGE were scaring the public and resulting in calls for budget cuts in government-supported science, scientists retreated in the professional competition. They now claimed to be interested *not* in the means to design the genes of the species, but in the more limited means that could cure genetic diseases in individual bodies through somatic "gene therapy." It would take almost twenty years for the scientists to gain back the ground they retreated from in this era.

Chapter 4 is a case study of the two-and-a-half-year process by which a U.S. government advisory commission created the most influential text on HGE—a process and text that I consider to have been the fulcrum of the entire HGE debate. I show how bioethicists and scientists were able to use the constraints of a government advisory commission to translate the substantively rational claims of theologians into the formally rational type of argumentation that they preferred. However, by not acknowledging that they were translating, but simply saying that they were "clarifying" the theologians' "vague" arguments, the commission delegitimated the input of theologians in future debates by making it appear that bioethicists were more qualified to debate HGE. The bioethicists would indeed be more qualified after that point, given that the commission had gone a long way toward successfully institutionalizing formal rationality as the only acceptable form of argumentation. The decline of the professional challenge from theology was particularly rapid after this point.

Chapters 5 and 6 cover the years after the report of the commission was issued. This was a time of increasing institutionalization of formally rational types of argumentation and increasing dominance of the bioethics profession. I describe how the institutionalized formally rational type of argumentation used by bioethicists and scientists allowed them to gain back the ground lost to the theologians in the 1970s and early 1980s. Chapter 5, covering the debate between 1985 and 1991, shows how the bioethicists' arguments about somatic human gene therapy became accepted, an acceptance that eventually allowed scientists to conduct the first somatic human gene therapy procedures. Having secured somatic gene therapy as part of their jurisdiction, the bioethics and scientific professions used the institutionalized form of argumentation to argue that germline HGE should also be under their jurisdiction.

By the final years in this study, 1992–95, somatic gene therapy was solidly under the jurisdiction of bioethicists and scientists, and more than a hundred experimental protocols had begun. Chapter 6 shows the further strengthening of the bioethicists' and scientists' jurisdiction, the further institutionalization of their formally rational type of argumentation, and the further marginalization of theologians and their substantively rational arguments. By this time bioethicists had reduced the number of ends to be pursued to four: autonomy, beneficence, nonmaleficence, and justice. However, while beneficence had been the end sought in most arguments, there was now a turn toward the pursuit of autonomy. This has allowed some authors to turn back to the vision of the eugenicists who started this debate, by arguing that if parents want to engineer superior traits into the germline of their progeny, then this should be their autonomous choice to make.

Chapter 7, the conclusion, begins with a description of events in the debate that have occurred since 1995. Specifically, indicating that debate has indeed come full circle, there has been increased discussion of the ethics of the genetic enhancement of desirable characteristics, as well as a push by scientists for the government advisory commission that effectively controls the ethics of HGE to allow germline HGE. I also discuss the 1997 debate over human cloning and how the process of the cloning debate provides further evidence for the claims made in this book. I revisit and evaluate theories of why debates become more formally rational, and discuss extending the insights of this book to other debates. I close with a prediction of the future of the various professions in the jurisdictional competition and a discussion of what this book has shown about how people should make collective moral decisions in pluralistic societies.

Chapter 1

Framework for Understanding the Thinning of a Public Debate

If one were to ask a focus group of people unfamiliar with the debate about germline human genetic engineering (HGE) to come up with arguments against the practice, they would probably not think to raise the argument that germline HGE is wrong because persons not yet born at the time of the research had not given their informed consent to risks to their own genome.[1] Yet this was one of the few arguments against germline HGE considered *legitimate* in bioethical debate in the 1990s.[2] I suspect that my imaginary focus group would think it odd to conceive of intergenerational responsibility through the language of consent, rather than responsibility, and they would likely find it pointless to enter a debate premised on getting impossible consent from not yet existing people. Surely the focus group would come up with another way of talking about our obligations to future generations.

My imaginary focus group has the freedom to apply the myriad ends or values available in society. They could use values such as these, for example: "there should be limits to human control of our environment," "we should perfect ourselves and our environment," "we should follow nature," "we should control nature," or "we have an obligation to shape future people regardless of their interests." These ends and many more could be used to argue for or against germline HGE.

The professionals who debate this topic today are not so free. They are operating with a very constrained list of universal, commensurable ends that have become institutionalized by the dominant profession in the debate. By the 1990s there were just four ends that could be legitimately used in arguments, and the informed-consent argument uses one of them: individual autonomy of decision making. Thus, people not yet born must give their informed consent to have their genes changed, in order to forward the end of autonomy.

This is just one example of why the recent debate is a very thin one. When the ends or values that can be considered in arguments are limited, something seems to be missing. One is left with the feeling that important points are not being raised. The debate, however, has not always been this thin.

In the 1960s there was a debate about the ends that HGE should forward. For example, in 1969 a distinguished molecular biologist and then proponent of germline genetic engineering, Robert L. Sinsheimer, was hopeful that his end, which I will call species perfection, would be forwarded through the emerging genetic technologies of his time. These new means were "a new eugenics" that, unlike the old, would not require huge social programs or pervasive social control, but rather would be conducted on an individual basis. He wrote that "the old eugenics would have required a continual selection for breeding of the fit, and a culling of the unfit. The new eugenics would permit in principle the conversion of all of the unfit to the highest genetic level . . . for we should have the potential to create new genes and new qualities yet undreamed" in the human species.[3] Sinsheimer was also arguing for an even higher end—that humans should be radically in control of themselves. He saw the new genetic technologies as "the turning point in the whole evolution of life. For the first time in all time, a living creature understands its origin and can undertake to design its future. Even in the ancient myths man was constrained by his essence. He could not rise above his nature to chart his destiny. Today we can envision that chance—and its dark companion of awesome choice and responsibility."[4]

This end was not agreed upon, but became the subject of debate. Theologian Paul Ramsey reacted against Sinsheimer's ends, which he described as attempts to make "man" "his own self-creator." The scientists' ends, he thought, reflected their attempts to form a new theology of science, "a cult of men-gods, however otherwise humble." Cloning, germline HGE, and the like were "proposal[s] concerning mankind's final hope," a new source of "ultimate human significance."[5] As Ramsey would reiterate often in this debate, we should follow God's ends, and humans are not God.

Twenty-five years after Ramsey and Sinsheimer's exchanges, the debate is no longer about ends. Rather, the current debate largely assumes a few institutionalized ends, and thus is limited to discussions of whether the means will forward the given ends. For these and other reasons I will discuss below, the earlier debate seems thicker than the current one. How then did the debate change from thick to thin—that is, how did it become reduced to a few ends, and why to such ends as autonomy?

Substantive and Formal Rationality

First, to explain how and why this change occurred, more precise terms are necessary. The change from thick to thin in public debates is akin to Weber's description of the rationalization of the economy, but applied to forms of argumentation (that is, the common patterns used in arguments). As an example of these different patterns, we may consider those commonly used by philosophers and sociologists in making arguments. Sociologists typically make claims based upon empirical data, while philosophers make claims based upon logical deduction from premises. My use of the terms *rational* and *rationality,* however, requires much more explanation. The best advice I can give to readers who are not familiar with social science treatments of rationality is to try to forget everything you know about the word and start over.

In common language, *rational* is used as the opposite of *crazy.* Among social scientists, *rational* is used to mean a belief that can be held legitimately, while *irrational* refers to illegitimate belief. Social scientists also demarcate many different types of rationality, based on the logic of this legitimacy. "Instrumental rationality" is the best known among social scientists and most closely articulates with common uses of the term *rational.* An individual is being instrumentally rational if he or she is maximizing the means toward given ends. Thus, a student selecting biology as her major is being instrumentally rational if her goal is to become a physician. Her choice is less rational if her goal is to become, say, a musician. Instrumentally rational is also what is meant by the term *rational* in economics and in rational choice theory in the social sciences.

The best known theorist of rationality is undoubtedly Max Weber. Weber discussed many forms of rationality, but his primary concern was with the rise of human *action* motivated by formal rationality at the expense of action motivated by substantive rationality. A pattern of action is *substantively* rational if it applies the "criteria of ultimate ends" or "ultimate values" to acts or means.[6] Therefore, a substantively rational argument about HGE asks whether the means of HGE are *consistent with* the ultimate ends or values. That is, ends and means are debated as a piece. A pattern of action is *formally* rational if it is *calculated* to be the most efficacious means for achieving *predetermined or assumed ends.*[7] A formally rational argument about HGE asks whether the means of HGE maximize, compared to other possible means, predetermined ends. (Clearly, formal rationality is closely related to instrumental rationality.)[8] The two different forms of argumentation are derived from this subtle yet basic distinction. The substantive is thick, and the formal is thin.

Weber did not apply his insights to public debates. In this book I build on the work of scholars influenced by Weber who are less interested in human action than in human discourse. For example, the Frankfurt school theorist Jürgen Habermas has put a discursive spin on the Weberian thesis, stating that the "system" (institutions governed by formal rationality) is colonizing the "life-world" (governed by substantive rationality).[9] The "life-world" is the location of free debate about ends, unencumbered by the biases introduced by formal rationality. The ability to have a substantively rational debate in the "life-world" has become limited because of this colonization—a phenomenon that Habermas, like Weber, bemoans. Similarly, American sociologists studying the voluntary sector have agreed with Habermas's description of the problem, finding that public debates in advanced industrial societies, because of "social and cultural forces," are increasingly limited to "questions of efficiency, practicality, cost-effectiveness, instrumental rationality, and expedience."[10]

Unfortunately, this literature offers little help in parsing arguments in a particular debate into substantive and formal, thick and thin. Weber's writing is notoriously unclear in its explanations of the types of rationality, and Habermas's writing is meta-theoretical, far removed from empirical examination. Moreover, even if there were an extensive empirical literature on this topic, each empirical topic is most likely different enough to justify the fleshing out of the details for each case. Therefore, I take Weber's and Habermas's distinctions in types of rationality in public debates as my basic inspiration, and fill in the details for the case at hand. This is what is required, I believe, to apply Weberian insights to the modern world.

Substantively and Formally Rational Forms of Argumentation in the HGE Debate

I return to Weber one last time to retrieve another helpful insight. The calculability required in formal rationality was so important in Weber's overall framework that all the different ways he describes the rationality of modern life in general are closely related to this calculability.[11] It has been pointed out by others that the critical feature of formal rationality that accounts for its thinness is the calculability required in the link between means and ends. However, this desire for calculability in ethical argument also results in the selection of particular types of ends in formally rational arguments. It is then not only the link between means and ends, but the features of these preferred ends, that make the debate feel thin. With this insight in mind, I further specify the ideal-type components of substantive and formally rational arguments below. These components were not derived from theory or from others' work a priori, but were created through a mix of insights from other scholars and inductive examination of the case at hand.[12] There are five components of arguments: the link between means and end, the extent to which ends are debated, the number of ends considered legitimate in the debate, the commensurability of ends, and the universality of ends.

Link between Means and End

The basic distinction raised by Weber and others regarding substantive and formal rationality is in the link between possible means and ends. The form of this link is recognizable by the role of the *consequences* of means in the argument. In early substantive debates, arguments were evaluated by determining whether the means were *consistent or inconsistent with* ends, regardless of the bad or good consequences of the act. For example, Ramsey expressed this link in substantively rational argumentation quite well when he argued against some of the schemes of the reform eugenicists by saying that a person making a correct decision would "*not* begin with the desired *end* and deduce his obligation exclusively from this. He will not define *right* merely in terms of conduciveness to the good end; nor will he decide what *ought to be done* simply by calculating what actions are most likely to succeed in achieving the *absolutely imperative end* of genetic control or improvement."[13] Despite the desirability of the ends some means may forward, in the substantive argument some means are inherently wrong because they are inconsistent with other ends.

Consequentialist reasoning is much more calculable, and is therefore a hallmark of formal rationality. Using this link between means and end, means that are the most efficacious for forwarding an end are ethical. A recent example of a consequentialist argument for germline HGE is that, given the implicit end of forwarding human health, the means of "germline intervention is more efficient than repeating somatic cell gene therapy generation after generation."[14] Unlike substantive rationality, which does not even purport to calculate, consequentialist reasoning requires calculation and selection of the more efficient means by examining their predicted consequences.

Debating and Defining the Ends

Substantively and formally rational arguments can be distinguished by the extent to which the ends are considered to be a part of the debate. In a substantively rational debate, means such as HGE technology and the end it furthers are of a piece—they are considered simultaneously—so the ends must be defined and argued for. Indeed, in the early days of the HGE debate, there was a substantively rational debate regarding what should be the ends of society, and whether HGE was consistent with them. Some scientists thought that HGE technology was consistent with the human value of "bettering mankind," and explicitly debated the nature of the "ideal" genetic man. (Albert Einstein was often used as an example.) Others tried to convince people that following God's will was the ultimate end, and that the end of the betterment of the species was the wrong end.[15] Yet others argued for the necessity of a challenge for the human species. HGE was consistent with this end because it would be a challenge to "produce a Man who can transcend his present nature."[16] Put simply, a substantively rational debate is about ends.

In a formally rational debate the ends are not open for debate, but are assumed, either explicitly or implicitly. For example, in recent years one of the few ends allowed into the debate over HGE is "beneficence," especially in its specific form of improving the health of people. Moreover, the possibility that the means may have implications for other ends is not considered.

Debating ends makes ethical debates less calculable. The reason for this is, as Weber himself pointed out, that there is no way to use calculation to select one end over another. Put differently, ultimate ends are incommensurable.[17] For example, the ends of beneficence and of individual autonomy cannot be adjudicated without appeal to higher level ends.

Number of Ends in the Debate

Because a substantive debate is about ends themselves, there is no limit to the number of ends that can be forwarded. For example, it is hard to describe in any condensed manner the numerous and myriad ends that were put forward in the debate over HGE that occurred from the 1950s through the early 1970s. Arguments were made for species perfectionism, beneficence, obeying God, creating meaning for human society, the pursuit of knowledge as an end itself, and many others.

In the later years, when the debate had become more formally rational, the ends argued for had been reduced to four: autonomy, beneficence, nonmaleficence, and justice. These ends were present in the earlier debate; it is the reduction in the number of ends that is critical for the case at hand. The ultimate force behind this reduction (as I shall argue below) is the need to make arguments more calculable. With only four ends, it is easier to evaluate the extent to which the means of HGE does or does not advance these ends. Of course, the debate would be even more calculable if there were only one end to be considered. Although there has been suggestive evidence in recent years that one end is coming to dominate all others—a topic that I will review in the conclusion—so far this debate has not yet become quite this formally rational.

It is important to add that in a *purely* formal debate, there would be no ends at all. The ideal-type of a formally rational institution in the Weberian framework is a set of institutional processes or means that simply increase calculability toward any and all ends.[18] The institutionalized process known as the "scientific method" is a good example of a purely formally rational institution. Scientific activity simply accumulates facts about nature that allow for a better calculation of the consequences of action, without any place within the method itself for a determination of the ends to which this knowledge should be put. Means (such as experimentation) then become the ends. The debate over the ethics of HGE does not become purely formally rational, but it does come closer to this ideal-type over time.

Commensurable Ends

A debate is more calculable if the selected ends can commensurate disparate ends. Commensuration, the process of transforming different ends into one common metric, is also a method for throwing away information to make decision making more calculable.[19] The classic commensurable end metric is the "utility" of utilitarianism, where all ends are translated or reduced to the scale of the "greatest good for the greatest number." If a commensurable end dominates a debate, then the debate is more calculable for two reasons. First, ends that can commensurate other ends are described using metric language. There can be more or less of the same property, such as "beneficence," instead of more or less of two different properties, such as "protecting nature" and "doing justice to our descendants." This makes calculation even easier. An early advocate of commensurable decision-making metrics in the HGE debate, Joseph Fletcher, displayed his use of a commensurable value scale when he wrote in 1971 that "if the greatest good of the greatest number (i.e., the social good) were served by [HGE], it would be justifiable not only to specialize the capacities of people by cloning or by constructive genetic engineering, but also to bio-engineer or bio-design para-humans or 'modified men.'"[20] "Greatest good of the greatest number" seems not only metric but almost quantifiable, and thus quite calculable.

Second, any end that can commensurate other ends is going to be limited—or, to use the metaphor I have adopted, it is going to be thin. Put differently, deeper ends resist translation into commensurate ends—more information must be thrown away to translate them. For example, how would one translate "fidelity to God" to utility? Authors assuming substantive rationality resist commensurable scales because their ends cannot be commensurated with other ends without distorting their meaning.

Perhaps the best known and most used commensurable values scale is money. Cost/benefit studies translate all ends to this most common of scales to reach a highly calculable outcome. The commensurable ends that became institutionalized in the HGE debate are far less precise than a money scale permits. "Justice," for example, allows for many readings. However, even these somewhat vague ends allow for a heightened degree of commensuration and calculability.

Universality of Ends

The final category is related to the previous one. Substantively and formally rational arguments are different because of their assumptions about the universality of the ends. There are two senses of *universality,* which enhance calculability in different ways. The first is that a commensurable, calculable end is defined in such a thin way that almost no one could disagree with it (e.g., "it is better to do good than to do harm"). For example, when entering a hospital, you are not asked whether you value your decision-making ability; rather, this is assumed to be universal, and a policy and procedure have been created by which all patients are asked to exercise their decision-making ability by giving their informed consent to a medical procedure. In the HGE debate many authors assume only that everyone would prefer a healthy body to an unhealthy one—a fairly thin, safe assumption that is probably universal, or nearly so. Using ends like these enhances calculability because the ends actually held by the individuals in question do not have to be consulted, only assumed.

A second sense in which an end is universal is across various means. Once again, this encourages the ends to be very thin because only thin ends could be applied to every potential debate about a technological means, such as HGE, abortion, or the cessation of life support. Universal ends in this second sense of the term enhance calculability because once a method is agreed upon for evaluating a means such as abortion, it does not need to be changed for a different means, such as HGE. For example, some persons in bioethical debates hold that the end of personal autonomy in decision making is the only end to be considered in evaluating the ethics of abortion. Because they also feel that personal autonomy is an end that is universal across various means, they can apply the same method to questions such as HGE, thus rendering decision making more calculable.

Universalism in both senses is foreign to authors who make substantively rational arguments. Their ends are thicker and thus not simple enough to be considered universal. Indeed, the entire point of a substantively rational argument is to convince others to adopt the end, so it cannot be considered to be universally held a priori. Since means and ends are of a piece, the same ends are unlikely to be important for any and all means. Authors making substantively rational arguments are therefore more likely to apply different ends to different means. The dimensions of

Table 1 Components of Debates Dominated by Substantively or Formally Rational Forms of Argumentation

Dimensions	Substantive Rationality	Formal Rationality
Link between means and end with which to evaluate argument	Means consistent with system of ultimate values or ends are ethical. Means that are inconsistent are not.	Any means that maximizes ends is ethical.
Debating ends	Ends are defined and defended.	Ends are implicit or considered outside the decision-making process.
Number of ends considered legitimate in the entire debate	Unlimited	Limited in number
Commensurability of ends	Ends are not commensurable.	Ends can be translated into a common end or a limited set of ends. Ends that cannot be translated are bracketed as unimportant.
Universality of ends	Ends are not believed to be universally held by individuals, although the author may want them to be. Different ends may be relevant to different means.	Ends, particularly when translated into a commensurable metric, are considered to be universal ends of all persons in society or even the world. The same ends should be applicable to all problems.

substantively and formally rational arguments about HGE are summarized in table 1.

The Effects of Substantive and Formal Rationality on Descriptions and Conclusions about Means

The differential need for calculability in the two forms of argumentation will lead actors to describe controversial means such as HGE differently, and to reach different types of conclusions regarding what should be done about the means under debate. Further parsing these effects will show additional ways in which the concern with calculability leads to a thinner debate.

Typical Descriptions of the Means

Since substantively rational authors are ultimately arguing about ends, they are typically not interested in the means themselves (such as HGE). They are interested in specific means only to the extent that the means point to a larger or deeper problem in the ends.[21] This assumption can often be recognized in a text by the sheer volume of different means being discussed. For example, some substantively rational authors discussing HGE simultaneously consider abortion and in-vitro fertilization to be "part of" the same problem—that is, all these means are consistent or inconsistent with the same end that they are debating. The debate then seems thicker, more fundamental or important, because one means seems to be connected with many other means.

One reason why formally rational debates seem very thin is that means are evaluated separately. For example, a formally rational text will *not* discuss abortion and in-vitro fertilization in the discussion of HGE. This is because an ethical debate defined as encompassing the means of abortion, in-vitro fertilization, and HGE has too many variables, and thus too much information is needed to calculate whether the means is the maximal solution to the defined ends.

A second effect of the form of rationality on the way a means is described is the definition of *who* is affected by the defined means. Ethical arguments assuming substantive rationality either portray individuals in the context of larger social groups or are not concerned with individuals except insofar as they are members of larger groups. For example, a substantively rational author might claim that germline HGE will change society's notions of family obligation. Authors assuming formal rationality will likely not describe means as existing for collectivities; rather, they will define the means as involving individuals. Formally rational authors tend to ask, What should a scientist or patient do in this situation, given the universal ends attributed to them? The move in recent years from what society should do to what individuals should do is one of the reasons why the debate appears to be thinner: acts of individuals seem less momentous than decisions for humanity. Considering the means of HGE to affect only the individual is a requisite to calculation, because it removes from consideration the complexity of social relations.

Finally, recent debates seem thin because they define the effect of the means in the very short term and are unwilling to debate means that may be possible in the future. This is a requirement of calculability, because the effect of the various means that must be weighed to reach an optimal decision cannot be predicted. Put differently, if the effect of the means on ends cannot be calculated, the debate must then be about the ends themselves. Moreover, formally rational authors also point out that the ends held by future people may be different as well, changing the calculations. As one formally rational text put it, "a working rule for ethical debate on new technologies" should be that authors "refrain from moral judgement on unverifiable possibilities."[22]

Assuming substantive rationality, the time frame is irrelevant. The effect of an act does not need to be known, because the substantively rational author asks only whether the act is *consistent with* our ends, not whether it will be the best way to achieve the ends. Indeed, for the substantively rational author, speculating about technologies that may not be operative for hundreds of years is a good way of arguing for ends. For example, in 1963 a paper was written titled the "Biological Possibilities for the Human Species in the Next Ten Thousand Years."[23] In a debate *about* the ends to which humans should be working, it is relevant to write papers with titles such as this because we could consider ten thousand years to be some sort of teleological endpoint—the end to which we as a species should be striving. In a formally rational debate about *means* with fixed ends, no calculations can be made about the effects of acts such as HGE ten thousand years into the future, so debates can only consider shorter time spans.

Typical Conclusions Regarding What Should Be Done

Following substantive rationality, means are right or wrong for a priori reasons—for their consistency with certain ends—not because of their consequences. This can lead to the conclusion that there are some means that should never be developed, despite the consequences. Using a metaphor popular in this debate, substantively rational authors are capable of arguing for a "red light" to further research in a given area.

For the formally rational debater, no means is inherently wrong, but is only wrong if it does not maximize the end pursued. There is then no means that should not be brought to the point where its consequences can be adequately calculated. When a means reaches this point, people may find that the ends have also changed. Thus, to continue the metaphor, these authors—in contrast to those who would enact preemptive bans—tend to see only green or yellow lights for research on HGE. A summary of the effects of using substantive or formal arguments can be found in table 2.

Understanding Change in the Form of Rationality in the HGE Debate

Given this more fine-grained description of the substantively and formally rational forms of argumentation in the HGE debate, we are still left with the question of why a debate would shift from substantive to formal. Put differently, what social forces select for calculability in ethics, the linchpin of formal rationality? There are two common explanations that I shall criticize in this book, both of which are deficient because they lack attention to interests, agents, and power. The first and most prevalent is the "expanding democracy explanation," put forward by the profession of

Table 2 The Effect of Substantively and Formally Rational Forms of Argumentation on the Descriptions and Conclusions about the Means

Dimensions	Substantive Rationality	Formal Rationality
	Description of means	
Range of means discussed	The means in question is representative of a deeper or larger problem. The means is therefore often linked to other means that are argued to be consistent or inconsistent with the same end.	A dispute is over only one discrete means, with related means intentionally bracketed so as to focus on the means at hand.
Whom does the means affect?	Society	Individual
Time frame for the effect of means	Time is irrelevant.	A means will only be considered a problem if it is imminent.
	Conclusions about what should be done	
Possibility of banning means	Anticipatory and permanent bans are permissible.	Anticipatory bans are impermissible. We must develop means first to determine what their consequences will really be. Moreover, the ends of society may change over time.

bioethics, which now dominates the HGE debate. The second is the "macro-historical process explanation," which holds that the shift to a formally rational, thin debate is part of a natural or inevitable process. This explanation is found in different forms in different fields of study.

The Expanding Democracy Explanation

As background for understanding the expanding democracy explanation, I will offer thumbnail sketches of two different theories of legitimate democratic decision making. The first of these recognizes that a decision based upon a pluralistic consensus is "that which survives a competitive process of debate and compromise, during which significantly affected parties have an opportunity and a right to modify, if not to veto, any particular conclusion or decision that they find unacceptable on grounds of their own well being."[24] This perspective, also called proceduralism, is the basis of the interest-group conception of American democracy. Consensus about legitimate ends to pursue is generated in the legislatures by the give and take of organized citizens who, operating with the same rules and opportunities to make their case, negotiate and compromise with each other. For example, in this model, the argument that abortion cannot be restricted because to do so would violate women's rights to make their own autonomous decisions is legitimate if all the interest groups that agreed and disagreed with this position were given a fair hearing by the decision-makers (in this case a legislature).

A second democratic model is called "overlapping consensus."[25] This position holds that the ends to be forwarded by any policy must be shared by all of the particularistic groups in society. These shared values—imagine the overlapping circles in a Venn diagram—are not determined through the give and take of politics, or through deliberation among the citizens, but typically through some process of academic reflection on past debates. So, to continue the abortion example, the argument that abortion cannot be restricted because to do so would violate women's rights to their own autonomous decisions is legitimate if bodily autonomy is one of those ends that is considered to be shared by all citizens.

A subset of the characteristics of formal rationality—assumed, commensurable, and universal ends—describes the "overlapping consensus" model. As will become clear below, I do not use this language of political theory to *describe* the change in the HGE debate because although it fits formal rationality fairly well, it cannot be used to describe substantive rationality. Moreover, while the change in operative political philosophy is one part of the explanation, there are other causes of the rise of formal rationality. Therefore, I treat the type of democratic authority assumed by a writer more as an explanatory variable than as what needs to be explained. More colloquially, the change in political philosophy is part of the story, not the story itself.

As I will argue below, the order of events in the expanding democracy explanation is the inverse of what actually happened. Consider the account of the rise of assumed, commensurable, and universal ends (the overlapping consensus) in this explanation. Reacting to a hegemony of establishment-oriented Protestants in the interest-group type of democracy used to determine public ethics in the 1950s and 1960s, in the following decade people began to question authority, and particularly the authority of dominant value systems. This gave rise to the recognition of value pluralism in America, which led to the establishment of this thin, neutral, common language that different groups could use in forums of debate about their collective problems, such as HGE. Or, as described in an account of the rise of government commissions to apply these new ends, "over the past two decades, a desire for mechanisms to articulate common values and foster consensus about biomedical advances in the face of cultural and religious heterogeneity resulted in the creation of Federal bioethics commissions."[26] After offering my theory of how these debates change, I will review why this explanation is not convincing.

The Macro-historical Process Explanation

The second explanation for the rise of formally rational, thin debates—that this change is part of a natural or inevitable process—takes several different forms. Some of the authors trained in bioethics have the deep assumption that there is a natural progression of human reason, a progressive enlightenment away from emotional, often religiously based arguments and toward more "rational," calculating, scientific, "neutral" arguments. In the same way that someone could read thousands of empirical articles in sociology and never encounter the statement, "this explanation is premised on the notion that human action is constrained by social structures," this "progress" assumption in bioethics and philosophy is so deep that it can only be read between the lines.

For example, bioethicist John Fletcher states that ethical issues "evolve generally in four stages: threshold, open conflict, extended debate and adaptation." This natural progression, described as agent- and interest-free, seems to track an evolution from substantive to formal rationality, with a progressive reduction in the number and particularity of the ends. In the second stage, for example, "differing moral convictions collide." In the third, "if emotions about the issue do not overwhelm debate," then clarification of "moral lines" can begin. In the fourth and final stage, "moral adaptation" occurs. In the remainder of his article Fletcher demonstrates how his favored "moral line" "had to be distinguished" for the debate to follow its natural course.[27] Or again, consider another bioethicist, Eric Juengst, who argues that ethical debates proceed through three stages: the stage of "romantic" inquiry, where knowledge is not "systematic," the stage of "precision," and the "generalizing" stage.[28] Notice how this also seems to flow, free of actors or interests, from less calculating and particularistic to more calculating and universal—akin to the transformation from substantive to formal rationality.

These bioethicists are simply tapping into one of the deep assumptions of the Western intellectual tradition, in which Weber played a critical part. Sociologists too have often assumed an extricable march toward the dominance of formal rationality over substantive rationality. The central theme in Weber's writings is the expansion of formal rationality into areas of social life once dominated by substantive rationality. Although methodologically Weber was not a believer in historical inevitability, he was extremely pessimistic about the outcome of this competition, believing that the "iron cage" of formal rationality would continue to tighten "perhaps until the last ton of fossilized coal is burnt."[29] Or, in the words of a Weber interpreter, Weber believed that a formally rational social life "is no mere possibility, but the inescapable fate of the modern world."[30] Later sociologists seemed to share this view, assuming in studies of formal organizations that organizational changes were "over determined by transcendental laws of rationality."[31]

These two explanations for the changing form of rationality in public debates over HGE are deficient because they do not examine the people who actually changed the debate, their interests, and the power relations among them. To understand the rise of formal rationality in the HGE debate, we must create theoretical constructs at a level much closer to the human agents who are, ultimately, responsible for social change.

The Production and Institutionalization of Formal Rationality

To build a theoretical edifice with which to conduct this investigation, we should begin by going to the root of the question. What is a debate? On the street corner, it comprises oral and body language. Among the professionals who dominate the HGE debate, the exchange occurs mostly through written texts. These texts are cultural products, created by actors with interests and social relations that have an impact on their content. Therefore, to understand the transformation in the form of rationality assumed in these products, we must look at the conditions of these texts' creation.[32] Most important, cultural products such as the arguments in books and journal articles do not simply rise from their social structural context, but are intentionally produced by people in competition with others in the same environment.

Producing Arguments from an Individual's Perspective

Three stages of the production of culture can be usefully demarcated as analytic categories.[33] The first is *production,* in which arguments are formulated by numerous persons. In the HGE debate, the produced arguments consist of every journal article or book written on the topic. As will become clear in the data given below, the barriers are low enough that almost any argument can be published *somewhere,* whether as a chapter in an edited volume or in an academic journal circulated to only dozens of libraries.[34]

The second stage is *selection.* Of all of these arguments, only a few are selected by the environment. Writers use some forms of argumentation over others, some fall by the wayside, others are supported. The critical explanatory variable is the resources garnered for the propagation of the argument, because writers are in competition to extract the necessary resources from an environment with limited resources.[35] On the most basic level, resources are needed to feed and clothe the proponents of arguments. For the producer of revolutionary tracts, the primary resource needed from the environment is money to buy food and shelter and run the press. For most of the writers in the HGE debate, the basic resources needed to continue making their arguments are the recognition and legitimacy to gain and keep access to their target audiences. From this perspective, not achieving tenure is a particularly large failure in resource extraction, because one may then lose the position from which to speak. In the field of ethics, where the point is to influence a decision, the people who actually decide whether to conduct HGE have the ability to provide the legitimacy for writers by using their argument in their decision making.

The third stage in this process is *institutionalization,* where routinized mechanisms for reproduction of an argument or form of argumentation come into being. The argument has by then become reified—it does not seem to have been created by anyone but is just "common sense" or even "fact"—and takes on a life independent of its creator.[36] For example, the argument that informed consent is necessary before experiments can be conducted on people has become institutionalized. It is taught in textbooks to medical professionals, rules in hospitals ensure its replication every day, and government regulations decree that its logic will be followed by researchers. Resources from the proponents of this argument are no longer required for its replication.

With this perspective, the rise of formally rational forms of argumentation can be seen as the cumulative result of the actions of many individuals who create competing arguments to extract resources from their environments. Explaining in a cumulative sense why the supporters of formal rationality "won" in this competition and why the supporters of substantive rationality "lost" requires examining the factors that allowed some actors to disseminate their arguments, while others could not.

The critical determinant of resource extraction—and thus ultimately of the ability to produce arguments—is then the degree of fit between the form of argumentation used by the writer and the form preferred by persons in the environment who have the power to decide what to do. These decision-makers, in ethics at least, have the ability to decide which arguments are "important" and "influential." This means that to understand the change in arguments about HGE, one must closely examine who the decision-maker is, for whom the writer was writing, and, in particular, the social context in which they are all embedded. For example, if the primary resource a young bioethicist at a Catholic college is competing for is tenure, to obtain that resource he or she must achieve a certain status in the field. Perhaps thirty-five years ago the pinnacle of status was to have your magnum opus used by the Church's marriage classes, but today the goal is to be cited by the government bioethics commission. This implies a difference in the preferred form of argumentation, which in turn will tend to change the form of rationality used by professors in similar structural locations.

This is not to say that people only consider which argument will maximize their resources when writing. Nor do I argue that people will produce any argument that they will receive resources for, or write only what the decision-makers want to hear. However, one would be naive to think that writers do not make compromises at the margins in order to obtain what is often the greatest resource—being listened to. Consider, for example, the entire field of discourse called "liberal Protestantism." Some scholars would describe liberal Protestant theology as an almost strategic compromise with the "forces of modernity"—a compromise made so that the theologians would continue to be heard.[37]

Consider also an account by a former theologian as to why he and others took up the form of argumentation used in the profession of bioethics. "We could have attempted to analyze the new science and the new medicine in the terms of the languages [of philosophy and theology] we had learned," he begins. "But we gradually realized that, if we did, our words would be uttered largely in vain" because the two languages would not function as needed with the target audiences. "We had to find an idiom that, at one and the same time, expressed substantive content and was comprehensible to many listeners. . . . Like strangers in a strange land, we had to devise new forms of communication among ourselves, with our scientific and medical colleagues, and with the public."[38] This theologian had a customary way of writing, based on theology, but given the new environment, this would not allow him to achieve his goal of influencing the debate on these issues. Therefore, he changed his idiom in order to be heard. In sum, people will change their discourse at the margins in order to forward their overall vision, and the accumulation of these small individual changes transforms an entire debate.

Although people are somewhat plastic, probably the stronger cause of change in a population of writers is that those who do not receive resources will drop out of the debate and those inclined to make arguments that will be supported are more likely to enter. Put somewhat differently, this is an explanation of the motivation of an individual as well as the social conditions that allowed an individual with a particular form of argumentation to take part in the debate.

Debating Communities

Tracing individual participants in the debate and their resource contexts is not only logistically difficult; it also distorts the actual influences on a debate. As mentioned above, it is important to move away from the notion that writers are isolated individuals free from constraint and toward a view of writers in their social context. Taking the most obvious step in this direction first, there is clearly not *one* HGE debate at any one time, but many. Individuals are embedded in these different communities of debaters where the members will not agree on the particulars—they are, after all, debating with each other—but they will share some assumptions. At the most basic level, one assumption that each community shares is what is to be debated. Empirically, for example, some of these communities share only the assumption that they are debating the relationship between the means of HGE and the means of genetic screening. However, most also share assumptions about whether they should use substantively or formally rational arguments and whether specific arguments within these general forms are legitimate. Because they believe their form of argumentation to be correct, each of these debating communities is competing with the others to institutionalize its assumptions as applicable to the entire debate.

For example, by the mid-1990s several distinct communities were debating the ethics of HGE. The most prominent framed the debate as being about health—they assumed that HGE should be considered a medical procedure. They concluded that HGE should perhaps be regulated a bit, but that research should continue apace. Most important, the form of rationality assumed by the authors in this community was resolutely formal. They did not debate the ends of HGE, but assumed only a few predetermined ends, which they considered to be universally held, and they evaluated arguments by the extent to which the means maximize the ends. At the same time a much smaller and less influential community, which described HGE using the language of eugenics, argued for solutions that included banning research, and, most important, assumed substantive rationality in their reasoning.

But we must also place these communities in their social context. The most striking finding about these communities is that the members tend to belong to the same profession. In the example I have given here, the most influential debating community was made up mostly of bioethicists, and the less influential, mostly of theologians. This serves as a reminder that when those debating the ethics of HGE vie for legitimacy with whoever is empowered to make the ethical decision about HGE, they are considered not as individuals, or only as members of communities with a particular perspective on the topic, but also as members of a profession.

Professional Competition and the HGE Debate

Consider all the professions and potential professions in a society as part of a competitive ecological system.[39] In the case of HGE, the professions are competing within this system for "jurisdiction" over promulgating the ethics of HGE. Jurisdiction may be defined here as the link between a profession and its work. It can be forged through state licensing (e.g., only licensed physicians can prescribe drugs), institutional affiliation (e.g., health maintenance organizations may pay for services provided by M.D.s, but not those provided by homeopaths), or public opinion (e.g., medical doctors are more effective than homeopaths). For the work of producing arguments about HGE, the relevant link is forged by the opinion of those who are empowered to decide about the ethics of HGE.[40] To link with the previous discussion, we may say that the people capable of providing the resources necessary for institutionalization are those who give jurisdiction. At one point in the debate, the public had the power to make decisions about HGE. The public provided the most important resource, legitimacy, to a group of theologians who were making arguments that resonated with the public. If the theologians had successfully institutionalized their arguments, then we would have said that the public had given them jurisdiction.

Some may question whether "doing" public bioethics can be called "work." The work of arguing about the rightness or wrongness of acts of HGE is less concrete than dentistry, but it is work nonetheless. Jurisdiction over similar work, such as helping individuals determine the "meaning of life," has long been the subject of great competition for jurisdiction among theologians, psychologists, philosophers—and recently, it could be argued, advertising executives.[41] In the same way that a psychologist's advice about a family crisis has greater legitimacy than the same advice given by a nuclear physicist, different professions in society have different degrees of legitimacy in the work of producing ethical arguments about HGE. Although people might listen to psychologists' arguments about HGE, it seems more likely that they would turn to biologists, theologians, or bioethicists when deciding whether society should engage in HGE.

In everyday understanding, a profession is defined by having an association or a special educational degree. However, it is now less common for academics to define a profession by its political form—an organization (such as the American Medical Association). Rather, a profession is defined by the abstract knowledge that it uses in its work. For example, physicians have an extensive system of how to diagnose and treat medical problems, and a method of deciding which treatment goes with which diagnosis. The key to professional competition is that this knowledge is *abstract*—that is, it cannot be automatically applied (although advocates of artificial intelligence programs in medical diagnosis are heading in this direction).

If this knowledge were set in stone—if there were no fuzziness over what is a "medical" problem—then the boundaries of jurisdiction between professions would also be clear-cut. However, because the knowledge of a profession is abstract, a profession can redefine its problems and tasks, defend them from competitors, and seize problems from competing professions—as physicians have recently seized alcoholism, mental illness, hyperactivity in children, and obesity.[42] In the particular case of the work of ethics, the system of abstract knowledge is the form of argumentation typically used by the profession. An ethicist's work is precisely to make these arguments. Thus, in the HGE debate, each profession will have its own form of argumentation that is abstract enough to claim jurisdiction over the ethics of HGE, and each will change its form of argumentation to account for changes in the environment and to compete against other professions.

Competition between Professions for Jurisdiction

Since the various professions' forms of argumentation are abstract, it is not obvious which should be used to evaluate whether the means of HGE is ethical. Members of each profession will use their form of argumentation to define the problem, suggest solutions, and apply the form of rationality the profession uses. With HGE, members of two different professions will hear of the same act—for example, a scientist's first insertion of a gene into the germ-cell of a mouse—and one will argue that this is the first step toward eugenics, while the other will argue that the same act is the first step toward gene therapy. One will call for a permanent ban on the procedure, another will call for continued bureaucratic oversight. The form of rationality they will use to make their arguments will also differ.

Most important, medical research scientists might look at the means of adding a gene to a human embryo to counteract a genetic disorder and describe the process so that the means appears to fall within the medical profession's previously established jurisdiction (e.g., HGE as "medical therapy"). They will evaluate the ethics of the means using the form of rationality institutionalized in their home jurisdiction. Theologians will look at the exact same means as the scientists and describe it so that it appears to fall within *their* home jurisdiction (e.g., HGE's use to treat disease as a fulfillment of God's desires for humanity). This method of claiming jurisdiction, in which one profession claims that a task is reducible to its already secure jurisdiction, is called *reduction*. It is the primary means used to legitimate an expansion of jurisdiction into new areas.[43] For example, the medical profession has recently argued that child misbehavior is largely reducible to the disease of hyperactivity, and can therefore best be addressed by physicians. The medical profession defines hyperactivity as a phenomenon of the inner workings of the body, and treatment of the body is the domain of physicians; therefore, the argument runs, the medical profession should have jurisdiction over the treatment of hyperactivity. In the HGE debate, jurisdictional claims are made using one or both of two analogies: first, that the means under debate are actually the same as the means over which a profession already has jurisdiction (e.g., transplanting genes in cells is the same as transplanting organs); second, that the ends pursued with the means are the same as those pursued in the home jurisdiction (e.g., transplanting genes in cells promotes human health).[44]

To understand why a profession would change the form of argumentation it uses, one must view professions in an ecological field, where each is trying not only to secure jurisdiction over new tasks, but also to defend its "core" jurisdiction from interlopers. Professions will change their form of argumentation, if this is necessary to enhance their profession's overall position in the ecological field. For example, it has been shown that scientists will change the form of argumentation that they use, depending on the professions they are competing with.[45] Similarly, in the HGE case, scientists began by assuming substantive rationality and then, as their competitors and resource environment changed, transformed their arguments toward formal rationality. Again, this collective professional change was the result of individual scientists changing their arguments and of scientists who entered the debate in later years using different arguments than those of the older scientists who were entering retirement.

The ecological perspective also provides the insight that professions cannot argue for any position that maximizes their position for each task, but must simultaneously defend their other jurisdictions. For example, scientists during the 1960s and 1970s retreated from an attempt at jurisdictional expansion because the modification of their form of argumentation designed for this purpose was threatening their core jurisdiction of laboratory research.

A shift in the beliefs of those granting jurisdiction—such as those people who are empowered to decide about HGE—can also result in changes in jurisdiction. For example, the rising legitimacy of scientific thinking in society due to the efforts of scientists and others has worked to the benefit of medicine—whose form of argumentation presupposes science—and against folk healers of various types, whose arguments are not based on science. In the case of the HGE debate, in the late 1960s people who were not involved with the HGE debate successfully institutionalized a generalized questioning of authority by the public, and specifically the idea that scientists do not always work in the people's interest. The institutionalization of this idea among the general public, who were the empowered decision-makers of that era, led to a delegitimation of scientific authority (i.e., jurisdiction) in general—leaving scientists open to challenge in the peripheries of their jurisdiction (such as HGE).

Types of Settlements between Competing Professions in the System

Professions generally vie for full jurisdiction over particular work. Medicine is the prototypical case of full jurisdiction, where persons not in the profession who conduct acts defined as "medical" are punished by the state. Often, however, professions reach settlements with other professions regarding the boundaries between the two. One type of settlement is *subordination,* where a dominant profession loses or gives up subsidiary acts to another. The classic case is doctors and nurse-practitioners, or doctors and X-ray technicians. Another form of settlement, which has been important in the HGE case, is that of an *advisory jurisdiction,* where one profession interprets, buffers, or partially modifies actions taken by another profession within the latter's full jurisdiction.[46] I will argue that bioethicists have collectively, through their individual interpretations of their social context, pragmatically decided not to compete with scientists for full jurisdiction, but rather have created an advisory jurisdiction to the scientists.

In sum, to understand the transformation in the form of the debate about HGE from thick to thin (substantive to formal), we must look not only at the different debating communities, but also at the professions in which those communities are embedded. The change over time from a substantively rational to a formally rational form of argumentation is thus seen as the result of a change in the profession with jurisdiction over this work or a change in the form of argumentation used by the profession with jurisdiction. The form of argumentation the profession uses is a function of what will earn it jurisdiction (legitimacy), while maintaining its present jurisdictions and undermining competitor professions.

Clarifying Comments

Macro-competition and Individual Motivations

The language of professional competition gives the mental image of secret cabals of professionals, perhaps meeting in darkened rooms, plotting the takeover of jurisdictions from weakened competitors. Worse, it paints the individuals involved as trying to compete with others to gain some sort of mercenary advantage. In this book I use the competition metaphor in a limited manner. First, competition was not the result of the collective decisions of professions to attack a competitor. Scientists never had a meeting where they decided to go after the theologians' jurisdiction. But, if there were no secret meetings planning jurisdictional attacks, how can I then defend statements I made above—for example, that professions will change their form of argumentation if necessary to enhance their profession's position in the ecological field? The defense lies in the commonsense notion, expanded upon by sociologists, that similarly situated individuals tend to act in similar ways. In trying to enhance his or her own position in a debate, a theologian may change the form of argumentation to make it more effective given the social context. Other theologians will be similarly situated and will behave similarly, and thus a profession acts collectively without central coordination.

Moreover, I do not believe that these professionals were competing for personal wealth or power. The literature on professions is often unclear in this area, leaving the implication that physicians compete against chiropractors for self-interested reasons. Although this was undoubtedly a factor in the HGE debate, I think that the main reason for competing was that each profession believed that its approach was the best way of addressing the problems at hand. More concretely, professions in the HGE debate decide to compete for jurisdiction because they feel their competitors are making bad arguments. Consider this account of some of the earliest meetings in the public bioethical debate where the different professions debated the issues: "One of my toughest problems during [these early years] was persuading the philosophers to sit down with the theologians and to take them seriously. The secular philosophers could not give a damn for what the theologians were saying and were even scornful."[47] In sum, professionals compete against each other because they think their competitors' form of argumentation is wrong.

An analogous case is competition between different types of academics. It is a common joke among academics that economists see the other social sciences as being a sub-specialization of economics, while sociologists view the others as a sub-specialization of sociology, and so forth. When one believes in one's form of argumentation, it is hard to not see it as applicable to a wider range of phenomena in the world than it is currently applied to. It is not hard to see why phenomena at the margins of a home jurisdiction might seem to be reducible to the home jurisdiction.

Disciplines or Professions

Many readers, upon examination of the data presented in the following chapters, will want to say that the authors are members of academic disciplines, not professions. They are indeed on the border between the two, but the term *profession* is appropriate for several reasons. First, what is an academic compared to a professional? An academic is one who refines the form of argumentation of the profession and teaches it to the professionals who carry out their jurisdiction. Thus, theologians are the ones who work out the form of argumentation that clergy members use daily, and professors of psychology do the same for practicing psychologists. Moreover, while many authors who debated HGE over the years have appointments in academic departments, they tend to be both working out the form of argumentation and applying it to problems in the world. At the margins of the data can be seen authors who are primarily concerned with fine-tuning the form of argumentation—typically described as theorists. Taking theology as an example, at the margins of the data are references to systematic theologians Karl Barth and Paul Tillich, but the debaters themselves are typically not systematic theologians. In Protestant circles, they are "Christian social ethicists," a branch of theology that not only applies the systematic theology of a Barth or Tillich to actual problems but also suggests modification of systematic theology in light of these applications.

Bioethics and the Bioethicist

It is extremely important to clarify here a source of confusion that results from the multiple uses of the term *bioethics*. Public debates of the type exemplified by the HGE debate have come to be known as *bioethical* debates. But there are at least three types of bioethical debates, which I will call *foundational, clinical,* and *public.*[48] Foundational bioethics discusses how the debates about issues such as HGE are related to broader societal concerns, such as systems of ethics, democratic practice, and the like. This book will conduct very little analysis of this debate. Indeed, this book itself could be considered part of this debate. Clinical bioethical debate concerns the ethics of interactions with patients in hospitals or in research studies. A typical question here is, Should life support be ended in this instance? I will have little to say about this type of bioethical debate either, except where it infringes upon my primary interest, which is in what I will call public bioethical debate.

Public bioethical debate is where societal elites—in this case, professionals—debate over what society should do about a problem such as HGE. In the language of social theory, this is the debate in the "lifeworld" that people like Habermas are so concerned about. Of course, in an ideal sense, there is no reason why the public cannot debate the issues without elites. However, given the specialization that has occurred in modern society, people now rely upon experts at least to lay out the various ethical arguments. The purpose of the public bioethical debate among the professionals is to influence the beliefs and values of the public, to come to some modicum of consensus, or in some cases to represent public opinion to policy makers.

This bioethics debate is to be distinguished from the bioethics profession, whose members I call *bioethicists*. Although all those who engage in bioethical debate are popularly identified as bioethicists, it is important for this analysis that the only professionals considered to be bioethicists are those who use the profession's form of argumentation.

It is particularly difficult to empirically classify debaters as bioethics professionals—as opposed to theologians, philosophers, or scientists—in the early years covered by this study. However, my method, described in the appendix to this book, errs on the side of caution and identifies people who could be called "pure" bioethicists. My method also allows the professional identity of authors to change over time, which occurs for a number of people in the data. Despite the measurement difficulties, this empirical research finds that those whom I identify as bioethicists do use a different form of argumentation than the members of other professions.

In short, members of other professions participate in public bioethical debate, but they are not bioethicists unless they use the system of argumentation of the bioethics profession. This distinction will help to clarify a common confusion these terms tend to create for readers who also participate in these debates, regarding whether there actually is a recognizable bioethics profession. The confusion arises because the earliest individuals involved with public bioethical debate, such as Daniel Callahan, did not set out to invent a new profession. Their dream was that the public bioethical debate would have representatives from multiple professions, explicitly combining the arguments of the social sciences, theology, philosophy, medicine, and science to solve ethical problems in society. Despite these intentions, a group of people who called themselves bioethicists—rather than theologians, philosophers, or members of any other existing profession—created a distinct form of argumentation for this purpose. In so doing they also created a profession, making distinctly different arguments than other professions did. As we will see, by the mid-1990s bioethicists involved with the broader public debate about HGE were engaged in a debating community quite distinct from the other professions.

The Explanation for Change in the HGE Debate

We can now see my argument in its full complexity. The HGE debate became more formally rational as a side-effect of a complex interaction of professions competing in an environment for jurisdiction. Specifically, as the narrative begins, a debating community dominated by eugenicist scientists had begun a quest to expand its jurisdiction beyond the lab benches toward explaining, through genetics, the meaning and purpose of human life to the public. HGE would allow human beings to design themselves into the perfect species, the scientists argued, providing meaning in human self-creation. They adopted a fairly substantively rational form of argumentation for this task—one that placed their arguments very close to the jurisdiction traditionally held by theologians. The theologians thought that their approach for explaining the meaning and purpose of human life remained superior. When the scientists faltered, the theologians reasserted themselves, establishing themselves as the primary competitor to the scientists for jurisdiction.

The theologians, with others, had whipped up public concern over HGE, and legislators began to react to this concern with proposals to regulate the scientists' experiments. Scientists felt that their core jurisdiction was under threat, as legislators considered banning some types of research. Scientists retreated from their jurisdictional aspirations, and in the process created a safer metaphor with which to claim jurisdiction. They stopped claiming that they were going to forward the end of providing meaning for humanity and switched to safer, metaphorical links to means and ends from their home jurisdiction. They claimed that they should have jurisdiction over HGE because they were interested only in the more limited means of somatic "gene therapy" (the transplanting of genes in the bodies of presently living people) and wanted to forward the end of healing disease. Thus, "gene therapy" was born, not out of scientific discovery but as a defensive maneuver to a jurisdictional threat.

Scientists, in their struggle to avoid severe damage to their jurisdictions, were able to bring about a change that would ultimately—and unintentionally—result in the dominance of formal rationality in this debate. In a critical moment, scientists were also able to deflect public regulatory control over their work in HGE and their home jurisdiction by advocating the creation of government advisory commissions that would "advise" the scientists of the broader ethical implications of their work. These advisory commissions eventually became the ultimate decision-maker for the ethics of HGE because the federal government—which at the time funded almost all HGE research—tended to follow the advice of these commissions in deciding what experiments were "ethical" and thus fundable. By establishing what arguments were legitimate, these government advisory commissions would grant jurisdiction to some professions over others.

To grasp the radical transformation in the debate caused by these advisory commissions, consider the alternative that beckoned before their appearance: citizens would make ethical decisions and express their conclusions about HGE to their elected officials, who would make laws to regulate it. That is, the pluralist model of democracy was in force. This was unacceptable to the scientists, who feared that an "excitable" public would shut down not only HGE research, but other research in their home jurisdiction that the public did not understand. In place of elected representatives, unelected representatives of the public on advisory commissions—who are much more distantly accountable to the public than elected officials are—would make ethical decisions for the public. They would discern the overlapping ends of the public a priori, and apply them to controversial issues.

The rise of this new provider of jurisdiction ultimately would result in the gaining of jurisdiction by the bioethics profession. At the same time as the emergence of commissions as decision-makers, an intellectual movement was begun by a group of theologians and philosophers to create what was called a "principle"-based system of ethics. This might have resulted in another dusty book in an academic library, except that this system turned out to be exactly what an unelected commission would need as a working system of knowledge, and it came at the right time. "Principlism," as it would later be called, borrowing heavily from philosophical ethics, was based on four universal, commensurable ends that could be universally applied to any problem: beneficence, nonmaleficence, autonomy, and justice. This form of argumentation also articulated well with the overlapping consensus theory of democratic practice, which could be used by government advisory commissions. It was, in the words of one of its proponents, "a philosophy of the people."[49]

In sum, the system of knowledge that these ends were embedded in was resolutely formally rational. Professionals in the debate began to adopt this new form of argumentation largely because they thought it was the best way to make ethical decisions *in light of the new decision-makers,* the commissions. People who entered these debates and who adopted this new form of argumentation—or people who converted to it—began calling themselves "bioethicists" and not scientists, theologians, or philosophers. A new profession was born.

Government advisory commissions legitimated the new form of argumentation, and thus gave immediate advantage to the new profession of bioethics. However, like all processes of institutionalization, in time the form of argumentation took on a life of its own, somewhat independent of the social conditions of government advisory commissions that gave it birth. Formal rationality spread not only because of the increased use of government advisory commissions, but because the profession of bioethics was taking over jurisdiction in this and other debates.

The top frame in figure 1 shows the profession of the authors of texts that debate HGE, for four different periods, with roughly equal numbers of texts in each time period. The bottom frame shows the professional distribution of the most influential authors in the HGE debate in the same periods.[50] We can see here that scientists are in decline throughout both series, although the decline results not from a decrease in influence, but from the rise of the allied profession of bioethics. More important, while theology was the primary competitor of the scientists until the mid-

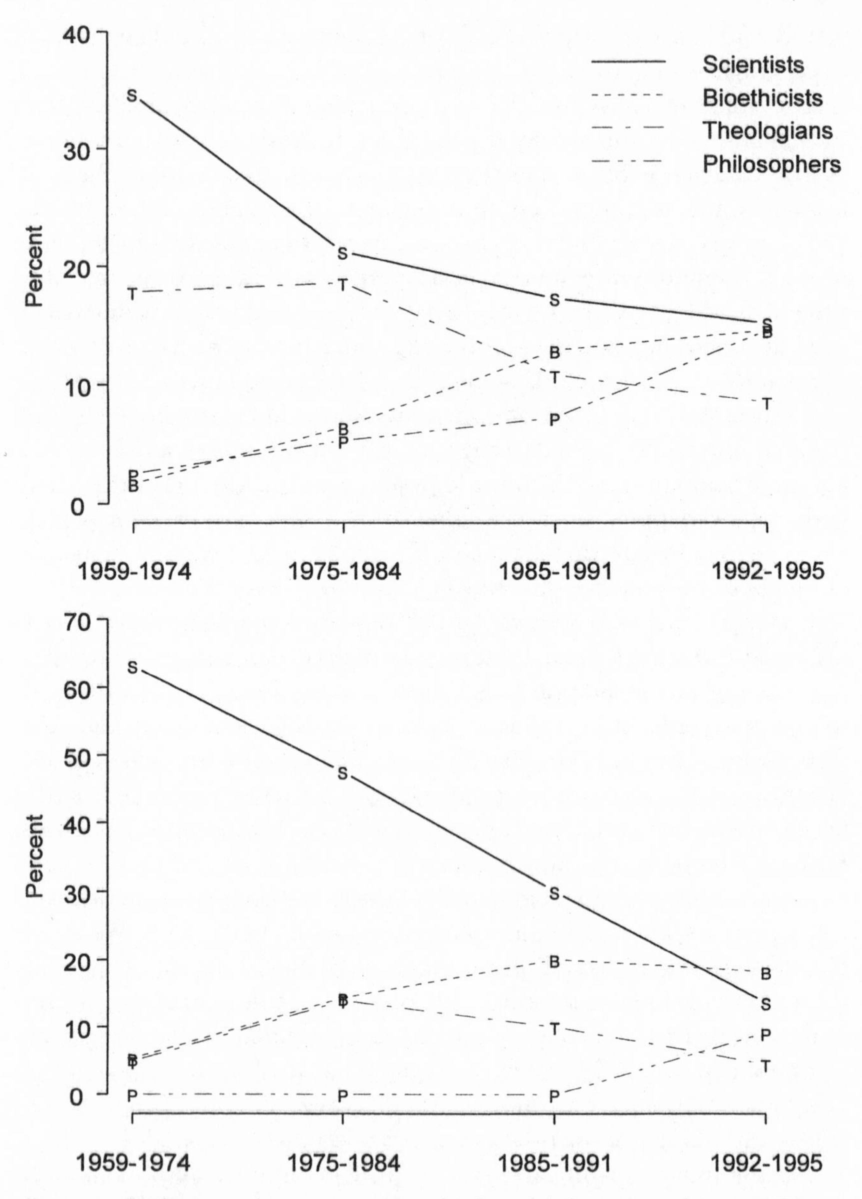

Figure 1 Professions of common (*top*) and influential (*bottom*) authors in HGE debate

1980s, after that point it was in decline, being replaced in the runner-up position by the bioethics profession and, to a lesser extent, philosophy, a profession from which bioethics has drawn much of its form of argumentation.[51] It is this increasingly strong jurisdiction of the bioethics profession that explains the rise of formal rationality in the HGE debate.

I shall also show how the professional competition affected the actual acts of HGE that have occurred or are being considered. With the institutionalization of the four already named ends as the only legitimate ends to pursue, scientists and bioethicists used one of these ends, beneficence, to justify regaining part of the scientists' jurisdiction that they had retreated from earlier. Somatic gene therapy had earlier been connected to the scientists' jurisdiction through the metaphor of means *and* ends. They had argued that the changing of the genes in a body was analogous to other medical manipulations of the body, and that these analogous means were to forward the same end of healing disease. However, this metaphor limited them to manipulations of the body. With the growing strength of the profession, they began to make their jurisdictional metaphor solely on the ends pursued: human gene therapy—be it of the body or of the species—pursued the end of healing disease. The only reason they could make this jurisdictional challenge is that the jurisdictional claims of their competitors had by this time been severely weakened.

With this new analogy, scientists and bioethicists claimed that they should have jurisdiction over the ethics of *germline* human "gene therapy," defined as the eradication of disease in the descendants of the present generation. By the end of the time period considered in this book, it was beginning to be argued that these same four ends used by the bioethics profession allowed for the genetic enhancement of children to make them stronger, taller, and the like. The debate had come full circle back to the claims of reform eugenicists of the 1950s—but with a formally rational form of argumentation.

Given the full exposition of my micro-level theory of change in the debate, it is much easier to reveal the weaknesses of the two competing explanations for this change. The "expanding democracy" explanation is incorrect because it treats a change in political philosophy as what is to be explained, whereas my data suggest that it is part of the explanation. Moreover, this explanation ignores the role of professional competition in spreading formal rationality, independent of government advisory commissions. Its greatest fault is that it has the order of events backward. This explanation would say that the new recognition of pluralism led to the desire to use the overlapping consensus model, which led to the use of government advisory commissions which applied overlapping consensus. To repeat one of these historical descriptions, it was "a desire for mechanisms to articulate common values and foster consensus . . . in the face of cultural and religious heterogeneity [which] resulted in the creation of Federal bioethics commissions."[52] In later chapters I shall show that the order of events was the inverse. First came the decision to move the debate about HGE to government advisory commissions; when writers learned that these commissions tended to use formally rational arguments, the debate changed accordingly.

Not only does this explanation have the order of events wrong, but it ignores an important reality: there have been many thick debates about ends since the 1960s, and such debates are not antidemocratic. For example, political theorists Amy Gutmann and Dennis Thompson demonstrate that debates about abortion in Congress are still about "fundamental values" (a term akin to my use of *ends*).[53]

In the conclusion to this book we will see that the "expanding democracy" explanation is not so much an explanation of the change in the debate as the Genesis narrative of the bioethics profession. This narrative retells the profession's founding as following the central themes of American democracy, thus legitimating its jurisdiction.

The macro-historical process explanation will also be found wanting for its mechanistic, agent-free form of explanation. In this critique I join other scholars who have recently refocused on Weber's insight that the spread of formal rationality is not a teleological or natural process, but rather a cultural form that, much like religion, must be constructed, maintained, and spread among a population, often to the advantage of the group propagating it.[54] For example, in Espeland's study of the rationalities employed in a dispute over the construction of a dam, she shows how the formally rational maximization of means to ends is not an inherent characteristic of humans, as rational-choice theorists suggest, but a cultural form that promotes the interests of a particular group in the dispute.[55] In this book, I too will provide evidence suggesting that formal rationality is not an inevitable by-product of modernization, but a by-product of professional competition for jurisdiction. There was nothing natural about it, although professions that obtain jurisdiction tend to create narratives of historical inevitability to further legitimate their claims.

The Relation between State Bureaucratic Authority and Formal Rationality

Although I will explain the relationship between state bureaucratic authority and formal rationality in greater detail in later chapters, pausing for a moment on this critical component of the explanation is crucial.[56] The critical question is, Why would a government advisory commission select for formally rational arguments? We must distinguish from the outset between representatives of the state whose decisions are directly legitimated by the public, and those whose decisions are very indirectly—if at all—legitimated by the public.

The authority of democratically elected officials is relatively unquestioned, and they often make substantively rational claims. If people do not like the ends pursued, they will vote the officials out of office. Moreover, legislative arenas are one of the locations where ends are determined through deliberation. Elected officials make substantively rational arguments all the time—as seen most recently in the human cloning debate, when some senators wanted to ban the procedure because it was inconsistent with what they perceived to be the ends that humanity should pursue.

The story is quite different among unelected government officials. Put simply, it is part of the U.S. political culture not to trust authority, especially government authority, and the authority of bureaucrats in particular. The result of this distrust is that while in other countries government officials are generally trusted to "exercise judgment wisely and fairly," in the United States "they are expected to follow rules."[57] Given this mistrust, for the government official "the temptation to substitute supposedly impersonal calculation for personal, responsible decisions . . . cannot but be exceedingly strong."[58]

In my terms, unelected officials cannot be seen as setting or debating ends, but must be simply calculating the maximal means for achieving rulelike ends that they receive from the public. Most decisions in the bureaucratic parts of the state receive their ends from an elected official who is clearly responsible to the public. For example, the president may tell the Environmental Protection Agency to protect endangered species.

Consider the challenge facing the decision-makers who attempt to follow rules on decisions about issues like HGE, and how this would lead to formal rationality. They are rarely told the "rules"—the ends to pursue—because for these difficult problems elected officials intentionally avoid making decisions about ends, to avoid the political controversy.[59] A government advisory commission could debate and set the ends in a substantively rational fashion, but this would be akin to having democratically unaccountable decision-makers creating the rules. Instead, they attempt to determine ends that are universal and thus shared by all citizens (such as "it is better to do good than to do harm"). Avoiding ends entirely would be even better. A debate would then become purely formally rational, with ends considered either outside the decision-making process and not open for debate, or in the extreme, with means treated as if they were ends.

Consider how an official demonstrates that he or she is *following* instead of *creating* the rules. Scholars have found that decision-makers in these settings attempt to keep their decision making transparent.[60] The first feature of transparency would be to make your calculations as simple as possible. Thus, as historian Theodore Porter has shown, bureaucrats have been the foremost proponents of decision methods that rely on commensurable ends scales, such as cost/benefit analysis, because they purport to be based on the rules of "impersonal calculation."[61] As I will argue below, commensurable metrics in ethics have the same appeal for government commissions in the HGE debate because they give the appearance that democratically unaccountable government officials have not set ends or used their judgment, but rather have applied preexisting rules to potentially controversial ethical debates.

Transparency also selects for formally rational links between means and ends. Decisions where the consequences of various means are compared to calculate which most efficiently maximize the ends purport to show the public the decision-making process. Decisions can then appear to be matters of facts—of consequences—such as mean X produces more of the good end than mean Y. There is no safer ground for an unelected government official than to be making decisions based on "facts."

In contrast to the method of comparing the efficiency of means in maximizing ends, consider the substantively rational method of deciding whether means are *consistent with* ends. Here there is no "more or less" comparison of different means, but only a yes/no decision of whether the one means is consistent with the end. While the argument that a means produces *more* rather than *less* is simple and transparent, an argument that a means is inconsistent with an end is neither simple nor transparent. In sum, the need for transparency also selects for formal rationality.

On Method

Flowing from this theoretical framework are the methodological decisions that I made to conduct my research. Most sociological research has to decide between depth and generalizability. Ethnographers of a single hospital provide great depth and richness without being able to make generalizations, while survey researchers cannot offer richness but rather settle for generalizability of their more modest findings. Ideally, a project should have both depth and generalizability, and I have attempted to have both through the incorporation of multiple methods.

In addition to the pursuit of depth and generalizability, the question under examination in this book presents additional methodological challenges. The difficulty is that I want to conduct a micro-level study of individuals and the groups of which they are members, yet I want to do so over many decades to observe subtle changes in the form of argumentation. The reason such a research design is rarely initiated is that the comparable micro-level data across many years are rarely available.

For example, one method would have been to look at public opinion polls across years. However, the first such polls on HGE were conducted in the early 1980s, and they are very limited in the questions they ask. Moreover, as we will see in the following chapter, by the time of the first survey the debate had already become highly structured. Another method would have been to interview participants who were active in each era. However, although I have interviewed participants when necessary, this method has the problem that the persons and professions that rose to prominence will tend to write the history of the debate in a way that makes their rise seem somehow natural, and potential respondents will have learned this history. On a more practical note, many of the participants in the early debate are no longer alive to tell their story.

It was therefore necessary to develop a more objective method of determining the content of debates in different eras, without compromising on depth or generalizability. Through a method explained in more detail in the appendix, I created a population of 1,465 texts representative of the professional debate about the ethics of HGE from 1959 to 1995 using bibliographic sources. Then, using methods developed by sociologists of science, I demarcated debating communities in each era by noting which groups of authors are influenced by the same authors, as determined by citation patterns.

I also have data from two types of participants in each community: average and influential members. Specifically, for each community there is a wealth of in-depth data on the *influential* authors: interpretive analysis of their texts, their career histories, and professional identities. For each community there are also less rich yet more generalizable data on the numerically larger group of *average members* of the community who write about the ethics of HGE. Since there are hundreds of these authors—as opposed to only dozens of influential writers—a form of content-analysis data from their texts is used, as well as data on each member's professional identity. All of this information about a community, quantitative and qualitative, is used simultaneously. To enhance readability, I have moved the details of these methods and the quantitative analyses of each community off stage, into the notes and appendix.

Finally, because the case of the advisory commission of the early 1980s called the President's Commission was such a crucial turning point for this debate, the analysis goes into as much depth as possible for that case, relying not only on the text the commission produced but on drafts of the text, intra-office memoranda, correspondence, analysis of the transcripts of the commission meetings and in-depth interviews with important participants in the process.

One feature of these methods should be clarified for those readers who have not been trained in the social sciences. Academic research about an author, particularly in the humanities, evaluates everything the author has written. For example, an analysis of the theories of Durkheim would include everything written by Durkheim. Moreover, two authors would not be placed in the same category unless what they wrote was substantively similar. Here, however, everything an author has written is not relevant. Rather, I have considered only what the broader members of the community thought was relevant as indicated by their citation of it. Moreover, trying to examine intellectual trends in hundreds of texts over decades involves the use of statistical methods, which requires the tolerance of imprecision of measurement at various stages of the analysis. Like any public opinion poll, the sampling of authors results in a degree of error, as does the technique of examining citations and aggregating authors into communities. The result of this error is that not every author in every community will seem to "fit" with the others, and not everything the authors in a community write will share precisely the same assumptions. Communities dominated by one profession, such as theology, will not be composed exclusively of theologians; some theologians will argue like bioethicists and some bioethicists will argue like theologians. Just as members of society at large are embedded in multiple status affiliations, such as occupation, religion, race, and so on, members of these debates also are involved with multiple communities. As in other sociological research, the extent of embeddedness in each community can be estimated statistically. This book will discuss the central tendencies of each author's relevant work, and of each community's form of argumentation, professional affiliation, and the like. I hope to convince readers unfamiliar with this approach of its utility.

Chapter 2

Setting the Stage: The Eugenicists and the Challenge from the Theologians

What ends should we seek? This was the question during the earliest years of the debate over human genetic engineering, from 1959 to 1974.[1] Debate was vociferous, but it was largely conducted assuming substantive rationality with few tendencies toward formal rationality. By looking at this period, then, we can glimpse the form of argumentation that will become more and more difficult to see as the debate progresses.

Over this period scientists, who began with jurisdiction, increasingly found their jurisdiction under threat. Some of the ends scientists were pursuing through HGE were extremely close to those pursued by theologians in their jurisdictions, and this eventually enticed theologians into challenging scientists for jurisdiction, as the public increasingly began to realize that it was the ultimate decision-maker in the debate. Theologians confronted the scientists on the issue of ends, by challenging both the ends that some scientists were explicitly advocating and the claim of not having any ends at all, made by the first proponents of formal rationality. The debate in these first years set the stage for more momentous and rapid changes, discussed in subsequent chapters.

Reform Eugenicist Scientists and the First HGE Debate

Analysis of the data reveals three distinct debating communities in these early years, the most influential of which contained a disproportionate number of scientists. I will briefly outline the evidence here, relegating details to the notes and the appendix. Figure 2 is a tree diagram of the *most influential* authors in this early debate. In the figure, the closer to the left two authors merge into a single branch, the more highly related

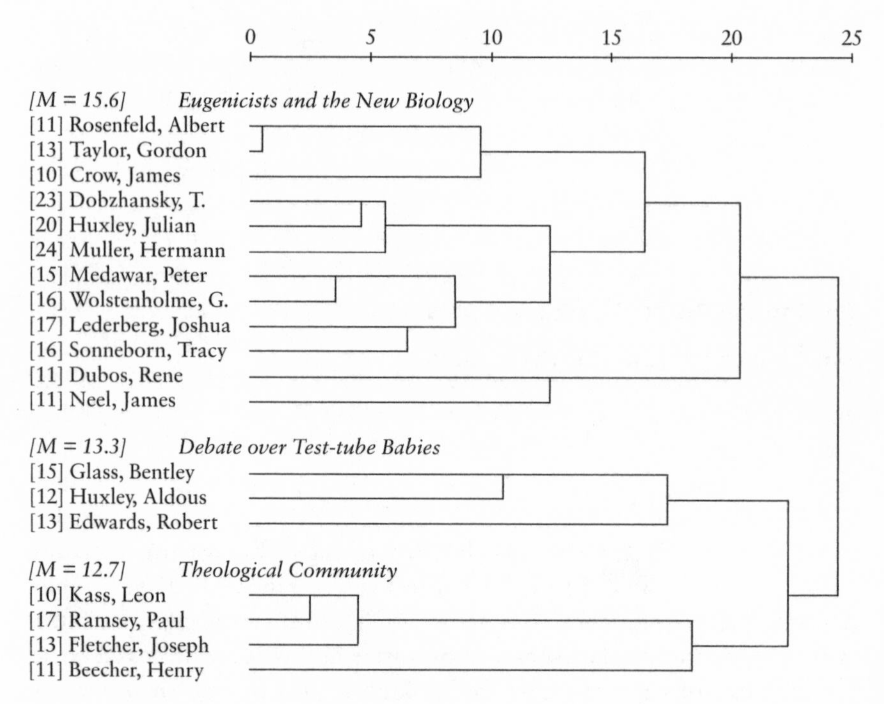

Figure 2 Clustering of most influential authors, 1959–74
NOTE: The number of texts that cite the author is given in brackets. *M* indicates the mean citations per author in the community. The scale at the top is the rescaled distance at which the clusters combine. $N = 69$.

their work was believed to be by the *average* members of the debate during this era. For example, the average debater in this era thought that the work of Kass and Ramsey was more closely related than the work of either of these authors was to that of René Dubos. In this tree diagram, influential authors are grouped into communities. The most influential texts by these most influential authors are listed in table A-2 in the appendix.

The claim that one community was more influential than another is based on a comparison of the mean number of citations that influential authors in the community received from the average authors in the debate. The top community received an average of 15.6 citations per influential author, more than either of the other two, so I consider it to be the most influential community in this era. The claim that this community is made up of scientists is based upon the profession of average members of the community who differentially cite the influential members of the different communities.[2]

The most influential scientist in the top community was population geneticist Hermann J. Muller (1890–1967).[3] A specialist in fruit-fly genetics who worked primarily at Columbia and Indiana universities, he later won a Nobel prize for his work on radiation-induced gene mutations.[4] As a lifelong leftist, he also spent some years in the Soviet Union during the 1930s until he became disillusioned with the Stalinist regime.[5] Most important, Muller was a lifelong advocate of eugenics. To understand Muller, as well as the other scientists in this community, we must understand the arguments of the earlier eugenics movement, upon which they built.

The Roots of Human Genetic Engineering: The Mainline Eugenics Creed

The idea of controlling the genetic evolution of the human species was not invented in the 1950s. Rather, the intellectual roots of the debate lie in the mainline eugenics movement of the late nineteenth and early twentieth centuries in both Great Britain and the United States.[6] The word *eugenics* itself was coined in 1883 by Francis Galton, an English scientist and cousin of Charles Darwin, whose goal was to set up a science of improving the human species by giving "the more suitable races or strains of blood a better chance of prevailing speedily over the less suitable."[7]

A number of ideas converged to make such a goal seem necessary. At this time Darwinism, and particularly the social Darwinism of Herbert Spencer, had begun to have great influence in England. This can be seen in the fear expressed by Galton that the "survival of the fittest" was no longer operative due to the conveniences of modern society. "Such measures as the minimum wage, the eight-hour day, free medical advice, and reductions in infant mortality encouraged an increase in unemployables, degenerates, and physical and mental weaklings," he claimed. The result was, in the words of Galton disciple Karl Pearson, that natural selection had been suspended and replaced by "reproductive selection," which gave advantage "to the most fertile, not the most fit."[8]

All of this was predicated on the notion that the "qualities" of an individual were genetically determined. Eugenicists thought that one could look to the leaders of society—intellectuals, artists, musicians, scientists, and members of the upper classes—to see the result of good breeding. Criminality, "feeblemindedness," alcoholism, and sexual immorality were to be found in the genetically unfit. Not only did the mainline eugenicists believe that all sorts of behaviors were genetically determined, they also believed that genetics was destiny. Pearson, for example, argued against the expansion of access to schools in Great Britain because "no training or education can create [intelligence]. . . . [rather] you must breed it."[9]

The patterns of these inherited qualities were assumed in England to be based on class lines. Perhaps not surprisingly, in the United States they were based on "racial" lines, often conflated with what people today would consider nationality or ethnicity. American mainline eugenicist Charles Davenport, the country's leading eugenicist of the time, wrote in 1911 that the genes of the Poles made them "independent and self-reliant though clannish," that the Italians tended toward "crimes of personal violence," and that the "Hebrews" were predisposed to "thieving" though rarely to "personal violence."[10]

The American mainline eugenics movement of the early twentieth century benefited from and encouraged the anti-immigrant feelings of the era. Immigrants, disproportionately poor and destitute, were seen as the source of great social problems, to which the eugenicists were quick to attach a genetic explanation. Davenport, for example, believed that the new immigrants during this era from eastern and southern Europe would make the American population "darker in pigmentation, smaller in stature, more mercurial . . . more given to crimes of larceny, kidnaping, assault, murder, rape and sex-immorality."[11]

These were not radical views at the time. Indeed, genetic explanations of behavior were the norm of the day—particularly on the part of "good Protestants" who were concerned about those "unfit" immigrants. Eugenics was a progressive social reform, and many clergy and others with good intentions were avid supporters of the movement. Even Theodore Roosevelt told Davenport that "someday we will realize that the prime duty, the inescapable duty, of the good citizen of the right type is to leave his or her blood behind him in the world."[12]

Given the connection with the supposed genetic traits of "inferior races" of recent immigrants, it is not surprising that the predominant solution to the eugenic "problem" was to try to restrict immigration from these populations. In 1924 the mainline eugenics movement had its most notable success with the passage of an immigration control act designed to dramatically limit the immigration of persons from eastern and southern Europe. The bill, which both the House and Senate had passed by overwhelming majorities, was signed by President Coolidge, who had earlier stated that "America must be kept American. Biological laws show . . . that Nordics deteriorate when mixed with other races."[13]

Eugenicists also had success in getting eugenic sterilization statutes passed in the states to keep the "feebleminded" from reproducing. One of these laws, which eventually passed constitutional muster, was the case of Carrie Buck, a seventeen-year-old "moral imbecile" who was committed to the Virginia Colony for Epileptics and Feebleminded. In a now infamous ruling in *Buck v. Bell* (1927), Supreme Court Justice Oliver Wendell Holmes, writing for the majority, concluded that sterilization was justified because "three generations of imbeciles are enough."[14]

Soon after the zenith of success marked by the anti-immigration statute and the *Buck v. Bell* decision, the notion that behaviors were directly carried by genes, and that these genes were found patterned in "races," increasingly came under attack. A growing number of critics recognized that the mainline creed was essentially hidden race and class prejudice. Indeed, the very notion that there were separate "races" came under question. Hermann Muller himself was one of these critics, writing at the time that eugenics had become "hopelessly perverted" into a pseudoscientific façade for "advocates of race and class prejudice, defenders of vested interests of church and state, Fascists, Hitlerites, and reactionaries generally."[15] After the eugenic policies of the Nazis had been exposed—in Kevles's words, "a river of blood would eventually run from the [German] sterilization law of 1933 to Auschwitz and Buchenwald"—the American mainline eugenics movement lay dead.[16]

One might think that the death of the mainline creed would be the death of the idea as well. But many people remained enthralled with the dream of genetic improvement. Those who were later called the reform eugenicists believed that biology revealed, in the words of Julian Huxley, "the inherent diversity and inequality of man."[17] They were "reform" eugenicists because they thought that the "valuable" characteristics to be encouraged in the human species were found across all class and racial groups. The reform eugenicists replaced the mainliners' concern about racial groups with concern about the "genetic quality"—or, for the pessimists, the "genetic deterioration"—of the entire population. Whether they wanted to improve or save the species, the first means proposed was what would today be called germline HGE—attempts to change the distribution of genes in future people. This earlier movement also provided two of the prominent ends that the influential scientists in the data examined here explicitly argued for in a substantively rational fashion: reducing the genetic load and providing a meaning and purpose of life.

The More Explicit End: Reducing the Genetic Load

The 1953 discovery by Francis Crick and J. D. Watson of the structure of DNA is often heralded as the beginning of the genetic age. Examination of the texts in the HGE debate reveals that a possibly more important event occurred eight years earlier: the bombing of the Japanese cities of Nagasaki and Hiroshima in 1945, which eventually led to concerns about the genetic load.

Muller introduced the concept of the "genetic load" in his 1949 address to the American Society of Human Genetics, titled "Our Load of Mutations," and reiterated it in most of his later work. Muller believed that modern society was heading toward a genetic apocalypse.[18] The release of radiation from nuclear weapons and other sources was causing mutations, as postwar Japan demonstrated. More consistent with the beliefs of earlier eugenicists, he also thought that because of developments in medicine and other forms of cultural advancement, such as the improvement of agriculture, the selection of the genetically fittest at the hand of nature was no longer occurring at a high enough rate to breed out undesirable traits. For example, diabetes was being propagated in the human gene pool because diabetics were surviving to reproductive age, while once they would have died in youth. The sum of these mutations he called the genetic load, and he and many others saw it threatening the future existence of humanity.

As the load increased, according to Muller, the average human "would have to be given a superlatively well-chosen combination of treatments, training, and artificial substitutes just to get by. The job of ministering to infirmities would come to consume all the energy that society could muster for it, leaving no surplus for general cultural purposes."[19] Ultimately, the "germ cells of what were once human beings would be a lot of hopeless, utterly diverse genetic monstrosities." He concluded that by then society would wish it had done something earlier because the "refashioning of these pitiful relics into human form would be a far more difficult task than the synthesis of human beings out of raw materials."[20]

Muller's paper had "more or less immediate effects" on the field of population genetics, prompting a plethora of studies to classify the loads, estimate mutation rates, and determine the effect of altered mutation rates on estimations of fitness.[21] By 1970 it was claimed that "population genetics today is largely a science dealing with the 'genetic load.'"[22] Avoiding the load, a specification of the more general end of beneficence, thus became one of the ends for which scientists argued.

The More Implicit End: Providing a Meaning and Purpose of Life

If reduction of the genetic load had been the only end debated among scientists, the history of the debate over HGE would have turned out much differently. Determining the extent of the genetic load is clearly in the jurisdiction of scientists—who else would have the skills to determine the rate of mutation and its effect on the population? Moreover, few members of the public would have objected to the end of saving the human species from genetic destruction. However, scientists at the time were trying to expand their jurisdiction beyond the discovery of facts about nature to a more active role in public affairs.

Following the horrors of World War II, these geneticists were part of a broader community of scientists attempting to find the meaning and purpose of human existence in evolution and biology, to create a secular "'scientific' foundation upon which to reestablish our system of ethics and to rest 'our most cherished hopes.'"[23] If human beings could no longer look outside nature for purpose and direction—as most theologies had done—the foundation for ethics was to be found in the "objective" facts of evolution, such as "greater complexity, biological efficiency and adaptive flexibility."[24] Society could no longer use a discredited traditional religion for its base: a new, human-based scientific religion was needed to save society. HGE would be a means to this end.

Sir Julian Huxley (1887–1975) was a British-born biologist, grandson of the famous defender of Darwin, T. H. Huxley, and brother of the novelist Aldous Huxley (*Brave New World*).[25] A lifelong advocate of eugenics, he was a leader in the shift from the mainline eugenics of his British forebears to the reform eugenics of the 1950s.[26] Like most of the other scientists in this community, he had particular scientific interests (in his case, birds), but also aspirations far beyond the lab bench. He was, in the estimation of one of his biographers, seeking "to create a religion of evolutionary humanism based on biology, and to bring these efforts to fruition through popularization and liberal political action."[27] A large part of "evolutionary humanism" was the rejection of traditional religion, and its replacement with science and rationalism:

> Evolutionary man can no longer take refuge from this loneliness by creeping for shelter into the arms of a divinized father-figure whom he has himself created, nor escape from the responsibility of making decisions by sheltering under the umbrella of Divine Authority, nor absolve himself from the hard task of meeting his present problems and planning his future by relying on the will of an omniscient but unfortunately inscrutable Providence. . . . More immediately important, thanks to Darwin, he now knows that he is not an isolated phenomenon, cut off from the rest of nature by his uniqueness . . . he is linked by genetic continuity with all the other living inhabitants on his planet.[28]

Thus, for Huxley, biology restores meaning to a world where theology has lost its relevance, for biology "reinstates man in a position analogous to that conferred on him as the Lord of Creation by Theology," offering "hope and meaning to human existence."[29] Muller had extremely similar views.[30]

Another influential member of this community, Theodosius Dobzhansky (1900–1975), a Russian-born population geneticist, also "saw himself as aiding the evolution of religious thinking in a scientifically evolving world."[31] He believed that "although a biologist may do his research on mice, Drosophila flies, plants, or bacteria . . . the ultimate aim should be to contribute toward the understanding of man and his place in the universe."[32]

Like Huxley, Dobzhansky was a follower of the unconventional French Jesuit theologian Pierre Teilhard de Chardin.[33] Teilhard de Chardin, whose work was banned by the Roman Catholic Church until after his death, believed that all biological evolution had a cosmic purpose and was evolving toward the "omega point," essentially the Christian eschatological kingdom. Biology and evolution, in his view, make meaning for humans—and not in a particularly traditional Christian manner. Following this type of theological notion, Dobzhansky wanted to create an entire worldview that would provide meaning. In his 1967 work *The Biology of Ultimate Concern,* Dobzhansky takes liberal Protestant theologian Paul Tillich's definition of religion ("religion, in the largest and most basic sense of the word, is ultimate concern") and mixes it with Teilhard de Chardin's evolutionary worldview to come up with an ultimate goal or direction for evolution.[34]

Dobzhansky saw the scientific control of evolution as a solution to the problem of meaninglessness inflicted by the death of a transcendent God. "Man and man alone knows that the world evolves and that he evolves with it," he wrote in 1962, and "the hope lies in the possibility that changes resulting from knowledge may also be directed by knowledge. Evolution need no longer be a destiny imposed from without; it may conceivably be controlled by man, in accordance with his wisdom and his values."[35]

Debating the Ends of the Human Species

Toward what ends should humans biologically evolve? It is hard to imagine a more substantive question, yet it has been largely ignored in more recent debates about HGE, despite the increasing ability to actually forward these ends. The reform eugenicists almost all believed that human beings control our genetic destiny, and that we should determine the ends to which we would use HGE technology. Even geneticist James F. Crow of the University of Wisconsin, a member of this community whose work tended to shy away from social or ethical claims, wrote that "the issue is not whether [humans are] influencing [their] evolution, *but in what direction.*"[36]

One might say that the debate about ends in this community involved those who wanted to work toward the end of stopping genetic load and those who wanted to also provide meaning to humanity through human self-perfection following some image of the "ideal man." Although they generally justified HGE by reference to the end of preventing the genetic load—an end that was probably universally held among these scientists—many of these reform eugenicists simultaneously pursued the ends of encouraging the propagation of "ideal man" traits. They knew that this end was not universally held, but hoped to make it so. For example, Muller was reported to have chosen the term *load* in his 1950 paper to avoid the term *eugenics* and thus to surreptitiously pursue his primary end of species improvement.[37]

The list of positive human traits that the reform eugenicists believed should be pursued reflected a strong faith in the reductionistic genetic determinism of their mainline forebears. Even in 1963, after much debunking of crude genetic beliefs about genetic determinism had occurred, Huxley would still say that he wanted to use eugenics to increase "man's desirable genetic capacities for intelligence and imagination, empathy and co-operation, and a sense of discipline and duty."[38]

For Muller, Crow, and Huxley, positive eugenics took the form of encouraging the reproduction of just a few of the most genetically gifted people. "Since society owes so much to a small minority of intellectual leaders," argued Crow, "a change in the proportion of gifted children would probably confer a much larger benefit on society than would a corresponding increase in the population average."[39] Reflecting a broader view of what is genetically determined than the "intelligence" implied by Crow, Huxley thought that "the great and striking advances in human affairs, as much in creative art and political and military leadership as in scientific discovery and invention, are primarily due to a few exceptionally gifted individuals," such as "Newton and Darwin, Tolstoy and Shakespeare, Goya and Michelangelo, Hammurabi and Confucius."[40]

The First Means

Assuming that the debate on the ends of humanity could be settled, how could these ends be achieved? At this point the only technology available was the selective breeding advocated by the mainline eugenicists, but with a voluntary twist. The success of a voluntary eugenic program depended upon the "genetically defective" persons in the population believing that they had an obligation not to reproduce in order for humans to seize control of their own evolution—and to save humanity from the apocalypse of the accumulating genetic load. People would have to realize that they had a "duty . . . to exercise their reproductive functions with due regard to the benefit or injury thereby done to society."[41] People would also have to give up other values, such as having children genetically related to themselves, and replace this with the "justifiable pride in accomplishment of a far more exacting and laudable kind than that of procreation: namely, having made children of especially high endowment possible and having brought them up."[42]

Muller began advocating a new means beyond encouraging certain people to forgo having children. His "germinal choice" plan relied on the newfound ability to freeze human sperm. Men who recognized their own inferior status would not insist on reproduction; their wives would instead receive artificial insemination of sperm from men with the "most outstanding" genetic qualities—men who, it was assumed, lacked the multiple mutations that were causing the load. This new means offered a new degree of precision: instead of assuming that any member of the upper classes was "outstanding" as the mainline eugenicists had done, Muller proposed that a donor's sperm would be frozen for twenty years to see how the donor actually turned out.

Perhaps even more optimistic than Muller, Huxley thought that people would flock to this technique, given the chance. If people were educated about heredity, and if legal restrictions on receiving information about sperm donors were lifted, people would naturally use germinal choice to have the "best" children. A "blind and secrecy-ridden A.I.D. [anonymous insemination by donor]" would be superseded by "an open-eyed and proudly accepted E.I.D. where the E stands for Eugenic."[43] Muller eventually founded a sperm bank in California named the Foundation for Germinal Choice to further his plan, despite criticisms that this would not achieve his ends.[44]

In sum, the scientists were conducting a largely substantively rational debate about the ends that should be pursued with HGE. With the exception of species self-preservation through the avoidance of the genetic load, no one thought the large number of other ends suggested were universally held or commensurable. Indeed, they all thought part of the problem was to convince the public to share their ends. These scientists, however, were using a generally formally rational link between means and ends, arguing for different HGE techniques on the basis of whether the techniques would maximize their favorite ends. This too would change in a substantive direction as the debate evolved.

The First Cracks in the Scientists' Jurisdiction

Scientists had essentially claimed jurisdiction based on an analogy of means: the determination of what were "good" genes was a matter of scientific expertise. A partial analogy of ends was also implied, with scientists attempting to develop means that would forward human health by eliminating the load. The pursuit of the end of providing meaning and purpose through the genetic perfection of humanity was more of a stretch, and was thus vulnerable to challenge. It was here that the first cracks in their jurisdictional claim would form.

Germinal choice is not HGE as I have defined it; it is more akin to assortative mating than to the chemical changing of the genome. Even proponents such as Muller believed that change in the human germline using these methods would be extremely slow because selecting "better men's" sperm was an imprecise business. By the early 1960s, however, members of this community had begun to see that there might be chemical means for changing the human genome, suggesting a new degree of precision in engineering. Instead of breeding people with "desirable" traits, they hoped that genes with known functions could be "spliced" or "surgically repaired." With this new precision, defining the "ideal genetic man" became even more important. Also at this time the scientists began to question whether they could determine the ends of society without consulting society more broadly. But instead of facilitating a debate about ends in the public, they quickly realized that such a discussion would not be in their interest.

New Means and the Recurring Question of Ends

The scientists discussed so far were mostly population biologists and geneticists whose techniques were observational. However, the new fields of molecular biology, cell biology, and biochemistry offered a reductionist approach to the questions addressed by Muller, Huxley, Dobzhansky, and the others. In the words of British eugenicist Lionel Penrose, speaking to a conference on human genetics in 1959, the observational methods of people like Muller were in decline, and "biochemical methods are now in the ascendant."[45] Although Muller and his contemporaries had defined the ends, they were quickly shunted to the scientific wayside in the development of the technological means to achieve these ends.[46]

If the reform eugenicists reduced the ultimate cause of human behaviors to a person's genome, the new breed of scientists took this one step further and reduced the human genome—and implicitly behaviors—to the genome's parts, the genes, and to the chemical composition of each gene. Given this new view, and with some advances in biochemistry, a new chemical manipulation of genes might be possible—opening up the possibility of new techniques of which the reform eugenicists could not have dreamed.

This new type of scientist rapidly entered the debate as the older scientists had defined it. In 1963 a meeting took place that had a strong impact on the HGE debate, bringing together "the now aging eugenicists and the younger generation of molecular biologists."[47] Tracy Sonneborn, a colleague of Muller's from the biology department at Indiana University, organized a conference in April at Ohio Wesleyan University, with funding from the National Science Foundation, that was focused on the new technological possibilities for more precise human control of evolution.[48] At first glance, this debate appears unchanged from that of Muller, Huxley, and the other observational scientists. The title of the conference proceedings, *The Control of Human Heredity and Evolution,* speaks to the ends pursued, as does Muller's keynote address, "The Means and Aims of Human Genetic Betterment."[49]

Throughout the conference, however, speakers emphasized the precision and power of the new techniques, which, compared with previous methods of genetic control, offered the possibility of a "direct attack on the human germ plasm."[50] A geneticist stated that "it is generally agreed among biologists that the means of [molding human nature] in a very minor way have already been available and in use for some time, but that it is reasonable to expect developments of these means on a colossal scale." The only thing that remained in dispute was "the time required."[51]

Although the fears of the genetic load generated a sense of urgency, the effects were not even hypothesized to occur for hundreds of years. Moreover, Muller's germinal choice was the baseline technology in this community, and it is easy to see why it would not have a rapid effect in the ends pursued. For one thing, it is not very efficient, since only half the genes (the sperm) would be chosen. For another, even if one accepts the eugenicists' view of the heredity of intelligence, artificially inseminating a woman with the sperm of Einstein would tend only to produce a more intelligent child. But using the radically new techniques, it was hypothesized that the genes that made Einstein intelligent could be engineered into others, suggesting a new precision.

With this new precision came a newfound sense of urgency. Sonneborn prefaced his volume with reference to Aldous Huxley's *Brave New World* in such an innocent manner that it is possible that he was the first person to make the connection between these new molecular biological techniques and the famous novel that became a hackneyed reference by the 1990s. Not only are many of Aldous Huxley's predictions apparently coming to pass, wrote Sonneborn, but they are coming at a totally unexpected rate. Noting that fifteen years after the publication of *Brave New World,* Huxley himself changed his estimate of the time of its occurrence from six hundred years to one hundred years, Sonneborn stated that "so rapidly is the pace of science accelerating that some of the newer prophets proclaim that the new and more extreme applications to man may be just around the corner."[52]

The new urgency raised an unresolved debate in this community over the proper ends of evolutionary progress. Given the diagnosed problem inherited from their predecessors—that humanity had lost meaning and purpose—and the general end that humans should control their own evolution, it suddenly occurred to these scientists, who had the means to do something about it, that the direction of "progress" remained unclear. Moreover, looking at the possibility of this radically new power, the scientists lost their confidence that they could lead society. For the first time in these data, scientists acknowledged that how this technology "will be used obviously will not be decided by scientists alone." But the question was one that should not be "decided alone by professional politicians or by theologians or by philosophers or by moralists. It should be decided by an enlightened and broadly based public opinion." The point of the symposium was then to contribute to the formation of public opinion and to "urge the public to assume its responsibility of contributing as wisely as it can to the formation of public policy."[53]

This democratic impulse was unusual enough that viral geneticist Salvador Luria felt he had to remind the panel (exclusively scientists) that to "claim the right to decide alone" would be "to advocate technocracy" and would be unacceptable. Luria thought that a solution would be for the "United Nations as well as the National Academy of Sciences of the United States to establish committees on the genetic direction of human heredity."[54]

Public attention was beginning to increase, and interest in the debate was growing outside the scientific community. By the late 1960s two influential members of this community had produced books for the general public that summarized the debates among scientists to that point, including the similar claim that human beings inevitably control evolution, so we might as well plan and do it better. Regarding the direction of "progress," they concluded that these issues were too important to be left to scientists, and that society should collectively decide the ends that science should pursue.[55]

Although it was not clear at the time, the suggestion that the public should determine the ends of genetic science would pose the first threat to the scientists' jurisdiction. Alerted to the magnitude of the decision they would have to make, the public began paying attention to the debate, and increasingly became the body that would grant jurisdiction. Scientists quickly realized that direct public involvement was actually a threat to their jurisdiction. No influential scientist in the data after 1963 echoed Luria's call for the ends of HGE to be determined by *unmediated* public opinion. While the debate about HGE began these pressures, it was a related debate over the technique of in-vitro fertilization (IVF) that reinforced them.

Test Tube Babies and the Growing Challenge to Scientific Jurisdiction

In 1969 embryologists Robert Edwards and B. D. Bavister and obstetrician Patrick Steptoe described an experiment they had conducted:

> Oocytes were released from Graafian and smaller follicles into a medium composed of equal amounts of Hank's solution containing heparin, and medium 199 . . . and buffered (pH 7.2) with phos-

> phate buffer.[56] Penicillin (100 iu/ml.) was added to these media. . . . After 38 h in culture many of the oocytes had extruded their first polar body and reached metaphase of the second meiotic division (metaphase—II). Ejaculated spermatozoa were washed once with Bavister's medium to remove seminal plasma, and were then re-suspended at a concentration of 10^6/ml. in more of the same medium. . . . Oocytes were washed through one or two droplets of Bavister's medium and then pipetted into the sperm suspension. . . . Fifty-six human eggs were inseminated.[57]

More than thirty-five years earlier, Aldous Huxley, brother of Julian, had begun his most famous novel with a scene in which a factory director gives a tour to a group of students:

> "These," he waved his hand, "are the incubators." And opening an insulated door he showed them racks upon racks of numbered test-tubes. "The week's supply of ova. Kept," he explained, "at blood heat; whereas the male gametes," and here he opened another door, "they have to be kept at thirty-five instead of thirty-seven." . . . Still leaning against the incubators he gave them . . . a brief description of the modern fertilizing process; spoke first, of course, of its surgical introduction . . . continued with some account of the technique for preserving the excised ovary alive and actively developing; passed on to a consideration of optimum temperature, salinity, viscosity; referred to the liquor in which the detached and ripened eggs were kept; and, leading his charges to the work tables, actually showed them how this liquor was drawn off from the test-tubes; how it was let out drop by drop on to the specially warmed slides of the microscopes; how the eggs which it contained were inspected for abnormalities, counted and transferred to a porous receptacle; how . . . this receptacle was immersed in a warm bouillon containing free-swimming spermatozoa . . . and how, after ten minutes, the container was lifted out of the liquor and its contents re-examined; . . . how the fertilized ova went back to the incubators; where the Alphas and Betas remained until definitely bottled; while the Gammas, Deltas and Epsilons were brought out again, after only thirty-six hours, to undergo Bokanovsky's Process.[58]

It is easy to see why the common members of this community thought that when discussing Edwards's research, they should be discussing Aldous Huxley's *Brave New World* at the same time (see fig. 2).[59] It was not just the similarity of the reproductive processes that suggested the connection, however: Huxley's book is also a warning about totalitarian control of society as well as of reproduction. Although Edwards's articles mentioned that IVF would help infertile parents have children "of their own," he was embedded enough in the dominant arguments in this era to see the implications of this work for allowing humans to control the direction of their own evolution. For example, noting recent research in mice, he pointed out that if his research proved successful, combining the genes either from two different animals or from different persons might soon be possible in vitro, leading to precise genetic control, the possibility of which had begun to stir the public.[60]

If Edwards's remarks on the possible uses of his new technique for implementing the control of evolution were brief and suggestive, others quickly expanded upon their significance. H. Bentley Glass (b. 1906), an influential member of this community, was an internationally distinguished geneticist.[61] Glass's writings are first and foremost solidly consistent with the attempts by his fellow geneticists of the time to forward a new meaning and purpose for humanity. "Ethical values do grow out of the biological nature of man and his evolution," he stated at one point. "The sciences may contribute to the resolution of our crisis of values," he claimed at another.[62]

Like the other scientists of this era, Glass was concerned with the genetic load, control of human evolution, and debate over the "good man" that we should engineer for the future. Like Muller's, Glass's "good man" would have "freedom from gross physical or mental defects, sound health, high intelligence, general adaptability, integrity of character, and nobility of spirit."[63]

Glass did not see IVF as primarily a means of helping infertile couples have children. Rather, like his fellow scientists in this era pursuing the end of species perfectionism, he anticipated that IVF methods would provide "a continuous, inexhaustible supply of the germ cells derived from selected male and female donors, carefully chosen on the basis of the demonstrated high quality of the children produced during their own lifetimes."[64] Exemplifying the diversity of the ends pursued in these early debates, Glass also argued that human beings would then have to genetically engineer themselves because "man requires a challenge and a quest if he is to avoid boredom."[65]

The linking of Edwards's research with Huxley's *Brave New World* was not a way of praising Edwards. References to *Brave New World* were a warning against the ends and means of the eugenicists. Indeed, the book was initially written in reaction to a work by eugenicist J. B. S. Haldane.[66] With eminent scientists such as Glass making claims that Edwards's work provided the means for the furtherance of the ends pursued by the eugenicists, it was not surprising that the IVF issue came to be seen as part of a broader controversy.

The Growing Public Threat to the Scientists' Jurisdiction

Sociologist Amitai Etzioni wrote in 1973 that "no other recent development in biological engineering has raised as much doubt among the public as that involving experiments in which conception has been carried out, and gestation fostered, in a test tube."[67] The enormous public outcry over IVF, as well as over other technologies, caused scientists to adopt a new defensiveness about their ability to determine the ends to pursue without public input—what they referred to as scientific freedom. ("Scientific freedom" is simply, in my terms, full jurisdiction over experimentation given to scientists by the public.) It may be hard to perceive from public debates in the present, but there was a time when the public, enamored with scientific progress, did not question the implicit jurisdiction they had given scientists over the ethics of their own research. By the 1960s, however, the public had determined that scientists' ends did not necessarily match their own. This was true not only in genetics, but in other areas as well. According to sociologist Kelly Moore, "by the middle of the 1960's, scientists were being blamed for, among other things, the war in Vietnam, alienation, a decline in the quality of life even as material prosperity increased, and a multitude of environmental problems."[68] It is clear that the anti–Vietnam War movement and the New Left were partly responsible for the disenchantment with scientific involvement in public affairs, as were the consumer and ecology movements.[69] A younger generation of molecular biologists, seemingly influenced by these movements, began to question the morality of the science itself.[70] By the mid-1960s "public confidence in technological development as the key to social progress gave way to disenchantment."[71]

It is in the public's questioning of the scientists' jurisdiction over the ethics of IVF and HGE that Glass's other contribution to this debate can be found. Glass at the time was a preeminent defender of the prerogatives of scientists, which were beginning to be challenged in the wake of Edwards's experiments. In general, he had a technocratic perspective, believing that the public had no role in controlling scientific research.[72]

Scientists such as the participants in the 1963 meeting on HGE had blithely called for public involvement in setting ends. Now to the great shock of the scientists it was discovered that, given the choice, the public might not give jurisdiction to scientists. Indeed, it appeared to the scientists that the public was questioning the wisdom of the entire scientific project. Ward Madden, chair of the lecture series that Glass's book was originally written for, displays the scientific community's shock at this disenchantment with its ends:

> Science is exuberantly flooding the world with knowledge. . . . transforming both our minds and our daily existence. . . . revolutionizing our concepts of the universe, life, man, and man's destiny. . . . [and] producing a thoroughgoing technological reordering of our habits and activities.
>
> And yet, strangely, there is a countercurrent to the scientific tide . . . a "counter culture" [that has placed science] on the defensive, even in the midst of its success.[73]

Suddenly, it appeared to the scientific establishment that these criticisms were threatening the laboratory-based core of their jurisdiction, which had been unchallenged for decades. "Science and technology [are] under attack," stated the president of the National Academy of Sciences, Phillip Handler, in 1970, and "with that attack has come a drastic decline in the scale and scope of our national scientific endeavor, a fall of perhaps 30% since FY 1968."[74] In the same speech Handler reiterated the expansionist ends defined by Julian Huxley: that science should provide meaning to life and that the public should adopt this new faith. "If our spiritual faith is somewhat shaken, whence do we turn?" he began. "Science may be an inadequate substitute—but it is the substitute we have. . . . Who can help but find deep satisfaction in contemplating the ingenious elegance of the manner in which the complementary double-stranded structure of DNA solves the most profound of all questions concerning the nature of life?"[75]

It is ironic that in a speech bemoaning the threats to the core jurisdiction of the scientists, Handler would continue the efforts at expansion that had made the public question the core jurisdiction. It was not simply these technological inventions that had sparked a public debate over the ethics of their use, but the ends to which the scientists intended to put them. Like it or not, the public was questioning the full jurisdiction it had implicitly granted to scientists, and this provided an opportunity for other professions to have their arguments heard and thus to compete with the scientists. If the scientists had restricted the end pursued to saving the human gene pool from the genetic load, the HGE debate would have been much different, because no other profession would have challenged this end. However, right up to the 1970s scientists were pursuing the end of creating meaning and purpose for humanity, which sounds like an end theologians pursue in their jurisdiction. Some theologians perceived it as such.

The Theological Challenge

The final community under consideration is disproportionately made up of theologians (see fig. 2).[76] These were liberal theologians, not traditionalist Catholics or evangelical Protestants.[77] By the mid-1960s liberal theology was at a high point in societal influence, as liberal Protestant and Catholic clergy took an increasingly strong role in the civil rights and antiwar movements.[78] Like scientists, liberal theologians were gaining jurisdictions, so it was reasonable for them to feel that they too had something important to say about the ethics of HGE, especially because the scientists had made their arguments almost theological. The average theologian in the debate was claiming jurisdiction based on a metaphorical link between the ends of HGE as defined by the scientists and the ends in the home jurisdiction of the theologians. This effort can be observed by looking at the writings of the influential members of this community.

Paul Ramsey (1913–88), a Methodist theologian and professor of religion at Princeton University, recognized that the scientists' arguments posed a challenge to theology's core jurisdiction.[79] As the scientists themselves had noted, theology had traditionally had under its jurisdiction work such as helping people define the meaning and purpose of humanity, and now the scientists were impinging on that work. For Ramsey, it was not simply a matter of opposition to the emerging technologies. Throughout his work he stated that he was actually opposed to what he called a "surrogate theology" of the "cult" of "messianic positivism" led by scientists such as Muller. Using the phrase "playing God" to summarize the form of argumentation used by the scientists—what he called their worldview—he recognized that the scientists' end was to provide the meaning of life:

> Taken as a whole, the proposals of the revolutionary biologists, the anatomy of their basic thought-forms, the ultimate context for acting on these proposals provides a propitious place for learning the meaning of "playing God"—in contrast to being men on earth.
>
> [The scientists have] "a distinctive attitude toward the world," "a program for utterly transforming it," an "unshakable," nay even a "fanatical," confidence in a "worldview," a "faith" no less than a "program" for the reconstruction of mankind. These expressions rather exactly describe a religious cult, if there ever was one—a cult of men-gods, however otherwise humble. These are not the findings, or the projections, of an exact science as such, but a religious view of where and how ultimate human significance is to be found. It is a proposal concerning mankind's final hope. One is reminded of the words of Martin Luther to the effect that we have either God or an idol and "whatever your heart trusts in and relies on, that is properly your God."[80]

The average members of this community looked to others besides Ramsey to question the ends of the scientists. Leon Kass (b. 1939) was during this period the executive secretary of the Committee on the Life Sciences and Social Policy of the National Research Council of the National Academy of Sciences. Trained as a biochemist and as a physician—and despite his employment during this era at the paragon of the scientific establishment—he was one of the most forceful critics of what he called the "intellectual foundations of the modern scientific conception of the world."[81] Like Muller and Huxley, he thought society was in moral crisis, but unlike them, he thought that the crisis was due to the unwillingness of scientists who argued for scientific freedom to determine standards (ends) to guide technological development (means). He stated, "We are witnessing the erosion, perhaps the final erosion, of the idea of man as something splendid or divine, and its replacement with a view that sees man, no less than nature, as simply more raw material for manipulation and homogenization. Hence, our peculiar moral crisis. We are in turbulent seas without a landmark precisely because we adhere more and more to a view of nature and of man which both gives us enormous power and, at the same time, denies all possibility of standards to guide its use."[82]

Similarly, Henry Beecher (1904–76), an esteemed professor of anesthesiology at Harvard Medical School, critiqued the end of the greatest good for the greatest number used by many medical research scientists at the time. The pursuit of this end had justified experiments on people without their consent or knowledge, in the name of forwarding beneficial knowledge for humanity.[83] Beecher's text "Ethics and Clinical Research" (1966) focused quite narrowly on convincing scientists to also consider the end of promoting individual autonomy of their research subjects by requiring their informed consent. Ramsey and Kass had both written on the same subject—following Beecher's lead.[84]

When an influential debater such as Ramsey appears in a community, it should be no surprise that his antagonist is there too. Protestant theologian Joseph Fletcher (1905–91) was a long-term antagonist of Ramsey's in Christian Social Ethics debates.[85] Fletcher had been dean of St. Paul's Cathedral in Cincinnati for five years in the late 1930s and had spent twenty-six years as a professor of social ethics at Episcopal Theological School in Boston. There he became an influential and prolific author, and served as the president of the American Society of Christian Ethics. Ramsey and Kass saw Fletcher as one of the "theologians-turned-technocrats" who believed, in Kass's words, that "man, with the dead God as his co-pilot, is to fly off into the wild blue yonder of limitless self-modification."[86] That is, to Ramsey and Kass, Fletcher was an apologist for the scientists' ends. Fletcher believed that "laboratory reproduction is radically human compared to conception by ordinary heterosexual intercourse" because "man is a maker and a selector and a designer, and the more rationally contrived and deliberate anything is, the more human it is."[87]

Why would the average theologian writing about HGE find these authors particularly influential? First, and most obviously, three of these four influential authors either were theologians themselves or engaged in debates with theologians, so the common authors would have been familiar with their texts and their arguments.[88] Second, the assumption using my method is that these influential authors' arguments either gave the average author something to build upon or were symbolically powerful. They were "important" in some way. If we look at the commonalities across the influential members of this community, we can see what was important: to push the debate, which had become somewhat formally rational with the younger scientists, back toward a more purely substantively rational form.

Defining Ends

Reform eugenicists such as Muller and Huxley had been debating ends. Later scientists, such as Joshua Lederberg and Edwards, had stopped being explicit about the ends of their research, which had raised so much controversy and had brought people like Ramsey and Kass into the debate in the first place. Through arguments about scientific freedom, where the ends were set by each scientist if they were set at all, the more recent scientists seemed to be advocating a purely formal rationality: letting their means (the technological possibilities) drive the ends.

Defining ends, not means, has been the strong suit of theology. The efficacy of theological means (such as prayer) almost by definition cannot be demonstrated, particularly by the modern canons of proof, so theology has focused on what the ends of society and individuals should be. Consistent with the strength of the theological jurisdiction, common theologians would select these influential authors because they concentrated on bringing into the debate a discussion of the ends that the means of HGE would forward. Kass, for example, recognized the lack of ends in the scientists' arguments and described what was wrong with this form of argumentation with a story: "Good afternoon, ladies and gentlemen. This is your pilot speaking. We are flying at an altitude of 35,000 feet and a speed of 700 miles an hour. I have two pieces of news to report, one good and one bad. The bad news is that we are lost. The good news is that we are making very good time."[89]

Society was "flying" toward perfecting the means for modifying the human genome, but it did not know why it was doing so. That is, scientists were seemingly on the verge of inventing HGE, cloning, and IVF, yet the end that these means served had not been established beyond satisfying human curiosity. "We are told that new technologies are coming and that we should attend, consider, and adjust to the social consequences," Kass wrote in 1972. However, "this formulation treats new developments as automatic, as insensitive to human decision and choice. But only a slavish mind and a slavish society let the means dictate the ends."[90]

Kass's statement here is essentially a reiteration of the Weberian thesis: that the maximization of means, as if they were ends, is characteristic of the formal rationality inherent in modernity. However, unlike Weber's seeming resignation in the face of modernity, Kass believes that the science can be redeemed, if the ends can simply be defined, because "to the extent that we are or seek to be rational men, we must insist, at least in our thought and discussion, on beginning with a serious deliberation about our ends and purposes. . . . [because] it is indeed the height of irrationality triumphantly to pursue rationalized techniques while insisting that ends or purposes lie beyond rational discourse."[91] Similarly, Ramsey concluded that Lederberg's arguments in favor of HGE and cloning were "simply an extrapolation of what we should do from what we can do. [For Lederberg] there are really only technical questions to be decided."[92]

What ends would the theologians argue for? There was a tension running through the theological community on the appropriate approach to take here—a tension found even in the writings of some theologians themselves. During this period theologians would examine an issue such as HGE, and determine an end that should be pursued. How that end was expressed, however, was an important issue. Some writers expressed ends in theological language, and some translated them into secular language, in hopes of reaching a broader audience.

Ramsey's work reflected this tension. For example, as stated above, he was one of the promoters of the argument that society should forward the end of autonomy in issues of human experimentation, through pursuing the consent of the participants in the experiment—a translation of his pursuit of the ends of agape and covenant with God. According to one reviewer, he made use of consent because he "sought to find a language accessible to as many people as possible despite their theological convictions and 'consent' appeared to cross over communities and traditions."[93]

The basic method was to examine each issue in a theological way, determine the ends relevant to the argument, and then debate these ends. In the words of one of the Protestant theological consultants to the President's Commission (the topic of chapter 4), "a rather typical, methodological approach for Protestants" consists in "isolating a number of general themes that can be supported biblically and then suggesting the tendencies or directions for decision making that seem appropriate in the light of these themes."[94] This approach led to myriad ends because every issue—abortion, HGE, human experimentation—would lead to different theological debates and different ends. Note that this translation is not commensuration. The richness of the theological end could generally survive translation to secular language because, at this point in the debate, any end that could be expressed in the English language could be used. Commensuration of theological ends, which would occur later on, involves the throwing away of information that does not fit into predetermined ends. To use a spatial metaphor, the theologians could create any (theological) shape they wanted and put it through a (secular) hole of the same shape. Later, theological square shapes would of necessity be stuffed into (secular) circular holes, losing part of their form as they passed through. These secularly stated ends were not assumed to be held by the population; the point was to convince the population to adopt them.

Ramsey, like the other theologians, also pursued untranslated, explicitly theological ends. One of these ends, agape, is a critical example because it illustrates other ways in which the theological debate was substantively rational. The debate over how to achieve this end moved the debate in an even more substantive direction. Agape (love of neighbor or Christian love) is for Ramsey "the basic rule or principle of Christian ethics."[95] He did not consider this end to be universally held, though he would have liked to convince enough people to make it so. Fletcher was also explicit that his end was agape, but he argued with Ramsey for decades over how it could best be accomplished. After becoming a bioethicist, Fletcher would translate the particularlist end of agape to the secular utilitarian end of the "greatest good for the greatest number." However, he kept alive the debate over the link between means and ends.[96]

The Link between Means and the End of Agape

Although they had not explicitly articulated a method of decision making, the scientists used arguments that implied a formally rational link between means and end. HGE was good or bad depending on whether it maximized their end compared with other means, so means such as Muller's germinal choice were compared with forms of HGE based on the efficiency of reaching the end of genetic control. Moreover, the medical research scientists critiqued by Beecher for experimenting on people without their knowledge or consent clearly thought that there were few means that were not justified by the end of providing medical knowledge to humanity. In contrast to the scientists, the theologians pushed the debate in an even more substantive direction by calling for the evaluation of means such as HGE by the standard of whether they were consistent with ends, regardless of their good or bad consequences.

Theologians' attraction to debates about the relations between means and ends is not surprising, given that Ramsey and Fletcher had been debating the topic back to the 1950s.[97] Again, the debate was over how to forward the end of agape, which both Ramsey and Fletcher agreed was the fundamental priority in Christian ethics.[98] Critics accused Ramsey of advocating a "love deontology,"[99] a belief in virtually exceptionless rules that were consistent with agape. Thus Ramsey argued for a rule that the act of marital infidelity was inconsistent with his ends, and should not be tolerated, even if another good end were to be forwarded by it. Fletcher disagreed. Ramsey then accused Fletcher of being "simply a sign of our times," and argued that by ignoring rules, under Fletcher's system agape would become "a slave to 'other than uniquely Christian sources.'"[100]

Fletcher's view on the link between means and ends reflects the pure calculation of formal rationality used by the scientists, as well as the view that there is no means that should not be considered to see if it maximizes the end:

> If the greatest good of the greatest number (i.e. the social good) were served by it, it would be justifiable not only to specialize the capacities of people by cloning or by constructive genetic engineering, but also to bio-engineer or bio-design para-humans or "modified men"—as chimeras (part animal) or cyborg-androids (part prosthetes). I would vote for cloning top-grade soldiers and scientists, or for supplying them through other genetic means, if they were needed to offset an elitist or tyrannical power plot by other cloners—a truly science-fiction situation, but imaginable. I suspect I would favor making and using man-machine hybrids rather than genetically designed people for dull, unrewarding or dangerous roles needed nonetheless for the community's welfare—perhaps the testing of suspected pollution areas or the investigation of threatening volcanoes or snow-slides.[101]

Although Kass, Ramsey, and Beecher would calculate maximally efficient means to forward ends from among those means considered to be consistent with ends, in contrast to Fletcher they believed that there were some means you did or did not use because they were not consistent with ends, regardless of the consequences. Kass, for example, preferred "arguments from principle concerning intrinsic rightness or wrongness, arguments which abstract from the difficult task of predicting and weighing consequences."[102]

A typical example of Ramsey's substantively rational "love deontology" was made at the 1965 Nobel Conference, appropriately titled "Genetics and the Future of Man." Responding to Muller, he argues that his own ends would require a different evaluation of the morality of the means of HGE than that made by the scientists. If we were to stop "intending the world as a scientist" and begin "intending the world as a Christian or as a Jew," he asked, what would be our ethics regarding genetic control?[103] The first difference would be that one of Muller's ends—preventing the apocalypse of the genetic load—would be seen as less important because "religious people have never denied, indeed they affirm, that God means to kill [the human species] in the end, and in the end He is going to succeed."[104]

Given that saving the human species is not an "absolute command of nature or of nature's God . . . [humans] will not define *right* merely in terms of conduciveness to the good end; nor will [they] decide what *ought to be done* simply by calculating what actions are most likely to succeed in achieving the *absolutely imperative end* of genetic control or improvement."[105] The means of HGE could then be wrong in and of themselves, not just because they did not maximize ends, but because, as Ramsey concluded, the means were inconsistent with proper human ends.[106] He similarly argued in other contexts that it was always wrong to experiment on people without their consent, regardless of the good consequences, because to do so was inconsistent with the end of autonomy or respect for persons.[107] In sum, the theologians who entered this debate considered Ramsey, Kass, and Beecher to be important to their arguments, and one of the reasons they were important was this focus on substantively rational links between means and ends.

The Theologians' Description of, and Solutions to, the Proposed Means of HGE

The theologians, reacting to the jurisdictional challenge of the scientists, had brought a new focus on defining the ends of HGE as well as a debate on the proper link between means and ends. Their efforts pushed the debate back toward the substantively rational form of argumentation from which scientists had recently been moving away.

Lacking the need to calculate consequences, the majority of these authors defined the problem as broadly as had the eugenicists, retaining the sense of a thick debate. The problem was not HGE itself, but the ends that HGE purported to forward or the lack of articulated ends at all. To look at the texts of the influential authors in this period, particularly the theologians, is to be given a tour of almost every related concern at the time—abortion, HGE, IVF, single parenthood—all tied to the deeper problem of the ends these acts were forwarding. This would change in later periods as the debate thinned.

Both the scientists and the theologians were largely concerned with society, not with the individual. The strategy of making the individual the unit of analysis came later with the need for more effective calculations.

Moreover, the span of time considered in the typical argument would not be surpassed in future debates; both the scientists and the theologians were interested in the effect of HGE on the far future.

The one difference between the theologians and the scientists lies in their general solution to the HGE issue as they had defined it. The scientists, as well as Fletcher, were using an implicit or explicit, formally rational link between their means and their ends. Since means were not wrong in and of themselves, but only potentially wrong given historically contingent ends, there was no reason to ban the development of means such as HGE. As Fletcher pointed out, using this form of rationality, there might be situations in the future where the means might maximize the ends.

However, if there were means that were wrong because they were not consistent with society's ends—that is, for substantively rational reasons—it stands to reason that they should never be used, regardless of how effective they might turn out to be. This substantively rational position, which was strongly contested by later authors, was most famously held by Ramsey, who stated:

> A man of frivolous conscience announces that there are ethical quandaries ahead that we must urgently consider before the future catches up with us. By this he often means that we need to devise a new ethics that will provide the rationalization for doing in the future what men are bound to do because of new actions and interventions science will have made possible.[108] In contrast, a man of serious conscience means to say in raising urgent ethical questions that there may be some things that men should never do. The good things that men do can be made complete only by the things they refuse to do.[109]

Substantive Rationality and Ultimate Decision-Makers

This chapter began with a description of reform eugenicist scientists who saw HGE as the means of forwarding their end of restoring meaning and purpose to human existence. To whom were they speaking? That is, who was the ultimate decision-maker, the entity with the ability to decide what to actually *do* about HGE, and logically, the entity that could offer jurisdiction? That decision-maker was the public, and this is evident by the target audience of the writers in this era.

Certainly, these scientists spent a great deal of their time speaking only to other scientists through articles in specialized journals, and the earliest writers did not seem so concerned about what the public thought. However, many of the scientists were writing as popularizers of science for the general educated public, as were the influential science journalists in their community.[110] One of the reasons to speak to the public was that the scientists' solution to the defined problem depended upon the public accepting that they had an obligation to engage in techniques such as Muller's germinal-choice scheme.[111] In Muller's scheme, the public would have had the ability to decide what to do about HGE with every act of reproduction. Similarly, at least in his earlier writings, Glass wanted to educate the public so that people would see the wisdom of following the scientists' ends. He wrote that the public must accept that "it is right to utilize science to develop and regulate human social life, adjustment to change and rate of social transformation [and] it is wrong—morally and ethically wrong—not to do so."[112]

Before scientists had the means to achieve their ends, they were speaking to a public that was not paying much attention. This situation changed with the development of the new biological techniques, when some scientists attempted to warn the public of the imminence of weighty decisions that they thought too important for scientists to handle alone, arguing that the ethics of HGE should be decided "by an enlightened and broadly based public opinion."[113]

The theologians were similarly speaking to the general public. Fletcher, for example, wanted "across-the-board cultural consultation" regarding HGE, outside of scientific communities, while Kass wanted "broader public deliberation" and "public decision."[114] The back cover of the paperback edition of *The New Genetics and the Future of Man*—the proceedings of a symposium funded by liberal Protestant institutions, edited by Michael Hamilton, canon of Washington Cathedral of the Episcopal Church—summarizes the motivations and the intended audience of the theologians:

> The scientist often works in what amounts to a vacuum of public opinion, despite the fact that he wields potentially more power over our destiny than all the Presidents and Premiers combined. As [this book] makes clear, society runs a grave risk by granting carte-blanche to the scientific community, particularly in the area of genetics. Test-tube babies, cloning, and gene manipulation were awesome enough merely as elements of science fiction novels. Now that they have become imminent possibilities, communication between the scientific world and the lay world is imperative. The essays included in this volume make an important contribution to such a dialog. . . . It is the contention of this book that an informed and thoughtful public is the best insurance we have that genetic research will open the door to a better world.[115]

The implication in this early debate was that the public would decide what to do, informed by either the scientists or the theologians, and would instruct elected officials to carry out their wishes.

I will make the case in subsequent chapters that if the public had retained decision-making authority over HGE—albeit through elected officials—the debate would have remained substantive. Instead the scientists, seeing that they were losing their jurisdiction over informing the public as to the ethics of HGE, sought to avoid the public determining the ends they should pursue, and successfully argued for the use of government advisory commissions that would determine the ends that the public held without consulting with the public. This effort ultimately resulted in the rise of formal rationality in the HGE debate, as well as the invention of the profession of bioethics.

Chapter 3

Gene Therapy, Advisory Commissions, and the Birth of the Bioethics Profession

The most important development in the transformation of the HGE debate from substantively to formally rational was the reaction by scientists to the beginnings of public interest in public control of science. Decisions regarding the ethics of HGE were poised to come under more direct public control through the creation of publicly accountable regulatory mechanisms. Scientists were alarmed by this development and were extremely concerned that their home jurisdiction was under threat, that their scientific freedom was about to be constrained and their funding slashed. As every good political organizer knows, to counter a strong public sentiment, one must create an alternative that appears to assuage public concerns, even if it actually does not. In an environment calling for the creation of government *regulatory* agencies, scientists instead proposed the subtly different mechanism of government *advisory* commissions—groups that would "represent" the public's concerns without being accountable to or involving the public. The first government advisory commission in the area of public bioethics was limited to examining medical experimentation on humans, as scientists fought to keep issues such as HGE out of its purview.

As the experts on the advisory commissions became the ultimate decision-makers of whether HGE would occur, the ability to grant jurisdiction shifted from the public to the commissions. Yet the commissions were embedded in the part of the state that lies furthest from democratic legitimation and therefore selects for formally rational arguments. The rise of this new decision-maker, with expectations of a new form of argumentation, provided an opportunity for a group to implement their new vision for making publicly legitimate ethical decisions. They had created a form of argumentation, superior in their view to that used by other professions, that was portrayed as a "philosophy of the people"—a system that would represent how the public would make decisions if given the opportunity. This system, first adopted by converts from the fields of philosophy and theology, was the form of argumentation of a new profession called bioethics.

The influence of this new profession grew rapidly. Through the influence of the first government advisory commission, its form of argumentation was written into public law as the proper method of making ethical decisions about research involving human subjects. Henceforward all researchers at institutions that receive federal funds would have to learn and adopt this form of argumentation.

While the first jurisdiction of the profession of bioethics was human experimentation, the groundwork was being done for the creation of additional government advisory commissions that would provide an opportunity for the profession to compete for jurisdiction in the HGE debate. The Asilomar biohazard controversy of the mid- to late 1970s, like the human experimentation debate that had preceded it, resulted in genetic engineering falling under the aegis of government advisory commissions as a way to avoid more direct democratic control. This move was successful in further removing the public from decision making regarding HGE, and placing authority in the hands of various government advisory commissions. As I will show below, this meant that the formally rational form of argumentation used by the bioethics profession would come to dominate this debate.

In this chapter I also show how the dynamics of professional competition led to the development of a new type of HGE. Most reviews of the debate imply that somatic cell human "gene therapy"—the use of HGE technology to treat genetic conditions generally considered to be diseases in the bodies of the living, not their descendants—arose as an issue because of the growth of scientific knowledge. But the argument for somatic gene therapy was not initiated as the result of a scientific advance; rather, it was the result of a jurisdictional retreat by scientists from the attempts by Muller, Huxley, and the reform eugenicists to forward the end of creating human meaning through genetically controlling the species.

I will also show that under continuing pressure from their competitors, most notably theologians, scientists fell back to forwarding only the end of beneficence through which their home jurisdiction was legitimated. As a part of this retreat to the home jurisdiction they also defined a subset of HGE that forwards beneficence: "therapy" for "diseases" in people's bodies. This new means was intended to metaphorically link to the means and ends in their home jurisdiction that was not under threat. For the present they would abandon to the theologians their jurisdictional claims based on forwarding the end of providing human meaning through genetics.

The Continued Theological Challenge

In the previous era of the debate, geneticist Hermann Muller had been the most influential author. His ends, form of argumentation, and means had set the tone for the entire debate. In a telling change, by this time period (1975–84) the most influential author was Muller's harshest critic, theologian Paul Ramsey.[1] Moreover, of the five separate debating communities identified in this period, there remains a distinct group of theologians whose leaders were collectively the most influential authors in the entire HGE debate, compared to the leaders of the other communities (fig. 3).[2] Theologians were the chief challengers to the scientists, continuing their critique of the scientists' ends—and often their lack of ends—and the scientists were clearly on the defensive.

The scientists were already moving away from the expansive ends advocated by the reform eugenicists, and were focusing only on the end of beneficence. The members of the theological community attacked the scientists for this reduction in the number of ends considered, as well as for their formally rational link between means and ends—the claim that any means were acceptable as long as they maximized beneficence.

The theologians, in contrast, argued for a substantively rational link between means and ends, according to which some means were inherently inconsistent with ends, regardless of how much they forwarded beneficence. They held that an additional end that the scientists were not considering—"respect for persons," or autonomy—meant that means designed to forward beneficence but inconsistent with autonomy were wrong. It is important to note that the members of this community were not applying the end of "respect for persons" exclusively to the means of HGE. Rather, following a substantively rational approach, they discussed many means simultaneously as indicative of the use of improper ends.[3] The theologians' critiques, combined with the more general cultural criticism of scientists' involvement with public affairs, discussed in chapter 2, brought even more pressure on the scientists' jurisdiction.

The Scientists' Jurisdictional Retreat and the Invention of Human Gene Therapy

The scientists who had begun to take the reins from Muller, Huxley, and the reform eugenicists had a distinct community in this era, with a debate separate from that of the theologians and other communities.[4] The scien-

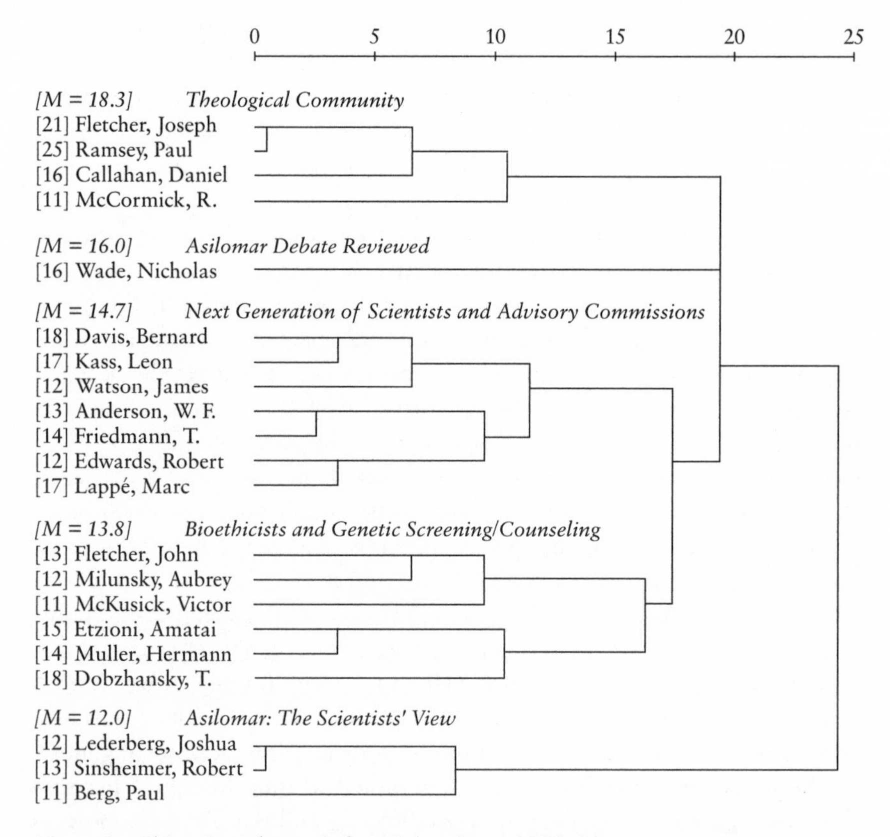

Figure 3 Clustering of most influential authors, 1975–84
NOTE: The number of texts that cite the author is given in brackets. *M* indicates the mean citations per author in the community. The scale at the top is the rescaled distance at which the clusters combine. *N* = 71.

tists in this community were shocked that the public could reject the scientists' "messianic positivism" (in Ramsey's terms), and thus their jurisdiction over HGE. The dominant discourse in this community was retreat from the jurisdictional aspirations of Muller and the reform eugenicists who had promoted the end of providing meaning for humanity through genetic engineering. As a strategic part of this retreat, these scientists began to promote a new, more limited means, which they called somatic "gene therapy." By limiting the claim to the application of HGE to the bodies of patients, they linked the new means with the means used in their safe home jurisdiction. By using the term *therapy,* they linked the intended ends with beneficence, which was also the primary end pursued in the home jurisdiction.

This move assumed that the public recognized the scientists' jurisdiction over making decisions about therapeutic techniques used on the human body. The scientists thus appeared to renounce any jurisdictional claims to the means of germline HGE, the technological dream of the earlier generation of scientists, which could be used to forward the ends that had been so controversial. (Somatic human gene therapy does not affect future generations because the sperm or egg cells are not changed.)

Fearful of the public threatening the home jurisdiction of laboratory research through funding cuts, the scientists chastised their colleagues who were still challenging the theologians through the use of ends other than beneficence, and urged a circling of the wagons around the home jurisdiction. For example, Harvard bacteriologist Bernard Davis, an influential member of this community who became a leading defender of scientific prerogatives in this debate, began a 1970 article by referencing nearly all of the most influential scientists in the previous time period. In that article he notes that "some of these statements, and many articles in the popular press, have tended toward exuberant, Promethean predictions of unlimited control and have led the public to expect the blue-printing of human personalities."[5] However, "most geneticists" have "had more restrained second thoughts." He argues that most of the spectacular claims of the previous generation of scientists concern polygenic behavioral traits, such as intelligence, which are not well understood. But the cure of monogenetic diseases is possible, because the problem is the result of only one gene, not of poorly understood combinations and interactions of genes.[6]

Did most geneticists have "more restrained second thoughts" because they had figured out by 1970 that intelligence and the like were polygenetic, while many diseases were monogenetic? Scientists in the previous time period had known that there was not one gene that determined "intelligence" or the other traits praised by eugenicists. Rather, claims about intelligence and the like by Muller and his contemporaries were descriptions of their ends of species perfection. The more likely reason for Davis advocating that scientists focus on monogenetic "diseases" is that the ends promoted by scientists like Muller were now threatening the home jurisdiction by making the public question the ends of all scientific activity. Davis writes that the "exaggeration of the dangers from genetics will inevitably contribute to an already distorted public view, which increasingly blames science for our problems and ignores its contributions to our welfare." This "irresponsible hyperbole" of the previous generation of scientists "has already influenced the funding of research. It therefore seems important to try to assess objectively the prospects for modifying the pattern of genes of a human being by various means."[7]

The mistake of the scientists of the immediately preceding era, it appeared to Davis, was to have challenged theology in the first place. Scientists had accepted a "naive" view that "failed to recognize the fundamental distinction between empirical questions, concerned with the nature of the external world, and normative questions, concerned with moral values."[8] In my terms, they failed to recognize the distinction between the formally rational argumentation used in their secure jurisdictions (empirical examinations of the consequences of means) and the substantively rational form of argumentation used in the secure theological jurisdictions (determining ends or "moral values"). This move was understandable, according to Davis, because it was science that had destroyed the traditional base of normative inquiry, with science having now "replaced earlier supernatural and animistic explanations of the universe by a coherent set of impersonal mechanisms . . . split[ting] the rock underlying Judeo-Christian morality." In his view, it is the "failure of science to provide a basis for a replacement [which] underlies much of the tragedy, anxiety, and rootlessness of the present age," precipitating attacks on science. These attacks then reflect "a long-standing public sense of spiritual loss, no longer suppressed now that criticism of science has become respectable."[9] Put differently, the pursuit by Muller and Huxley of the end of providing meaning through genetics was understandable, but ultimately untenable.

In sum, many influential scientists in this era perceived the decline of federal spending on science to be the result of the public's disenchantment with science. They thought this was the result of the extravagant claims made by such writers as Huxley and Muller, which instigated critics such as Ramsey, whose writings raised public concern over the issues.[10] When scientists retreated from their jurisdictional aspirations while under attack from theologians and other critics, they did not abandon all of the means that could fall under the rubric of HGE. Rather, they carved out a subset of means from HGE, one that they could claim to be "the same" as means over which the profession already had jurisdiction: inventing medical "therapies" for "diseases" to forward the end of beneficence. Earlier scientists, of course, had wanted to cure disease in addition to pursuing their loftier ends, but this new constraint upon their ends reveals the pressure they were under.

Theodore Friedmann, a molecular geneticist and pediatrician, was another scientist in this community. He took Davis's advice to focus debate on monogenic diseases, and described this new means as "gene therapy."[11] The limits of this new jurisdiction are suggested by how Friedmann begins. Instead of starting off with claims about controlling human destiny, as was typical of texts in the previous era, he begins with the statement that "at least 1500 distinguishable human diseases are already known to be genetically determined"—diseases such as phenylketonuria and cystic fibrosis.

Friedmann cites one of the first uses of the term *gene therapy,* a 1970 article by H. Vasken Aposhian, who argued that scientists should "prepare" (note that he does not say "seek the approval of") "the political and lay community for the moral and other implications of gene therapy."[12] Part of that preparation is the creation of terms that divorce the ends pursued by Muller, Huxley, and the reform eugenicists from the new end of the beneficent healing of disease. Aposhian claimed that the "fear of genetic research might be relieved to some extent if the term 'genetic engineering' was abandoned in favor of the term 'gene therapy' [because] 'genetic engineering' has the revolting connotation to many of impersonal scientific manipulation of the future of human life and offends the dignity of many."[13] While this is suggestive, the motivations for creating the means of gene therapy are most evident in a debate between a young biochemist, W. French Anderson, and theologian Paul Ramsey.

In 1971 Anderson was a biochemist employed at the National Institutes of Health. He later became the most influential author in the HGE debate, an outspoken advocate of gene therapy, and one of the most prominent researchers trying to develop the technique.[14] At a 1971 conference Anderson explicitly claimed gene therapy for the scientists' jurisdiction, out of the broader means called HGE, leaving for other professions the remainder of means that could be called HGE. He stated that "gene therapy is the name applied to the attempt to treat hereditary disease by influencing the genes directly." Several other terms had been used, he noted, including "genetic engineering, genetic intervention, genetic surgery, and gene technology." "Gene therapy is only a part of a broader area which has been called genetic medicine and which involves the diagnosis, prevention, and treatment of hereditary diseases."[15]

Ramsey and likeminded critics had apparently won their desired jurisdictional settlement: scientists would create means to forward beneficence as long as these means did not affect other ends that the theologians held dear. The theologians would worry about the means by which humans would learn meaning and purpose. Ramsey was invited to respond to Anderson's paper, and like the generals grabbing as much territory as possible before the ceasefire, he used the opportunity to push gene therapy as far away from the ideas of Muller and Huxley as possible.

"We have only a handful of crucial words standing between light and darkness," he began, and "to blur the meaning of even one is to hasten darkness." "'Therapy'—treatment—is one such crucial word standing between us and disaster."[16] Anderson, it appeared to Ramsey, was making a metaphorical link to the scientists' home jurisdiction in a way that would allow scientists to maintain their grander ends of controlling the genes of humanity. Anderson was using the analogy that medical research science forwarded the end of beneficence by relieving suffering, and that gene therapy could be justified as relieving the suffering of future generations. Anderson "[spoke] of gene therapy upon an as yet unconceived individual, upon germinal matter, [and] the gametes."[17] Thus the new means, as defined, could still forward Muller's various ends. Ramsey wanted to press the scientists toward metaphorically linking therapy with the means in their home jurisdiction. He noted that therapy, as traditionally understood, involves only the individual body, not the species; what Anderson called therapy included the engineering of human germ cells, an act related to the claims of Muller.[18] Ramsey pushed Anderson's jurisdictional claim back further by suggesting a more limited metaphor for the new means: "genetic surgery." With gene surgery as the metaphor, there could be no mistaking that only living bodies are eligible for HGE, because it is impossible to conduct surgery on a species. Ramsey's surgery metaphor never took hold, and although for many years after that point scientists did limit their claims of gene therapy to the bodies of existing people, the abstraction in the *therapy* term permitted a later reexpansion in jurisdiction (see chapter 5).

Having argued for the further limitation on the means under the scientists' ethical jurisdiction, Ramsey then outlined the different forms of argumentation to be used in the different jurisdictions. For HGE of the germline, there should be discussions of the "fundamental questions," such as the ends that scientific knowledge should forward; such discussion would rely on a more substantively rational form of argumentation.[19] The limited case of somatic human gene therapy, what Ramsey wanted to call genetic surgery, is "under accepted medical ethics governing dangerous therapeutic intervention, justified only when the consequences would be worse without the trial treatment and there is no other recourse." That is, under acts deemed "therapeutic," the "ethics of therapy calls for balancing judgements as to consequences"—essentially a formally rational link between means and end.[20] If limited to means of somatic human gene therapy, in Ramsey's view the debate could be a legitimately thin and formally rational one, focused on only the end of beneficence, because the means in question were not inconsistent with any other ends he thought to be important. If scientists should someday suggest means that had other effects, such as changing the germline, Ramsey would want a thick, substantively rational debate about ends.

As Ramsey predicted, the retreat to the secure jurisdictional ground of medical research through advocating the limited means of gene therapy came with the formally rational method of argumentation. For example, in the article introducing gene therapy, Friedmann noted that "the ethical problems posed by gene therapy are similar in principle to those posed by other experimental medical treatments." He then elaborated five "ethico-scientific criteria" for evaluating whether or not to conduct gene therapy, all of which concern whether enough knowledge is at hand to calculate risks and benefits, such as whether long-term tests have been conducted on animals.[21] Content-analysis data for the texts of the common members of this community (a form of data available beginning in this time period) show that the common members share these assumptions, discussing the ethics of human experimentation in "gene therapy."[22] In Weberian terms, this debate would be one of fact, not of value. The one end to be considered, beneficence, had been set, and it was no longer to be questioned in a reasonable debate, which was now about how to maximize this end.

If all had ended here, theologians could have registered a jurisdictional victory. But this formally rational method of argumentation did not remain restricted to discussions of when and how to engage in somatic human gene therapy. Much to Ramsey's dismay, it later became institutionalized as the only proper form of argumentation used for HGE more generally. This occurred because at the same time scientists were redefining their task as human gene therapy, they were also trying to derail proposals in Congress to create regulatory agencies, akin to the Securities and Exchange Commission, that would approve all experiments on HGE and other controversial means. What they proposed, instead of government regulatory commissions, was the creation of government advisory commissions, designed to be a buffer between scientists and an irrational, easily excitable public that wanted to set the ends for the scientists' experiments. It was these commissions that allowed formal rationality to dominate the entire debate.

Fear of the Public and the Creation of Advisory Commissions

The scientists' critics in this and the previous period share the implicit or explicit claim that because the scientists' experiments affect the entire population, the public should decide whether the experiments should occur.[23] The scientists were right to be fearful. In 1968 Senator Walter Mondale had proposed the creation of a government commission, citing genetic engineering as one of the areas of scientific effort that should perhaps be controlled.[24]

Even though the commission Mondale had in mind was advisory, perhaps due to its lack of definition—and the dominant public discourse of "controlling scientists"—many eminent scientists assumed that it would eventually have a regulatory role. The famous scientists who testified before Mondale's Senate subcommittee were largely hostile to the idea, causing Mondale to complain that one biochemist testifying before the committee "seemed reluctant 'to have persons other than the people in your laboratory look at the social implications.'"[25] Asked whether other professionals might help scientists address the ethical implications of their work, one medical professor replied, "If you are thinking of theologians, lawyers, philosophers and others to give some direction . . . I cannot see how they could help . . . the fellow who holds the apple can peel it best."[26] Mondale's effort failed in 1968, and he reintroduced his measure in 1971. Facing similar hostility, Mondale even responded to one scientist, "I sense an almost psychopathic objection to the public process, a fear that if the public gets involved, it is going to be anti-science, hostile and unsupportive."[27] This was indeed the belief of many scientists at the time.

Scientists soon realized that some sort of public oversight was going to occur, whether they liked it or not. Moreover, by the early 1970s public opinion had continued to go against them, and the most likely form of oversight seemed to be creation of a regulatory commission. While regulatory commissions are at least one degree removed from elected officials, it has been shown that elected officials are ultimately able to control regulatory actions through various signaling devices, such as bringing in the head of the agency to testify to Congress. Most notably, Congress often micro-manages regulatory commissions by adding directives to their budget authorizations.[28]

Instead of regulatory commissions, the scientists suggested the creation of advisory commissions. Such a step would give the public a forum in which to air their concerns, but without the ability to regulate scientists' actions. Elected officials could not direct such an entity to make certain decisions, because an advisory commission would have no formal authority. IVF pioneer R. G. Edwards himself was quite clear about the need for structures that cannot be influenced by legislators. Noting that biologists must "invent a method of taking counsel of mankind" or "society will thrust its advice on biologists . . . in a manner or form seriously hampering to science," he maintained that what was needed was an organization "easily approached and consulted to advise and assist biologists and others to reach *their own* decisions."[29]

W. F. Anderson similarly endorsed an advisory "commission on genetics, composed of physicians, scientists, and concerned laymen," which "might be of enormous value in providing a central and public forum for establishing goals and safeguards." However, he argued that "the establishment of a regulatory commission . . . could be most dangerous," because HGE "holds such promise for alleviating human suffering . . . that no individual or group of individuals should take it upon themselves to make the decisions."[30]

The advisory commissions as envisioned by the scientists in this community were seen not so much as advising the scientists as "preparing" the public for what the scientists intended to do anyway: that is, the commissions would help mold the ends of society to accept the means being developed. Anderson, for example, explained that his ideal "advisory" input from the public was in the case of the artificial heart, where "the country, and the world, quickly adjusted itself to this new, revolutionary procedure."[31] Here the scientists were proposing that these commissions advocate purely formal rationality. Technological possibilities (means) would determine the values (ends) that society should adopt, rather than the reverse. Ever alert to the encroachment of this form of argument, Ramsey, responding to Anderson, asked, "Is adjustment to inevitable developments what we mean by wisdom?"[32]

The scientists' vision was not perfectly implemented. The government advisory commissions that were created did not become merely venues where critics could vent concerns while scientists remained free to pick and choose what they agreed with. The commissions, although they lacked power in themselves, made recommendations to federal agencies, which did have regulatory powers. Still the scientists were largely successful in their efforts. One study of the history of the debates over the public control of science in this era concludes that "during this period in which the democratic approach to decision-making appeared to be gaining acceptance and impact, the political challenge it represented was successfully contained [by scientists], to such an extent that the technocratic approach—and the process of decision-making by elites that lies behind it—was never seriously threatened."[33]

The first government advisory commissions in the public bioethical debate were concerned not with HGE, but rather with human experimentation, which had raised greater public concern. At this point HGE was still theoretical, but unethical human experimentation was actually taking place. Moreover, as a result of critiques such as Beecher's, there had been increasing political pressure for the government institutions that funded most biomedical research to ensure that other ends besides the pursuit of medical knowledge were considered in the ethics of research on human subjects. The first move in this direction came in 1966 when the National Institutes of Health (NIH) required that each institution that received a grant set up what is now called an Institutional Review Board (IRB) that would evaluate research to make sure the informed consent of the subject of the experiment was obtained and that the benefits outweighed the risks in the experiment. By mid-1973 revelations regarding the unethical treatment of research subjects by government-funded medical researchers made some sort of a national commission nearly inevitable. One tragic case was the infamous Tuskegee experiment, in which African American men with syphilis were left untreated, in an effort to determine the effects of advanced stages of the disease.[34] The National Commission for the Protection of Human Subjects of Biomedical and Behavioral Research was created by Congress in 1973, and first met in 1974.

The Birth of the Bioethics Profession

To fully understand the case at hand, it is important to step back from the data and follow up on the effect of the 1973 commission.[35] One might think that if such a commission were to represent the public, it should contain, at minimum, a variety of members of the public from different communities. Instead, the 1973 commission was established on the basis that the public's interest would be represented if mediated through the variety of professions competing for jurisdiction with the scientists. Thus, the commission was mandated to have eleven members chosen from the fields of "medicine, law, ethics, theology, biological science, physical science, social science, philosophy, humanities, health administration, government and public affairs," with no more than five being medical researchers.[36] As the first government commission on medicine, science, and society—in other words, public bioethics—its formation was a momentous occasion. Most important, the commission selected and promoted a form of argumentation about human experimentation—primarily a reduced set of ends that should be considered—and these ends were subsequently adopted by the new profession of bioethics.

Why would a government commission define a set of ends? This is a particularly important question for my thesis since, as I explained in chapter 1, government advisory commissions are reticent to set ends because they lack democratic legitimacy. In this case the reason is clear: the commission was instructed by Congress to create a set of ends.

During the legislative debate over the creation of the National Commission, Senator Edward Kennedy wanted to create a permanent, regulatory commission, outside the influence of the NIH—one that, like the Securities and Exchange Commission, would have the power to enforce its decisions.[37] In a compromise with Rep. Paul Rogers, who supported the position of the scientists at NIH that the commission should be advisory, Kennedy agreed to the creation of a time-limited, advisory commission—provided that the Department of Health, Education, and Welfare published satisfactory regulations of its own for the recipients of NIH grants. Kennedy had compromised more direct public control, but he inserted a mandate into the bill creating the National Commission that to formulate these regulations, the commission would "conduct a comprehensive investigation and study to identify the basic ethical principles which should underlie the conduct of biomedical and behavioral research involving human subjects" and "develop guidelines which should be followed in such research to assure that it is conducted in accord with such principles."[38] That is, the advisory commission was to determine the legitimate ends for arguments about research on human subjects—ends that would be applied to make ethical decisions by the members of IRBs throughout the country who, by accepting government money, were implicitly making decisions for the government.

The commission had been given a classic task for a bureaucratic government entity: create rules that could be uniformly and ritualistically applied across situations. Senator Kennedy essentially guaranteed that the commission would arrive at a formally rational form of argumentation, given how the bill had defined its task.

Forms of argumentation from theology, philosophy, science, or the other professions that had been involved in the debate about human experimentation could not meet the implicitly formally rational requirements of the commission. However, at the same time that the commission was trying to create its principles or ends, a group of philosophers and theologians at Georgetown University had been working on a principle-based approach to ethical problems that could be applied to public policy. The commission tapped into this group of scholars, and called upon them to work out these principles for the concrete case of human experimentation.

The commission proceeded by calling for papers on the topic of principles, hiring as a consultant a philosopher who had been working at Georgetown on a principle-based textbook, and holding a retreat to discuss what these principles or ends would be. The commissioners thought they were creating a system that would meet "the need of public-policy makers for a *clear* and *simple* statement of the ethical basis for regulation of research."[39]

At the retreat the commissioners set to work deriving these "clear and simple" principles; at one point in the retreat they had arrived at seven. After complaints that the list was not "crisp enough," and that some of the principles were not universal, the group narrowed the principles down to three: "respect for persons, beneficence, and justice."[40] Having set the ends, the commission created three procedures by which to guarantee that the proposed means would not violate the ends: "informed consent" to guarantee respect for persons; "risk/benefit assessment" to guarantee beneficence, and "fair procedures for the selection of research subjects" to guarantee justice.[41] The ends were not determined by consulting the public, but were derived from academic reflection as overlapping, consensual ends—in the commission's words, as the ends "among those generally accepted in our cultural tradition."[42] These ends, with their associated procedures, were reported to the Department of Health, Education, and

Welfare through publication of the *Belmont Report,* named after the conference center where the retreat had taken place.

To understand why the commission selected a formally rational system, it is important to recognize that it was creating principles that could be made into fairly calculable rules, promulgated by the NIH to all recipients of grants. These rules would need to be applied by unelected and—by this point in the controversy over human experimentation—mistrusted government representatives. This is exactly the situation where rulelike behavior is expected. Moreover, lack of trust in the discretion of medical researchers was the reason for the commission's existence.

A formally rational, highly calculable system would solve these problems. First, the ends had to be portrayed as universally held by the citizens, but had to be applied without a method of determining empirically what the ends of the citizens were. Second, the ends had to be applicable to any problem that arose, since it would complicate decision making to have different principles for different types of research. Third, maximal calculability could be had by creating one commensurable, universal ends scale, such as utility.

This degree of calculability was not achieved, due to the pressures that had led to the creation of the National Commission. All the controversies that had finally pushed Congress to create the commission had to do with experimenting on people without their permission or with experimenting on people who were not free to consent (such as prisoners and mental patients), to further societal beneficence (i.e., medical progress). Thus, the purpose of the commission required that an end such as justice and/or autonomy (respect for persons) be included. No one involved with the debate wanted to stop pursuing medical research motivated by beneficence, so this end had to remain. Although the three principles represented a great improvement in calculability from other possible forms of argument, it turned out in practice that each served as a commensurable end to which other ends could be translated, but that each was somewhat incommensurable with the other two.[43]

The commission's creation of three ends resulted in a thicker and more substantive debate than would have occurred if the scientists had continued their jurisdiction unopposed by the theologians. Yet the debate was much less substantive than it would have been had the theologians gained jurisdiction.

Theology and the Search for Universals

It is important to clarify why a group of theologians would want to create a new form of argumentation, based upon a limited number of commensurable ends that were not ultimately translations of a particular theological position. This is especially important, given that other theologians were still arguing for secular translations of theological ends. To understand this development, one must first grasp the state of theology in the early 1970s. At that time there were movements in both liberal Protestant and Catholic theology toward a form of theology that was accessible to all people, whether or not they believed in God.

The 1960s were a tumultuous time for theology in the United States. I have already discussed the debate among Protestants between "norm and context" or "rules and situations," positions represented by Ramsey and Fletcher, respectively. From the "situationist" wing also came what has become known as "death of God" theology. In Harvey Cox's influential version of this thesis, God had freed humans through secularization, and they must now live with the implications of this freedom.[44] The "death of God" advocates and the situationists both aimed to "eliminate any exclusively Christian conditions or terms" from what was also called "the new morality."[45] To slightly oversimplify the argument, there was nothing unique about Christian ethics that could not be obtained through secular sources. There was therefore no need to dip back into the religious tradition to come up with secular ends that were translations of theological ends.

Among Catholics, a similar strand of thinking had emerged. Reacting to neo-scholasticism, by the early 1970s a variant of natural law theology had emerged that Vincent MacNamara calls "an autonomous ethic" theology.[46] It goes beyond present purposes to delve into the history of Catholic theology, but a quotation from one of the advocates sums up the view of this new school: "Christian morality, in so far as its categorical determination and materiality is concerned, is basically and substantially a human morality, that is a morality of true manhood. That means that truth, honesty and fidelity, in their materiality, are not specifically Christian but universally human values."[47] As one critic at the time stated, "whereas a few years previously theologians regarded it as natural to demand that the teaching of morality should be theological, i.e. 'Conceived in terms of scripture and of salvation history' things have been entirely reversed 'so that Christian morality is understood in rational, philosophical terms, i.e. in terms of empirical human science.'"[48]

This focus on philosophical rationality among Catholics, quite strong in the natural law tradition, also explains why it would not be unusual for at least a faction among Catholic theologians to look to a calculable connection between means and universal ends. Among Protestants, this impulse also long existed in the strand of thinking represented at this time by Fletcher and the situationists. For example, on the first page of his survey of (Protestant) Christian ethics since the late 1960s, Edward Long discusses the long debate between the two sides: between "the defenders of reason" and those who have "rejected the place of reason in making moral judgements."[49]

In light of these factions within theology, it is no surprise that a group of theologians would try to create principlism as a more effective way of influencing public debates, given the new environment. Consider Quaker theologian James Childress, one of the founders of principlism. According to a review of his work, his theological beliefs require a universally accessible form of ethics:

> The medical ethics of James F. Childress makes it clear that the boundaries of the community of ethical responsibility should not be drawn too narrowly. The prominent theological theme in Quaker thought (Childress's tradition) of "answering that of God in every person" carries a universalistic impulse, requiring the ethicist to take account of and be accountable to nontheologically informed positions. The responsibility to answer to a community constituted by "every person" may entail a less explicit, background role for theological convictions in moral discourse. As in Childress's case, there may be *theological* reasons for not doing medical ethics theologically.[50]

In sum, there was a faction within theology that believed in universally accessible ethics, stated in secular language, that did not have to return each time to the well of theological discourse for discernment. As one might imagine, once this turn has been made, it becomes less clear what one needs traditional theology for. It is therefore no coincidence that theologians in this faction showed a strong tendency to become bioethicists. The theologians who were not a part of this faction, such as Paul Ramsey, continued in their more particularistic, nonuniversal, end-seeking ways.[51]

Reaction against Traditional Theology

These theologians and philosophers pushed for their form of argumentation because they thought it was an improvement over the alternatives, and they set out to argue against its competitors. According to an early participant in this effort:

> We could have attempted to analyze the new science and the new medicine in the terms of the languages [of philosophy and theology] we had learned. But we gradually realized that, if we did, our words would be uttered largely in vain. The philosophical ethics of that era had very little to say about the substantial content of moral decision and action. Theological ethics used terms that were incomprehensible to many who were not believers or were believers of another sort. We had to find an idiom that, at one and the same time,

> expressed substantive content and was comprehensible to many listeners. . . . Like strangers in a strange land, we had to devise new forms of communication among ourselves, with our scientific and medical colleagues, and with the public.[52]

This form of argumentation had two parents, philosophy and theology, but theology was the only serious competitor in public bioethics debates at the time. As noted in the quotation above, philosophy was not at the time engaged in applied questions.[53] Theology was so engaged, and it was theology that this new group primarily argued against. At the opening in 1971 of one of the first bioethics centers in the United States—at Georgetown University, the birthplace of principlism—Senator Kennedy, speaking as the president of the Kennedy Foundation that had endowed the new center, spoke of the "future importance of bioethics for the formation of public policy." At the same press conference, the president of Georgetown University, when challenged as to how a Catholic university could reconcile the particularisms of Catholic theology with this new universalistic discipline, "tersely" replied that the new institute would be a "'truly ecumenical and catholic effort,' defining catholic in its classical meaning of universal."[54] The new "systematic, secular (principle-based) approach to bioethics" created at Georgetown was "a new platform from which to speak, displacing the suspicions of sectarianism and religious ideology."[55] Reflecting on this history, Daniel Callahan has concluded that the acceptance of the place of a public bioethical debate in society was gained by "push[ing] religion aside."[56] Or, as one of the textbooks that later institutionalized the form of argumentation of the bioethics profession described the history of public bioethical debates:

> The history of bioethics over the last two decades has been the story of the development of a secular ethic. Initially, individuals working from within particular religious traditions held the center of bioethical discussions. However, this focus was replaced by analyses that span traditions, including particular secular traditions. As a result, a special secular tradition that attempts to frame answers in terms of no particular tradition, but rather in ways open to rational individuals as such, has emerged. Bioethics is an element of a secular culture and the great-grandchild of the Enlightenment.[57]

What became known as either the Belmont principles, principlism, or, more pejoratively, the "Georgetown mantra," became the accepted form of argumentation in public bioethical debates about human experimentation. The institutionalization of the Belmont principles was due to several converging events. The most important was acceptance of the commission's advice, when the Department of Health, Education, and Welfare made the *Belmont Report* public law governing the research activities of federally funded scientists. This meant that every IRB at an institution that received federal money had to apply these principles.[58] Since journals refuse to publish results from research not reviewed by IRBs, the principles became the standard not only for federally funded research, but for privately sponsored research as well.[59] This was a huge resource given to the new profession of bioethics in its competition with other professions: the government was essentially requiring researchers at every research university and hospital in the nation to learn its form of argumentation.

The Expansion of the Bioethics Jurisdiction

Although at the Belmont retreat there was a substantively rational debate about what the ends should be, after the publication of the *Belmont Report* the ends became institutionalized. Like the "formal, abstract, general rules" of bureaucratic administration examined by Weber, they became assumed and therefore were not argued for or defended. In a hallmark of formal rationality, the ends came to be taken as outside the realm of debate, leaving only thin debates about whether technologies (means) maximized these given ends.

Here we had a formally rational form of argumentation and a new profession, bioethics, to forward it. If the Belmont principles, which were selected as those ends "particularly relevant to the ethics of research involving human subjects," had remained limited to the jurisdiction of human experimentation, there would be no reason to raise principlism in a book about debates over human genetic engineering. Ramsey, for one, probably would have approved of the *Belmont Report* had its application remained limited to human experimentation. However, shortly after publication of the report the principles stated in it "grew from the principles underlying the conduct of research into the basic principles of bioethics."[60] That is, bioethicists began to apply the form of argumentation of the bioethics profession to the ethics of all issues having to do with human beings, medicine, and science, as they tried to expand their jurisdiction.

Several related factors contributed to the spread of this form of argumentation to topics beyond human experimentation. First, the state became involved with more debates that were considered to be bioethical, requiring the same formally rational type of argumentation that had led to the rise of principlism in the discussion of human experimentation. For example, one account of the growth of the bioethics profession states that "there was a political urgency to many of the biomedical issues" at the time. "The media craved the biomedical controversies and federal and state policy makers wanted answers."[61]

Second, participants in these debates took the time to create a full-blown form of argumentation, applicable to all problems in science and medicine, which others could use to make arguments about issues other than human experimentation. For example, the academics from Georgetown who had helped the National Commission were at the same time writing a textbook called *Principles of Biomedical Ethics,* now in its fourth edition, which would further spread the form of argumentation.[62] Departing slightly from the commission, they derived four ends by splitting the principle of beneficence into nonmaleficence and beneficence. (In common language, *nonmaleficence* means refraining from harming a person; *beneficence* means contributing to a person's welfare.) Most important, in this textbook the form of argumentation was not limited to the problem of human experimentation, but was expanded to cover almost all problems related to ethical decisions in science, medicine, and society—abortion, euthanasia, medical rationing, and many more. Once the form of argumentation of bioethics had been enshrined in public law for human experimentation, and embodied in a popular textbook, it began to spread rapidly. There was an "enormous demand" for training in ethics, met by countless books, workshops, and courses designed to make "the theories and methods of ethics" "readily available to more people in a shorter time."[63] The "major strategy" in these educational products was the principlist approach embodied in *Principles of Biomedical Ethics.*

According to observers of the profession, this one textbook, more than anything else, "shaped the teaching and practice of biomedical ethics in this country. . . . [becoming] a standard text in courses and a virtual bible to some practitioners." The ethical framework provided by the book "shapes much of the discussion and debate about particular bioethical issues and policy, whether in the academy, the literature, the public forum or the clinic."[64] The institutionalization of this form of argumentation for human experimentation and increasingly for other problems was so strong that one set of critics would go so far as to begin their essay with the mocking claim that "throughout the land, arising from the throngs of converts to bioethics awareness, there can be heard a mantra '. . . beneficence . . . autonomy . . . justice.'"[65]

The principles are also often thought to be a compromise between two different ways of linking means and ends. Autonomy and justice are taken to be substantively rational ends, with which any means must be consistent (that is, they are what philosophers call deontological ends). Beneficence and nonmaleficence are taken to be formally rational ends—that is, means that maximize these ends are selected (what philosophers call consequentialist ends). However, since bioethicists have not determined how the principles relate when they conflict—and almost all issues involve more than one principle—the deontological ends are actually maximized. The typical conundrum would be: would X increase the amount of autonomy while minimizing the decrease in justice and beneficence?

Alternative forms of argumentation for public bioethical debates were suggested and could have been adopted. Most notable are the systems proposed by Robert Veatch, then a scholar at Georgetown University, in *A Theory of Medical Ethics,* and by H. Tristram Engelhardt in *The Foundations of Medical Ethics.*[66] Although experts will note the differences between these various approaches, they are all versions of principlism.[67] It is also worth noting that those who created principlism see principles as more than simply ends as I have defined them. In actual debate, however, nuances tend to be lost, and thus the principles function more or less as ends.

The creators of principlism also decry the overuse or misuse of principlist systems, claiming that principles are only a part of applied ethics. Still, they are all that the average bioethicist attends to.[68] The intentions of the inventors of principlism are not as important here as how principles are actually used.

The first jurisdictional expansion of bioethics was from the ethics of medical research to the ethics of medical practice more generally. According to historian David Rothman, after being institutionalized with medical research, "the new rules for the laboratory permeated the examining room, circumscribing the discretionary authority of the individual physicians. The doctor-patient relationship was molded on the form of the researcher-subject; in therapy, as in experimentation, formal and informal mechanisms of control and a new language of patients' rights assumed unprecedented importance."[69] This was quickly followed by attempts at expansion into any area that could be metaphorically linked to either medicine or science, as long it somehow involved human beings.

The institutionalization of this form of argumentation concerned some of the original advocates of interprofessional public bioethical debates. As early as 1982 Daniel Callahan, founder of the Hastings Center and an influential author among theologians in this period, was warning about the new hegemony in ethical thinking in these debates. Referring to the new hegemony as "some kind of ultimate moral big bang theory" or an "'engineering model' of applied ethics," he bemoaned the emphasis on one narrow form of argumentation. The interprofessional approach that he had championed in the early days of the field had had "the distinct advantage of leaving the way open for the insights of religion, of cultural observation and social analysis . . . and of concepts of human dignity and purpose that had a wider scope than mere autonomy." Opposing the creation of a separate profession of bioethics with its own form of argumentation, he hoped that "the other disciplines that are a part of applied ethics [would] . . . shout and scream" when they detected the "diversion from what was intended to be a richer agenda."[70]

Nonetheless, this new form of argumentation proved congenial to "the educational and policy making purposes being pursued" by bioethicists.[71] That is, it was well suited for extracting resources from the new environment where the bureaucratic state was the primary decision-maker. Reflecting on this success, a current Georgetown scholar reports that the first director of the Kennedy Institute at Georgetown University, André Hellegers, was able to "promote the success of the Georgetown vision [by] marshall[ing] hitherto untapped federal and private funding and university resources for operating expenses and endowed chairs in bioethics. He fostered the need for medical bioethics in the government agencies and biomedical research and clinical centers that were seeking to develop ethical and regulative policies, and he supplied that need."[72]

Just as the government advisory commissions became mediators between the will of the people and the scientists, so did the profession of bioethics. Reflecting on the history of the bioethics profession in 1993, historian Daniel Fox drew an analogy between nuclear weapons and medical research, noting that in both there has been a tension between professional and civilian control. "Arms control intellectuals and bioethicists have been critically important mediators between the ideologies and the technical fantasies of the professionals on the one hand, and the most adamant and uninformed advocates of civilian control on the other."[73] The "task" of bioethicists is the translation of arguments into what has been set as the commensurable, universal (and numerically limited) ends of society, so that the bureaucratic state or any other formally rational institution, such as a business or a hospital, can make legitimate decisions without directly consulting the public.

The rise of government commissions provided an exceptionally good environment for the growth of principlism, and as a result of the focus of the first bioethical concerns—and the focus of concern of the first government commission—the jurisdictional homeland of the bioethicists began as the interaction between medical researcher and subject. This is what their form of argumentation was originally designed for.[74] Yet as bioethicists have moved beyond their home jurisdiction, they have used this same form of argumentation to address new problems—that is, to gain new jurisdictions. Bioethicists at this point did not yet raise a challenge for jurisdiction over HGE, but they began their jurisdictional expansion with a related problem of genetic counseling where the analogy of the means and ends to their home jurisdiction of doctor-patient interaction could most easily be made.

Bioethicists and Genetic Counseling

Returning to the data, within the broader HGE debate during this era is a community of bioethicists who are primarily concerned with genetic counseling and screening.[75] While Muller was planning his grand scheme of germinal choice, other reform eugenicists had pressed on with a low-technology solution of setting up facilities devoted to what would eventually be called "genetic counseling," with the aim of improving the biological quality of the population. By the early 1970s the carriers of at least fifty genetic disorders could be conclusively, not probabilistically, identified through chemical tests. It began to be suggested that "genetic screening" be conducted on segments of the population more disposed toward certain diseases. In 1972 Congress passed the National Sickle Cell Anemia Control Act, which provided funds for the screening of African Americans, who are more susceptible to the disease than other groups. Other diseases followed, and by 1975 nearly half a million people had been examined for the sickle-cell trait, while tens of thousands more had been tested for Tay-Sachs or thalassemia.[76] Eugenicists assumed that parents, armed with information from the screening, would make the "right" decision about whether or not to have children or to abort fetuses that had genetic diseases.

Genetic counseling was being done by physicians and other paramedical professionals, and it involved the interaction between these professionals and people who could be legitimately called "patients" who might be suffering "diseases." Thus the analogy to medicine could be made, and bioethicists had been gaining jurisdiction over determining the ethics of medical interaction. Moreover, the paternalism of physicians and scientists experimenting on people without their consent for the good of society—a practice that the bioethicists had successfully challenged—was similar to the paternalism of the eugenically motivated counselor who would advise the patient to pattern his or her reproduction according to what was good for the human species.

Examination of the debate in this community shows that at this point it was still substantive, in that participants were arguing over the ends to pursue in the ethics of screening and counseling. On the one side were those, influenced by Muller and Dobzhansky, who promoted the end of reducing the genetic load and designing the genetic future of humanity. On the other were the first bioethicists, who wanted to forward autonomy—one of the ends later put into the *Belmont Report*—to restructure the way decisions were made about screening and counseling.

For example, consider John Fletcher (no relation to Joseph Fletcher). By the time most of the average members of this community were writing, he had moved from being a theologian to a bioethicist, and he makes many arguments against eugenic ends using an implicit version of the bioethical form of argumentation.[77] Like the other challengers to the scientists at the time, he argues for physicians and scientists to move back to their jurisdictional homeland of curing individual illness and away from Huxleyite schemes for the perfection of humanity: "to adhere to their therapeutic calling and resist any attempt to hasten the 'kingdom of God' through technical progress."[78] The ends forwarded by the means of prenatal diagnosis in the arguments of the reform eugenicists—"one more of those steps whereby man, consciously or unconsciously, has grasped the reins of his own genetic destiny"—are wrong because they are a threat to parent-child relations.[79] In the end, it is the needs of the family, as understood by the family, that should determine the morality of screening through prenatal diagnosis and any subsequent abortion decisions.

Fletcher's reasoning epitomizes the particular strand of the debate that appeared a few years later in the *Belmont Report*. First, the end of autonomy is pursued: parents are autonomous decision-makers, and no one else can tell them either to have amniocentesis or to abort or not abort a fetus.[80] Once autonomy is ensured, the decision of whether prenatal diagnosis should be conducted is made by means of a risk/benefit consequentialist analysis of whether it would maximize beneficence. Thus, when bioethicists cited texts on genetic screening and counseling, it is easy to see why they selected John Fletcher, who soon became one of the more influential bioethicists.

The profession of bioethics had been created, with its initial jurisdiction being the relations between medical researchers and research subjects. Moreover, the profession's thin, formally rational type of argumentation was optimal for instances in which the representatives from the bureaucratic state were the ultimate decision-makers—that is, cases in which decisions had to be legitimated as rulelike calculations maximizing ends that the citizens would have suggested had they been asked. Bioethicists had begun to appear at the edges of the HGE debate, applying their form of argumentation to genetic screening and counseling. Yet, to this point, the debate about HGE appeared to be still outside the bioethicists' jurisdiction, with scientists and theologians still in competition. However, at approximately the same time that the *Belmont Report* was being drafted, a separate controversy arose, closer to the HGE debate. This controversy made government advisory commissions the ultimate decision-maker for the HGE debate itself, suggesting that the bioethics profession could gain jurisdiction here as well.

The Asilomar Controversy

While the debate over the ethics of HGE had raged between such spokesmen as Muller and Ramsey, the basic research needed to make HGE a reality continued apace. Progress with microorganisms was rapid, and by the early 1970s techniques had been developed to cut DNA at a specific site through restriction enzymes, suggesting how this type of engineering in humans might eventually occur. Techniques that involved the use of plasmids and bacterial viruses to carry foreign DNA into cells, with the cells adopting the DNA, were in use, and the ability to synthesize pure DNA was at hand.[81]

At about the same time biochemist Paul Berg had been successful in combining DNA from a bacterial virus with DNA from a monkey tumor virus, and by 1973 other researchers spliced a piece of the DNA from a toad into the *E. coli* bacterium.[82] In 1972 Berg had planned to splice DNA from the monkey tumor virus into *E. coli,* which lives in many people's intestines, but he had had second thoughts about the potential hazard this might cause if some of the engineered *E. coli* were to escape the lab and come to reside in the intestines of the human population. Berg's competitors, however, continued their experiments, and in 1973 an internal debate erupted among these scientists, with some of the leaders calling for the National Academy of Sciences to establish a committee to examine the hazard that these hybrid molecules might present to laboratory workers and to the public.[83]

Consistent with the scientists' acknowledgment of the many challenges to their jurisdiction in this era, many scientists feared that if the public were to become aware of this internal debate about what became known as biohazards, the scientists' jurisdictional defense of HGE and even the home jurisdiction would be threatened by the public trying to control or limit scientific activity. Harvard bacteriologist Bernard Davis, the ardent defender of scientific prerogatives discussed above, wrote to Paul Berg of his concern about the "danger of public over-reaction to the presumed menace, which could lead to renewal of fear of genetic engineering in man, and perhaps to fear of molecular genetics in general." Similarly, Joshua Lederberg wrote to the director of NIH stating his concern that "carelessness about the details of enforcement and compliance [with possible controls] may result in a smothering blanket that goes far beyond any needs that are reasonable for the problem."[84]

Scientists attempted to limit the problem under debate to the possibility of hazards. This kept the debate on a technical level—whether certain procedures were risky or not, and how risk could be minimized—and therefore justified keeping decision making entirely in the hands of scientists.[85] That is, in my terms, they struggled to keep this debate in their home jurisdiction of evaluating consequences, which scientists do for all of their experiments, and away from contested jurisdictional arenas, such as whether DNA should be moved between species at all.

A committee of the National Academy of Sciences was formed, composed of leading molecular biologists and biochemists. Although it was suggested that Leon Kass and Jonathan Beckwith be invited to represent a "more radical or skeptical" view, they were not invited.[86] The committee eventually published a letter in *Science* magazine, as well as in other venues, calling for a temporary moratorium on particular types of research. Most important, the letter described safety as the only potential problem with genetic engineering, and called for an advisory commission at NIH, composed only of scientists, to propose methods for evaluating and addressing any hazard, and for a follow-up meeting of scientists to further discuss the issue.[87]

On the day of the release of the letter calling for a temporary moratorium and the establishment of a scientist-controlled government advisory commission, the NIH announced it was setting up such a commission.[88] Plans were being made for an international conference of scientists, to be held at the Asilomar conference center in California in February 1975, where scientific consensus could be reached on the extent of the hazards and possible safeguards, providing critical input to the new commission.

At the Asilomar meeting, the fear of public influence on scientific decision making was pervasive.[89] Berg and the other scientists who were organizing the Asilomar meeting "surmised that if the NIH did not act with dispatch in responding to the potential risks of gene-splicing, Congress might pass restrictive legislation," which would be "detrimental to the interest of biology."[90]

At the meeting, when consensus seemed remote, "participants were effectively warned of the invisible threat of legislative intervention." For example, Berg stated, "If our recommendations look self-serving, we will run the risk of having standards imposed. We must start high and work down. We can't say that 150 scientists spent four days at Asilomar and all of them agreed that there was a hazard—and they still couldn't come up with a single suggestion. That's telling the government to do it for us."[91]

At the meeting organizers continued to struggle to avoid debate about ends. Although persons like Sinsheimer had been arguing that the new techniques whose safety was being challenged actually raised more fundamental questions, such as whether humankind should "take into our own hands our own future evolution," organizers intended to avoid this discussion.[92] In the introductory remarks to the conference, co-chair David Baltimore spoke about ruling out the issue of "the utilization of this technology in gene therapy or genetic engineering" because it was "'peripheral to this meeting' and likely to 'confuse it in a number of ways.'" Like Davis distinguishing between fact and value, Baltimore stated that genetic engineering "leads one into complicated questions of what's right and what's wrong—complicated questions of political motivation—and which I do not think this is the right time [to discuss]."[93] That is, the scientists tried hard to make this a debate about the efficacy of means for forwarding set ends, not a substantively rational debate about ends themselves. Sinsheimer later concluded about Asilomar that "the eagerness of the researchers to get on with the work in this field was most evident," and that there was "no explicit consideration of the potential broader social or ethical implications" of the research.[94]

The final report of the conference called for a resumption of research using particular safety measures for certain types of experiments. All told, the Asilomar conference had been a success for the scientists in minimizing threats to their jurisdiction over HGE or their other jurisdictions. The debate had been constructed as being about whether the means of gene splicing threatened beneficence or nonmaleficence, not whether gene splicing would also forward species perfection or any other of the more controversial ends. More generally, scientists had succeeded in keeping the public, or anyone else, from being involved with decision making. This conclusion was reached by Senator Kennedy, who complained at a hearing after the Asilomar conference that the meeting was "inadequate because 'scientists alone decided to impose the moratorium and scientists alone decided to lift it.' The factors under consideration, however, extended far beyond their technical competence, said Kennedy. 'In fact they were making public policy. And they were making it in private.'"[95]

Although the most ardent advocates of total scientific freedom were dismayed at the creation of a government advisory commission to suggest regulations to NIH, the commission was to be composed entirely of scientists, the majority of whom were actual or potential users of the technology.[96] The first meeting of what would soon be called the Recombinant DNA Advisory Committee (RAC) of the National Institutes of Health took place right after the Asilomar meeting. The NIH released its first guidelines for conducting the research in 1976, based on the advice of the RAC.

Although figure 3 shows that the biohazard debate was a part of the larger HGE debate in this period, it moved along a separate track in future years.[97] Wright shows in her examination of the debate that through the skill of the scientists involved, the issue later became noncontroversial.[98] It was folded entirely into the home jurisdiction of scientists' everyday tasks of evaluating risk in their pursuit of scientific discovery. The critical issue for the HGE debate is that the Asilomar controversy institutionalized the government advisory commission as the solution favored by scientists, far preferable to elected legislators passing laws.

The RAC had become the ultimate decision-maker about the ethics of the genetic engineering of microbes. It had decided that nonmaleficence (safety) was the only end under discussion, and from its inception in the mid-1970s to the mid-1980s the RAC limited its discussions to evaluating the means toward this end. As we will see in chapter 5, the RAC later became the primary decision-maker about the ethics of *human* genetic engineering, after the efforts of yet another government advisory commission, the President's Commission. With these commissions as decision-makers, the form of argumentation of the bioethics profession became even more dominant in the HGE debate, and thus the debate became thinner as formal rationality became assumed in any "reasonable" argument.

Chapter 4

The President's Commission: The "Neutral" Triumph of Formal Rationality

In 1980 the HGE debate was very much in flux. The scientists had retreated to wanting to debate the more limited means of somatic human gene therapy, but this new jurisdictional settlement had not been institutionalized. The Asilomar debate had led scientists to be even more fearful of the public becoming the ultimate decision-maker. They felt, quite rightly, that their jurisdiction over the ethics of any form of HGE was still at risk. Bioethicists had yet to enter the HGE debate in great numbers—the competition remained between scientists and theologians.

As we have seen, the Asilomar debate had produced a government advisory commission on the genetic engineering of microorganisms (the Recombinant DNA Advisory Committee, or RAC). While the scientists had limited the scope of this commission to a consideration of research risks, the public apparently did not fully accept the strong distinction the scientists had drawn between the genetic engineering of microorganisms and the genetic engineering of humans. In 1980 the Supreme Court ruled in the *Chakrabarty* decision that one of these genetically engineered life forms could be patented. The decision drew public attention to the fact that scientists were creating forms of life so novel as to warrant attempts to garner patent protection for their inventions. In response to the *Chakrabarty* decision, the leaders of the National Council of Churches, the Synagogue Council of America, and the U.S. Catholic Conference wrote a letter to President Carter outlining their concerns about the emerging genetic technologies.[1]

The letter made arguments about genetic technologies of all types and assumed the substantive rationality that theologians had typically used in this debate. Since the letter's authors were not focused on calculability in ethics, they did not limit their concern to one narrow component of genetic engineering, such as HGE. Instead, they implored President Carter to start an investigation of "the entire spectrum of issues involved." They were not interested in individuals so much as in "the long-term interest of all humanity," and by pointing to experiments that were not yet possible, they implied that they were more concerned about the future than the present. Safety issues seemed less critical than deeper issues that could lead to a "fundamental danger" to humanity. The ends they were concerned with—defining the meaning of human life—placed the problem solidly in the theological jurisdiction. They stated that the problems with HGE were "not ordinary" but "moral, ethical and religious," and dealt with the "fundamental nature of human life and the dignity and worth of the individual human being." The writers pointed out that the ends to be pursued with genetic engineering were not as settled as scientists portrayed them, and they asked, "Who shall determine how human good is best served when new life forms are being engineered?" Referring to the reform eugenicists who had been competing with the theologians, the letter writers pointed out that "there will always be those who believe it appropriate to 'correct' our mental and social structures by genetic means, so as to fit their vision of humanity." Repeating a phrase used by Paul Ramsey in his response to eugenicist Hermann Muller a decade earlier, they pointed out that those who would plan to control the destiny of the human species through genetics now had better tools in their hands and that those persons "who would play God" would be tempted as never before. Like other theologians in the debate, they argued for a greater public role in decision making, and against control by scientists: "It is not enough for the commercial, scientific or medical communities alone to examine [the issues]; they must be examined by individuals and groups who represent the broader public interest." They did not seem interested in advisory commissions, but looked toward more direct public regulation, asking, "Given all the responsibility to God and to our fellow human beings, do we have a right to let experimentation and ownership of new life forms move ahead without public regulation?"[2]

President Carter did what presidents have long done with controversial issues. Instead of deciding, as an elected official, what the United States should do about HGE, or proposing legislation for Congress, he gave the issue to a government advisory commission.[3] President Carter requested that a report on genetic engineering be prepared by the President's Commission for the Study of Ethical Problems in Medicine and Biomedical and Behavioral Research, the successor to the National Commission discussed in chapter 3.[4] Two and a half years after the letter to President

Carter from the religious leaders, the commission released the report called *Splicing Life,* which was essentially a repudiation of the way the religious leaders had described the problems, the solutions they had proposed, and particularly their substantively rational form of argumentation. The report did not find "in the gene splicing now being planned or undertaken the 'fundamental danger' to human values, social norms or ethical principles that alarmed [the religious leaders]."[5] Given this conclusion, there was then no reason to limit the research of scientists.

Showing the strength of the rejection of the religious leaders' position, fully half of the analysis section of the report was dedicated to "clarifying" the concerns expressed in the "slogan"—"playing God"—that the religious leaders had used. Reflecting the tendency of government advisory commissions to focus on formally rational arguments about means, not ends, the executive director of the commission later wrote that one of the best features of the report was that "by carefully dissecting the complaint that gene therapy amounted to 'playing God,' the report was able to differentiate *important* concerns about means and consequences from *rhetorical* claims."[6] In terms of professional competition, the "clarification" of the "rhetorical" claims, such as "playing God," was a redefinition of the theologians' claims to make them fit the formally rational type of argumentation that scientists and bioethicists were using.

For many reasons, the report was the fulcrum of the HGE debate. First, it was the most influential text in the debate from the year it was published to the end of this study (1995).[7] This quantitative claim is backed by the judgment of the executive director, Alexander Capron, who claimed in 1990 that "the basic framework set forth in *Splicing Life*" had influenced the evolution of thinking on HGE during the seven years since the release of the report.[8]

Second, the *Splicing Life* report suggested that a "next generation" of the RAC at NIH should provide ongoing oversight of the problem of HGE.[9] Acting on a recommendation from the President's Commission, the NIH added a Working Group on Human Gene Therapy to review protocols for the first human gene therapy experiments.[10] The chair of the RAC at that time later went so far as to say that from the *Splicing Life* report and a congressional hearing on the report in 1982 "one can draw a straight line to the review process for gene therapy proposals currently in place in the United States."[11] Not only did *Splicing Life* justify the beginning of HGE in humans, it established which form of argumentation would be used in any decisions that would be made. This new subcommittee of the RAC was instructed to "consider explicitly issues such as those raised in the *Splicing Life* report," and it used the report to draft research protocols for scientists who wanted to conduct HGE experiments.[12]

Finally, perhaps the report's greatest influence was that it "quiet[ed]" the "fears" that HGE had generated among the public, which were threatening the scientists' jurisdiction.[13] The report was able to describe the HGE debate in such a manner that the elected representatives of the public in Congress apparently felt that they could safely turn over their decision-making authority to scientists and government advisory commissions—thereby further institutionalizing the formally rational type of argumentation of bioethicists and enshrining the government advisory commission as the ultimate decision-maker.

Executive Director Capron later wondered whether, without the President's Commission in place, concerns of the sort expressed by the religious leaders "would have been the 'great wave' submerging all work in human genetic engineering." However, referring to a later lobbying campaign by religious leaders, he continued that the "recent attention to another clerical letter suggests that considerable weight may be given to objections in the absence of a commission; in resisting the call for a ban on germline experiments, leaders in Congress pointed to the pending proposal to establish a new commission to continue with the analysis begun in our report on *Splicing Life*."[14] Although Capron felt that calming the fears of the public is an "honorable function" of a government advisory commission because it entails "peeling away mistaken ideas and rhetoric," one must wonder how "mistaken ideas" are identified and how it is determined what is "rhetoric" and what is argument. Here I take the position that these evaluations were influenced by the form of rationality assumed in a government commission. In sum, the commission served as a buffer between the concerns of the citizens and the scientists, reducing the threat to the scientists' jurisdiction and advancing the jurisdiction of the bioethicists.

Given the distrust of scientists by the public, how was this commission able to defuse the criticism and "calm" the public—essentially leaving the scientists to continue their research unmolested? There are two issues here: why the report of the commission was so influential, and why it had such a pacifying tone.

To respond to the first issue, the influence of the report is derived from the "neutral" appearance of the commission. Political scientists have suggested that the primary reason for the continuing use of presidential commissions in general is that they appear to be "non-political" and "free from the taint of partisan political advantage."[15] This particular commission was influential because it gave the appearance of not siding with any group involved in the jurisdictional dispute, as though it was the neutral arbiter between the theologians and the scientists. This neutrality was established through the use of formal rationality. As discussed earlier, government decisions that are only distantly legitimated by democratic processes, such as those of government advisory commissions, are only legitimate if they appear to be following rules. The President's Commission was then influential because it claimed not to be setting ends, but only calculating the consequences of alternative means for ends that were themselves outside the debate. When ends are introduced in the report of the commission, it is made clear that these were not set by the commission or by any one group in society (such as the scientists), but are rather the universal ends held by all citizens, and thus worthy of being followed with rulelike devotion. Without this "doctrine of liberal neutrality" the President's Commission and other ethics commissions "could be seen merely and inevitably as a manifestation of state interests, including the interests of powerful groups with special access to political authorities."[16]

The appearance of pursuing universal ends is also forwarded through the use of consensual decision making.[17] Consensus in American political culture is thought to be a process of enlightened derivation of the ends of the public by open-minded participants learning from each other. Compromise, in contrast, suggests that participants arrive with fixed ends in mind and thus are not open to the derivation of the "best" public ends.[18] In Capron's words, striving for consensus was beneficial because "it encouraged the commissioners to seek the common ground that best expresses the moral insights and values of Americans today, in light of our shared, albeit not uniform, religious and philosophical traditions."[19]

The perception of following rules was also forwarded in this commission by the formally rational link between means and ends. Since the debate was portrayed as *not* being about ends—which would have given the appearance of *setting* the rules—the commission was just evaluating what effect HGE would have on preset or assumed ends. Thus, the commission appeared to be evaluating facts about HGE, not setting ends—the most legitimate standing ground for a government advisory commission.

Finally, the commission influenced the debate by changing the myriad ends pursued in each of the particularistic communities in the United States to the thin, universal ends used by the commission. By deciding both to recommend policy that purportedly reflected the universal ends of the citizens and to "educate" these same citizens about the proper, thin ends to pursue, the commission set about changing the ends that it was supposedly following with rulelike precision. Thus, the commission was influential in calming public opinion because it set about changing public opinion.

To respond to the second issue, the report had such a calming tone because this was in the interests of the scientists on the commission. This of course contradicts the notion of "pluralistic neutrality": commissions that forward the interests of one group should not be influential, because their lack of neutral rule following would be exposed. The most telling insights about the transformation of the HGE debate lie in this tension between the need to appear neutral to all groups and the desire to forward the interests of one group over another.

Scientists were able to have their interest in calming the public forwarded because their position ended up being "correct"—but only after the arguments of their opponents had been translated from substantive to formal rationality, rendering the arguments senseless. Most critically, this translation—which was not done with animus—was not announced, but rather was the result of attempting to satisfy the "neutral" requirements of a government commission. When the problems identified by the theologians were examined by the commission, they seemed less pressing after their underlying arguments were translated, and thus the theologians were made to appear to be in agreement with their own irrelevance to the debate.

The commission appeared to have evaluated all the arguments of the theologians and to have found them wanting, although in actuality it had examined arguments in which the ends had been commensurated to the bioethicists' ends, and the link between means and ends had been similarly changed. The great power of the *Splicing Life* report for altering the substance of the HGE debate, and also for delegitimating substantively rational arguments in future debates, makes it critical to examine in detail the process of creating the report.

The First Commission Meeting on the *Splicing Life* Report

The scientists on the commission had the Asilomar experience vividly in mind at the time when the letter from the religious leaders to President Carter was forwarded to the President's Commission at its third meeting. From the very first moments of what would be a two-and-a-half-year process, it was determined that any report that would be issued should calm the public's concern so that scientists could continue their research. That is, the commission determined a priori that its report would defend the scientists' jurisdiction against the gains that theologians and others had made.

Although the commission had a list of topics to address in its original mandate, HGE was not one of them, although statutorily the commission was allowed to add topics at its discretion. According to Renie Schapiro, the staff member who worked the longest on the report, the staff was very interested in addressing HGE when the letter arrived.[20] The commissioners, however, were reluctant to do so, for fear of fueling public concern in the wake of the Asilomar controversy. The staff, perceiving the commissioners' reluctance, was able to convince them that a report could straighten out the confused public:

> When we brought this up with the commissioners, there was quite a lack of interest in it actually. . . . and there was truly a pivotal moment which made the difference in my mind, between whether we took it or not—and that was when we flashed a *National Enquirer* article, which talked about that there were going to be some kind of hybrid plant/animal things. . . . We showed this to the commissioners and said . . . the issue is not whether genetic engineering is a good or bad thing, or whether there are problems, but how the public perceives it. Because if this is the kind of stuff the public is seeing then, do we have an obligation to at least clarify for the public what the real issues are and to separate fact from fiction? And it was that argument that won the day with the commissioners. That the public was misinformed, was getting increasingly misinformed, and that our job ought to be to inform the public in a better way.[21]

The commissioner who was initially the most opposed to conducting a study of the ethics of HGE was medical geneticist Arno Motulsky of the University of Washington Medical School. In his first comments in this debate, in which he became a principal participant, he stated that his "general reaction [was] negative" toward conducting a study. "Maybe I am too close to it," he continued, "but my feeling about all these matters is that the public and the media . . . have been much too worried about some of these things."[22] He then raised the Asilomar debate as an event that had gotten the public overly excited and had led to calls for the public regulation of scientists, something he opposed. In a 1997 interview he reiterated:

> All these horror stories, possible horror stories, turned out to be groundless. So in the meantime, the whole public was really worried about it. The Asilomar conference was called early to come to grips with some of this, and again, it was publicized. In retrospect, again, people will say, well, it fulfilled a purpose because the scientists said we were responsible and responsive to the public, and we'll do everything to make sure things are correct. And we will get the public fully informed. But to keep the public fully informed about something that is a very, very low probability just raises fears, which in retrospect may not have been appropriate.[23]

In this first meeting discussing HGE, Motulsky eventually acquiesced to the study by saying, "It is conceivable we might do some good by . . . see[ing] where we stand, and try[ing] to cut down some of the hysteria and excitement that has sometimes happened in this particular area."[24] The operating guideline for preparing the report, according to one staff member, "wasn't quite the open-ended, 'Let's explore this issue' as much as it was, 'let's make sure the public doesn't have some crazy ideas.'"[25]

Motulsky clearly shares the interests and concerns of the broader scientific profession, as discussed earlier. According to a staff member, Motulsky "saw himself as a representative of the scientific, and particularly the research community," and was "very intent on the science going forward in this area."[26] Motulsky thought that the commission "should reassure the public that this was important science, would promote the public good, and should be able to go forward without excessive regulation or restriction; excessive as he and his scientific colleagues saw it."[27]

Motulsky was able to have his interests acted upon because, according to the staff, he had "a tremendous amount of power and authority." Schapiro said that she "was on the phone with Arno all the time—sort of getting his OK. . . . Not every report had one person who was so closely identified with the issue."[28] When asked in the interview whether Motulsky's disagreement with a draft would have swayed other commissioners, she stated, "In fact, I don't think I would have ever gotten to the point where he would have disagreed with the draft. He would have seen it before it got to the [other] commissioners and we would have smoothed out our differences."[29]

It is important to clarify here just what the public was perceived to be misinformed about that was generating fear, because the commission's perception of the misinformation changed as the report evolved. In this first meeting, the misunderstandings of the public were considered to be factual.[30] For example, in an exchange between Motulsky, Capron, and another commissioner about the distinction between the cumulative effect of somatic compared to germline engineering, after Motulsky corrects Capron for a scientific mistake, Capron concludes that "that is exactly the kind of distinction I think it is useful to make to the public."[31] For example, the public needed to be made aware of the fact that scientists were not currently creating plant-animal hybrids that pump gasoline, as suggested by the *National Enquirer* picture, and that this was not even conceivable in the near future. Later, after it was accepted that the substantively rational arguments of the theologians would have to be translated to formally rational arguments, the mandate to "calm the fears" resulted in calming the fears generated by using substantive rationality, not those resulting from factual misunderstandings.

The scientists and others on the commission, particularly Motulsky, were interested in calming the public that had been overexcited by the Asilomar debate. Motulsky himself was in a position to strongly shape the report. Yet for any report to have the desired effect, it could not appear to be serving the interests of scientists, but must appear to be based on the public interest and to follow the public's values.

It has become a sociological truism that the most effective power is that which is exercised invisibly.[32] The report could not serve its function of defending the scientific jurisdiction if this interest were stated baldly at the beginning. Rather, the interest in calming the fears of the public would be legitimated through another, apparently neutral claim: that government commissions had to make decisions that were calculable. The first change in public presentation was to stop saying that the commission needed to allay fears, and to begin saying that the commission had to dismiss speculations about HGE technology. This transition was introduced in this first meeting by Motulsky, who thought that "we probably should only talk about the foreseeable future. To talk about what happens 50 or 300 years from now I think is not too appropriate because there are so many issues that we need to face immediately that we should stick to the pragmatic." He argued that "one problem in this field is to talk and think about scenarios that are so futuristic that upset the public, that get into the media, in the sense that always the most sensational is being picked out, and that is what people lock onto." His conclusion was that the commission would "have to be careful and paint what is possible, what the facts are, and what the facts aren't."[33] At the end of the meeting the commissioners concluded that the value of informing the public was worth the risk of stirring up more fears. Accordingly, they directed the staff to prepare a limited study of the ethical problems with genetic engineering and to find out whether these problems were being addressed by other government entities.

The First Draft of the Report

If one does not consider forms of argumentation or issues of jurisdiction, the first draft of the President's Commission report, written in September 1980, was not in the interests of scientists and other proponents of genetic technologies.[34] It listed fifteen unresolved problems with genetic technologies, such as possible biological warfare—issues that assuredly would not have had a calming effect. But taking into account forms of argumentation and jurisdiction, we see that the first draft described the problem of HGE, and genetic engineering more broadly, so that scientists clearly had jurisdiction over the problem and so that the form of argumentation used by theologians was not relevant. For example, under the heading of "gene therapy," the first two major concerns are "the safety of the patient" and the "safety of future generations." Nonmaleficence is therefore set as the assumed end in these arguments, and the only question is whether the means of "gene therapy" maximizes the end. It is scientists who have jurisdiction over determining the safety of experiments.

At the end of the document, in the sections titled "control of evolution" and "human identity," it appears that a few problems that would be placed in the theologians' jurisdiction will be addressed. However, the introduction to these issues suggests that they are actually being defined in such a way as to place them under the jurisdiction of scientists. These issues "have a mystical flavor that makes them particularly amorphous, difficult to deal with, and exasperating to *scientists,* whose response is usually to brush them aside," the section begins. "Yet, stripped of some *quasi-religious rhetoric* about interfering with nature or assuming God-like powers, these concerns too, can be stated in quite *practical* terms."[35]

In "practical terms" the first problem with human beings controlling their own evolution is that it could be hazardous to future generations who might need what are now thought of as "bad" genes. Although I do not deny that this is an ethical problem, or one that concerns theologians, it asks whether HGE will maximize the preset end of nonmaleficence, which is a question of fact under the jurisdiction of scientists. If the debate had been described as determining to what end these "God-like powers" *should* be used, it would have fallen under the jurisdiction of theologians, who interpret what is "God-like." This draft was sent to the commissioners before the next meeting so that it could be discussed.

The Fourth Meeting of the Commission: Mobilizing State Constraints for Calculability

At the next meeting of the President's Commission, in September 1980, the commissioners heard extensive testimony from government officials, scientists, and philosophers about what they saw as the important problems with HGE. The commissioners then debated whether to conduct a more thorough study than had been done for the draft they had received, and if so, on what assumptions it should be based. Motulsky again took an active part in the debate, stating that there were many "areas that can be clarified and demystified" in HGE.

At the beginning of the afternoon session executive director Capron summarized the discussion among the commissioners. Assuming the scientists' interest in calming fears, he pointed out that many "ethical and social concerns" have, "by their very magnitude and somewhat elusive nature, evaded careful attention," but are now "getting popular attention, . . . and some of that popular attention is inflammatory, and I guess you could say *borders on science fiction.* The kind of treatment that *The National Enquirer* gave to the issue, which we distributed to you, illustrated with cartoons of walking cactuses . . . [and other] applications which excite the public, and perhaps frighten them."[36]

Clearly, any report that they would create would try to calm peoples' fears by not allowing any "speculations." Capron suggests that a study could "try to separate those issues which seem realistically close at hand, . . . separating realistic uses of genetic engineering which I call straight-faced questions . . . from those which are more farfetched."[37]

The commission was in something of a bind. Capron thought that although the "expression which we heard from the church groups" was a reflection of the "somewhat confused" public concern, it nevertheless "deserve[d] to be taken quite seriously."[38] The White House also wanted the commission to conduct a more thorough study. The commission could not just ignore the "farfetched" problems that the religious leaders had raised, simply to placate the interests of the scientists. It was therefore stuck between Motulsky's position of not wanting to address any topic that he considered a "speculation" that would lead to public fear, and the need of a government commission to be responsive to the citizen concerns that it perceived to be represented in the letter from the religious leaders.[39]

The first step toward satisfying the jurisdictional interests of scientists in avoiding fear-inducing speculations, but in a "neutral," responsive manner required of a government commission, was taken by commissioner Albert Jonsen, a bioethicist. Jonsen offered a replacement for the "no speculations" requirement that would ultimately be seen as more legitimate: a distinction between "vague" and "concrete" problems. There are vague concerns and there are those "which are really much more specifiable, definable, arguable kinds of questions," he began. Although both types of questions are important, there are "probably different ways of going after them. Some of the questions asked by the [religious leaders] are of that very large sort. Some of the things [in the first draft] are of that sort."[40]

Jonsen wanted the commission to structure its investigation of HGE along three levels: first, "the broad, *speculative* issues that are extremely difficult to define"; second, "much more focused issues, but issues which don't seem to fall within any clear means of controlling what people eventually do about them"; and third, "more focused issues that do fall within a fairly clear range of controls" (such as safety issues). "Speculative questions should be discussed in conferences that people have from time to time," he said, and should not be dealt with by the commission.[41]

Jonsen's suggestion, like Motulsky's, would ignore the "speculative" questions of the religious leaders that might raise fears, but for a much more legitimate reason than the self-interest of scientists: the requirement of a government advisory commission. He legitimated his argument through the commission's mandate from Congress to report its findings and "any recommendations for legislation or administrative action" to the president, to Congress, and to each federal agency to which a recommendation in the report applied.[42] To fulfill this part of the mandate, it could be argued, reports would have to be conducive to the formation of policy that could be used by government officials. Jonsen clearly took this charge quite seriously.[43]

However, the commission had "*chosen* to speak to many different audiences, depending upon the topic—not only the President and Congress, to whom it reports directly, but also the American people."[44] I emphasize the word *chosen* because, unlike the commission's role in recommending policy, its educational mission to the American people was not explicitly laid out in the authorizing legislation.[45] This created a tension between the commission's chosen educational functions—such as calming the public—and its policy-recommending role. The constraint to describe the HGE issue in terms amenable to policy and procedure—that is, in calculable terms—turned out to be a resource that people could use to shape the public education component of the report in the direction they desired. Describing problems in the thin, formally rational terms corresponded nicely to the educational mission of calming the public.

Jonsen's argument was powerful with the commissioners because it satisfied both the implicit educational task of allaying public fears and the explicit policy-making task required of the government commission. Jonsen connected all the levels of "vagueness" to the degree of control that a government entity could have. He argued against accepting issues on the second level, for example, because they "don't seem to fall within any clear means of controlling what people eventually do about them." The third level—issues that "fall within a fairly clear range of controls"—is appropriate for a government commission.[46]

Jonsen seemed to believe what historians have said about decision making in bureaucracies: that making policy and procedure requires legitimating decisions through calculable facts or rules and not setting ends. Put differently, if the consequences of means cannot be calculated, the debate becomes one of setting ends. To drive home the point that the commission should not set ends by discussing the vague issues in his first or second levels, Jonsen told the story of the European discovery of the New World in 1526. The king of Spain asked a group of theologians what the obligations to the inhabitants of the New World were—that is, what *should* be done. This commission of theologians produced a "splendid report, a remarkable modern document" about these obligations—which was, as history shows, ignored. Returning to the task at hand, Jonsen thought that if the commission addressed the vague topics, it could produce "a super paper about ethical obligations, but it wouldn't have anything to do with what happens from here on out."[47]

"Vague" concerns resist calculability, and are thus improper for a government commission. The concern for calculability is clear from Jonsen's criteria: issues at the second level are those for which "you can draw on a lot of data . . . to come to those kinds of determinations," while issues at the first level are those for which "no amount of data is going to answer" the question.[48]

Some of those present had misgivings about this approach. For example, Capron correctly anticipated what would result from this strategy. Summing up the commissioners' thoughts on the vague-concrete distinction, he stated that suggesting that the first level "is somehow very amorphous, is to diminish the importance of it. By not elevating it to the utmost importance, it almost says that these are things which are not likely to be taken into account in decisions that are made."[49] This concern was lost in ensuing debate.

The vague concerns raised by the theologians could now be ignored in the commission's effort to educate the public and change the HGE debate, not because addressing those issues was against the scientists' jurisdictional interests, but because it was a requirement of a government advisory commission not to do so. Agreeing with Jonsen's argument, the commissioners concluded that the first draft of the report should be rewritten using the vague-concrete distinction with "the larger speculative questions" at level one described but not evaluated.

Focusing on the Concrete Concerns

The staff did not immediately rewrite the report, but the results of the next step in the process solidified the intention to write only about concrete problems. Two months after the meeting, the draft the commissioners had debated was sent to a panel of expert consultants, selected to represent a range of views.[50] Following the draft report they had before them, the consultants focused on the concrete issues, such as biological warfare and Asilomar-type safety issues, and not the concerns that had animated earlier debates, such as whether human beings should seize control of their own evolution.[51]

The response of Sheldon Krimsky, professor of urban and environmental policy at Tufts University, and a critic of scientists during the Asilomar controversy, was influential for the creation of the next draft. Unlike the other consultants, Krimsky outlined an entire method of analyzing problems that was highly formally rational. Because of its consistency with

Jonsen's conclusions in the last meeting, Krimsky's outline was used to restructure the next draft.[52]

Krimsky's method included the "clarification" of questions and the "reduction of ambiguity and vagueness." Only problems "ripe for solution" should be considered. Speculative issues should not be examined because the calculations could not be made: problems must be "close enough to realization for our social institutions to comprehend the diverse consequences of one choice over another." His process then focuses on the "identification of institutions in place that can deal with" these issues. It appears that if a "clarified" issue cannot be addressed by a government institution, it should not be considered.[53]

In just three months, by March 1981, the relatively short preliminary document had become a fifty-one-page draft, which shows the effect of accepting the assumptions that the document must calm fears, and that it must contain no speculations or vague concerns that cannot be acted upon by government regulations. The report describes and addresses fourteen problems by first raising the issue, discussing it, and then, following Jonsen's and Krimsky's advice, describing the regulatory device that could be applied to it. For example, the first concern is "will the drugs and biologics produced by these methods be safe and effective?" After the discussion, it is concluded that the federal Food and Drug Administration (FDA) has "begun to address questions" involved with DNA technology. Similarly, the concern "do the risks of gene therapy to patients mean that it ought not to be done?" is dispatched by stating that the commission advocates an expansion of the Recombinant DNA Advisory Committee at NIH to address this issue. The result of requiring that the problems discussed in the report not be speculative or vague, and that they be suitable to government action, is failure to address the "fundamental danger[s]" pointed to in the letter from the religious leaders.

The constraints used to create this draft were so stringent that most of the problems addressed in it did not involve HGE, but rather the nonhuman applications of genetic technology—the only genetic techniques not considered speculation at that time. Moreover, using these constraints there is little discussion of ends, so all the issues can be resolved through scientific data. This became the thinnest draft of the HGE report and the one most amenable to policy formulation by unelected officials.

Refocusing on the Vague Problems

A second round of consultants received the March 1981 draft of the report for review. In late March and early April they met as three separate groups of four to seven consultants plus staff in daylong meetings at the commission's Washington, D.C., offices. (Some consultants participated by telephone.)[54] The consultants seemed concerned that the vague issues were those the public cared most about, and that to be effective, the report would have to address those issues.[55]

Although many participants at these meetings were calling for a refocusing on the vague problems with HGE, it was Leon Kass (see chapter 2) whose suggestions seemed to have been followed most closely.[56] Since Jonsen's strategy of ignoring the vague problems identified by the religious leaders apparently would not work, Kass offered a method to resolve the contradiction between the commission's constraint of having to seriously address these problems and its need to not address them because they were speculative and vague, and thus likely to cause fear in the public. In a letter to Capron summing up his concerns with the draft, Kass stated that he understood "that you mean to conduct the inquiry under the heading of 'technology assessment.'" (Technology assessment evaluates the consequences of available, not speculative, means.) However, "*for this reason,* you have trouble raising and addressing the more general and moral-political concerns that are on the minds of the religious groups and those non-scientists who worry about these things."[57]

Resolving this contradiction would require dropping the technology assessment focus, he argued. Addressing the concerns of the religious leaders would necessitate defining the ends that HGE forwards. That is, satisfying the religious leaders' concerns would entail adopting components of the substantively rational form of argumentation and rejecting the formal rationality inherent in the commission's assumptions:

> For to begin with "technology" . . . is to continue to give first place to power and derivative place to ends or goods or "values." It is of the essence of the modern view of technique . . . that it is neutral with regard to the uses to which it is put. Yet the human agents who practice or support science are not themselves neutral regarding the ends. Everybody acts on some notion, even if only implicit, of what they regard to be a good or worthy human life, for themselves or for their communities. The disjunction between the human view of the human and the modern technical view—"Now I have bricks, what can I do with them?"—is a deep cause of our problems with technology and with the relation between science and human affairs generally.[58]

The second of Kass's suggestions is that vague problems should not be dismissed, as they had been in the previous draft, but rather should be more thoroughly probed:

> On the so-called "softer" concerns, which some people may seek to dismiss *because* they are brought forward by religious leaders and

> theologians, I think you need to do much more careful and thorough probing. I am increasingly convinced that, properly analyzed and understood, the concerns expressed in such phrases as "playing God" or "dehumanization" are expressions of wisdom, even if passionately expressed and poorly argued for. What my fellow scientists call "emotional" may be more *reasonable* than the *rational,* coldly "objective" opinions with which they contend.[59]

Kass became a figure of irony in the evolution of *Splicing Life.* The commission ignored his first plea for the definition of human ends in HGE. To include ends in the commission's process, Kass had suggested reformulating the report following an earlier National Academy of Sciences (NAS) report that included discussions of "beliefs, ideas, and values" in each section. This suggestion was rejected by the consultant who was rewriting the report, partly because "it would have required major restructuring," but also because the NAS report "was so controversial" that it would not be "a good model to follow if we wanted a consensus document."[60] In other words, to create a report along Kass's lines, the commissioners would have to debate ends, and thus would be unlikely to reach a consensus. Consensus in the commission, which was necessary to show that universal ends were being pursued, actually required that ends not be discussed.

The commission did try to take Kass's second point seriously by thoroughly probing and looking for the wisdom in the concerns expressed in such vague phrases as "playing God," instead of ignoring them. But since the commission rejected his first suggestion—to begin with debating human ends or values—the probing was based on formally rational assumptions.

Indeed, the next draft was dedicated to this probing, the attempt to "raise, in as precise a fashion as possible, ethical and social issues that ought to be considered alongside the medical, scientific, technical and economic issues that already arise in the decision-making process."[61] This raising of vague arguments in a neutral light without translation into formally rational assumptions could not stand up under further scrutiny of the commissioners when the draft was distributed at the next meeting.

From Probing Problems to Translating Forms of Argumentation

The eleventh meeting of the commission took place in July 1981, nearly a year after the commissioners had last addressed the issue of HGE. In the meantime they had been working on many of the topics assigned to the commission. Reflecting the multiple tasks of the commission, the discussion of HGE took up only the morning of the first day of the two-day meeting, with the rest of that day devoted to discussions of access to health care and continued discussions about how to define death. The second day of the meeting was devoted to informed consent in medicine and the patient-provider relationship.

To summarize the evolution of the report, in previous deliberations and drafts the commission had begun by describing only the problems with HGE that would support the scientists' jurisdiction. Other issues (vague problems) were to be ignored, following Jonsen's earlier argument, on the grounds that addressing them would raise fears and would not lead to constructive policy proposals. This approach, as the consultants pointed out, did not address the issues that had been raised by the religious leaders. The commission then moved to explicate those vague problems in a neutral manner in the next draft.

During the discussion of this most recent draft, a critical event occurred. The emphasis in the report changed from probing the vague problems raised by the religious leaders to probing, and ultimately translating, the form of argumentation used to describe these problems. That is, the form of argumentation was translated from substantively to formally rational.

Jonsen stated in the July 1981 meeting his belief that the draft before them was "deficient in its treatment of the religious viewpoint" and that it did not address the questions posed by the religious leaders.[62] Essentially, he was arguing that if the vague issues were going to be addressed, they had to be thoroughly explicated. Yet, since the report was intended to allay public fears, it had to deal with these "big, looming, more incomprehensible questions . . . in a way that avoid[ed] the doomsday rhetoric and the great vague fears that are often suppressed." The draft before them did not do this, and the commission thus had a "very tough balancing act to do" between what I would describe as the constraints of adequately responding to the questions asked and of satisfying the desire of Motulsky and others to calm the public.[63]

Jonsen's solution to this predicament was to think in a different way about the problems raised by the religious leaders. Just as "the potential for the control of human beings that was raised by the discovery of the Americas [by the Spaniards] set up a problem that was quite new and quite different than any ethical problems that had been faced by civilization up to that time," the new challenges in HGE were "very similar to the problem that faced the Spaniards." The Spaniards assigned a professor to "set out a new structure for thinking about this problem," and the commission needed to do the same.[64]

This "new structure for thinking" was the formally rational type of argumentation, which Jonsen believed could answer the problems raised by the religious leaders in a way that would not cause fear. Jonsen began by arguing that the primary impediment to the commission's addressing the concerns of the religious leaders is that people do not understand the theologians' seemingly vague "language." "One of the basic problems with the religious viewpoint is that the theologians and religious ethicists use a language which is metaphorical and symbolic very frequently. . . . [It is a] theological language, and nontheologically trained readers have a tendency to feel it is extraordinarily vague. Well, it isn't vague. It is metaphorical and has very clear meanings to the people that use them. And we need a chapter that can appreciate the underlying meaning behind much of that language."[65]

This "underlying meaning," which is "clear to the people" who use it, should nonetheless be translated into a formally rational language because

> it is important, I think, to recognize that the churches and the groups that brought the original problem to us . . . represent what might be thought of as the mainstream Christian tradition, which has . . . very often approached ethical issues by using a certain language which comes out of their tradition, but then when it comes to an analysis of those issues has done it largely in terms that are quite similar to a secular analysis.
>
> And you will find that those churches . . . will have *concerns* quite similar to those that one might find in society in general. And therefore, part of the translation issue is to say, "Can you move from the *language that they use to express their concerns* to the *substance of those concerns*, which is quite similar to broad social concerns?"[66]

Jonsen is saying here that the commission does not have to ignore the vague problems, but can simply change the "language" used to express them. This quotation is ambiguous in that there are two ways of converting religious language. One method, which I have called translation, used by many of the theologians, is to refer to the richness of the theological tradition and then translate the ends to secular language, resulting in myriad ends. Another is the method of the bioethicists, which is to take the religious arguments and commensurate—that is, throw away excess information that does not fit into the preestablished ends. Metaphorically, instead of creating shapes that fit through holes, the latter method takes a theological square and shaves it off so that it fits in preexisting round holes.

Although the commission later effectively gathered a list of the problems related to HGE that the religious leaders identified, it did not consider the range of ends deemed important by the theologians, nor did it use the same connections between means and ends. By translating the theologians' vague language, or arguments, to more concrete arguments, the commission translated the theologians' arguments from substantive to formal rationality, while portraying its work as an accurate depiction of the theologians' views. Put simply, the commission had moved from describing the arguments that vague problems were indeed problems to translating the arguments themselves. It pushed the myriad shapes of the substantively rational debate into the preset holes of the formally rational one that both the bioethics profession and government advisory commissions preferred.

Seeking Clarification from Religious Leaders and Theologians

Taking seriously Jonsen's concern that the draft of the report needed to address the concerns of the religious leaders, in the fall of 1981 the commission asked the signers of the letter to President Carter to have papers prepared that explored "the ways in which religious traditions and doctrines contribute a perspective on genetic engineering different from those expressed in a secular context."[67]

The religious leaders in turn asked theologians from their traditions to respond to the commission, and two did so, writing on behalf of the National Conference of Catholic Bishops and the National Council of Churches.[68] The theologians did indeed describe concerns similar to those expressed by secular writers, although they tended to have more interest in the vague problems. Yet one of the two papers, as well as input from other theological consultants (see below), demonstrated that although the stated problems might be the same, the form of argumentation that underlay these problems was markedly different. As many of the theologians pointed out, the religious leaders who had written to President Carter had identified problems that could also be described in a secular manner, but their form of argumentation implied different methods for identifying problems: different ends and different connections between means and ends.

Consider the document produced by Methodist theologian J. Robert Nelson for the National Council of Churches.[69] It described many problems with HGE that the commission was already aware of—and many of which it had already dismissed as vague in earlier meetings. Nelson described the problems of eugenics, the engineering of animals and plants, the crossing of species barriers, and the social, economic, and political implications of these technologies. For example, in the case of "modifying germ cells," which the commission had debated extensively, Nelson concluded, in a passage that was quoted in the final report, that "in *principle* there is no reason why a prohibition is needed against modifying the genetic content of germ cells as distinct from somatic cells."[70] This was immediately followed by a risk/benefit analysis, implying that the debate was one of calculating whether the means of germline HGE maximizes beneficence as an end.[71]

Passages such as this give the impression that Jonsen was correct in thinking that not only were the problems identified by theological and secular authors the same, but that the two forms of argumentation were the same as well. Yet the remainder of Nelson's document contradict this impression. Nelson devoted ten of eighteen pages to outlining ten "theological affirmations" that define his "cluster of values": the beliefs or ends that HGE does or does not support. This is the common method of theologians, delving into the Christian tradition to determine what one's ends should be, and then expressing these ends in more or less secular language. This gives the paper a thicker, substantively rational tone because the ends are defined and argued for. They are not treated as universal, nor are they commensurable, at least with the common commensurate ends being used by scientists and bioethicists. To take one telling example of Nelson's "theological affirmations," in discussing the problem of genetic disease, he seems to be arguing for an end opposed to beneficence as articulated in the *Belmont Report:* "There are varieties of suffering: some are clearly evil, some are beneficial, some are morally ambiguous. The wisdom of Christian faith counsels neither resignation to suffering nor avoidance of it. It teaches patience and endurance where suffering is inevitable, and willingness to suffer when it is altruistic and vicarious. The prime symbol of faith is the cross, representing even the divine suffering on behalf of humanity in the person of Jesus Christ."[72] Nelson is not saying here that people with genetic disease should just suffer. Rather, in his analysis beneficence as an end has a more nuanced meaning than commonly expressed in HGE debates.

The papers from the theologians, along with a copy of the letter from the religious leaders to President Carter, were sent to a group of distinguished theological consultants.[73] They were essentially asked to evaluate Jonsen's earlier conclusion: "whether the theological concerns about genetic engineering differ from the concerns expressed in a secular context."[74]

Many of the theologians questioned the assumptions behind the request to identify differences between theological and secular views. The overall point of many was that the identified problems were indeed the same, and thus should be acted upon by the commission. However, the form of argumentation used to identify these problems as problems—and ultimately to suggest solutions—was distinct.

For example, Karen Lebacqz, associate professor of Christian ethics at Pacific School of Religion, stated that the answer to the commission's question about the similarity of secular and theological concerns is "both a 'yes' and a 'no.'" "The concerns are the same in many ways," such as the "level of harm permissible to human beings." Yet there is a difference, which "emerges primarily in methodological issues." The primary "methodological issue" is that "instead of looking for general principles such as respect for persons and beneficence that are consonant with the Christian tradition but also widely applicable . . . the crucial question is how we need to live in order to become the sort of people that our Christian story tells us we ought to be." Instead of looking for "widely applicable" ends—such as those used by bioethicists—the theological perspective would suggest that Christianity is an entire way of living, a whole series of ends, which cannot be commensurated to the (few) *Belmont* ends.[75]

Reflecting on the liberation theology strand of Protestant Christian social ethics, Lebacqz concludes that one end to be pursued is to fulfill God's desires for the poor, because "God is understood to side with the poor and oppressed—with those who have no power or who find themselves at the mercy of the dominant power structures." "In light of this perspective," she disagrees with one of Nelson's points, and translates her end—derived from her broader tradition—into secular language: "the primary criterion" for "judging the uses of genetic engineering and related biotechnologies" would be "the impact of those technologies on the balance of power and the life prospects of those who are poor, oppressed, or deprived of power in our society."[76]

As we have seen, theologian Paul Ramsey had been a defender of the theological jurisdiction in these matters, and as we would predict, he had a similar reaction. "You ask 'whether the theological concerns about genetic engineering differ from the concerns expressed in a secular context,'" he began. "My answer is that they do not; that, indeed, the concerns are disappointingly similar to worries the commission has heard before. But my answer *also* is that the question is not one to pose." He then challenged the distinction between "theological opinions" and opinions from "so-called secular sources." All the concerns are legitimate, he continues, but the theological issues are different in their "depth, grounding, reasons and perdurance."[77] He wanted the commission to recommend "anticipatory nevers" in light of those concerns.

Gabriel Vahanian, professor of religion at Syracuse University, saw the entire investigation as avoiding the central problem: that U.S. culture "has marginalized religion and in turn is itself being marginalized by its own scientific 'ideology.'" Asking whether salvation had been "somehow been reduced to health," he sounded much like a theological version of Kass as he went on to say that "unless we come up with a theologically pertinent understanding of technology and its utopianism . . . our do's and don'ts . . . sooner or later will give way to . . . a surrender of ethics to science, a subjugation of 'ought' to 'is.'"[78]

Ready to write the next draft, the commission's staff struggled with how to interpret the additional theological input. Since no one among them had theological training, the staff turned to Jonsen, the former Jesuit philosophy teacher, for guidance. Concluding that the theologians identified the same problems as the secular experts, he limited consideration of arguments to those that could be used to make "*practical decisions* about the various sorts of problems posed by genetics." Ramsey's input, for example, while "telling us that we should be able to definitively say no to certain sorts of things," unfortunately expressed these things "in somewhat general terms. It is hard to know ahead of time exactly what we are supposed to say no to."[79] Examining Ramsey's letter to the commission shows that, in typical substantively rational style, he does not explain how one links the means of HGE with the ends he spends so much time articulating. He is concerned with ends, and considers means such as HGE only insofar as they tell us about these ends.

Why did Jonsen limit the legitimate forms of argumentation to the "practical," even though the public could be educated using imprecise forms of argument, as theologians and others had done? As before, Jonsen emphasized that to recommend rules and regulations for the state, decision making must be calculable and practical to decision-makers. He concluded that the "*general perspectives* that arise out of the Christian faith are extremely *difficult to transform into policy and procedure*."[80] Thus, the problems identified by the religious leaders would be considered in the report, but because the ends—"the general perspectives"—of the theologians were difficult to "transform into policy and procedure," these ends would be commensurated into the formally rational ends used by bioethicists. The commission stated in its final report that "while religious leaders present *theological bases* for their *concerns*, essentially the same *concerns* have been raised—sometimes in slightly different words—by many thoughtful secular observers of contemporary science and technology. . . . Therefore, no attempt need be made to limit an examination of the content of various specific concerns to the *religious format* in which some of the issues have been raised."[81]

Writing the Final Report

The theological input gathered and analyzed, the staff continued work on what would essentially be the final version of *Splicing Life*. It was now early 1982, and recently hired staff member Allen Buchanan began to work on this report together with Renie Schapiro. His understanding of his task is informative. By the time Buchanan joined the staff, the function of the report had evolved from presenting facts to presenting a form of argumentation to the public. Reflecting on the assumptions the staff used for writing the report, in a later interview Buchanan stated, "What I took to be the main idea of the approach right from the beginning . . . [was a] combination of demystifying and debunking on the one hand, and clarifying on the other, and trying to bring out what were the really important concerns that needed to be addressed, while doing the service of trying to separate out the concerns that perhaps weren't so important or perhaps were misplaced."[82] He also stated that the debunking of bad arguments was undertaken to calm the public:

> I think there was also in the background, probably, a concern—certainly a concern of mine and I think it's a legitimate one—that though there are great risks, perhaps, in genetic technology, there are also potentially tremendous benefits and that it would be unfortunate if there were a kind of chilling effect on the scientific endeavor as a result of misplaced or confused concerns. . . . I think there was a desire on the part of at least some people on the commission and the staff to make sure that there wasn't an unwarranted chilling of the enterprise by misplaced concerns.[83]

Buchanan also understood that one of the tasks of the commission was to teach the public to use the new form of argumentation that was more amenable to policy debates:

> I think everybody's understanding of the work of the commission was [that] it was to try to elevate the public and policy-making discourse—that is, to perform an educational role and to prepare policy makers and political representatives and the public generally to be able to grapple with issues in bioethics, biotechnology, in a more fruitful informed way. So, on the one hand, it was important to sort of articulate what the real issues were; dispense with pseudo issues if there were any; articulate the concerns in a way that made them more manageable, make them more amenable to rational consideration and debate rather than just make slogans. "Don't play God" and things like that. So there was an attempt to articulate the issues—to make for a better public and policy-making debate.[84]

This was of course a worthy cause. But, once again, there is no objectively true interpretation of ethics, so the assumptions had to come from somewhere. These different assumptions are suggested by the difficulty the staff had in understanding the arguments of the theologians. Although Jonsen thought that the "metaphorical" language of theologians "has very clear meanings," to a staff untrained in theology and assuming a calculable, formally rational type of argumentation, the theologians made little sense:

> We looked at the letter, and I can remember a certain kind of frustration with what we took to be the evocative, but elusive character of the phraseology in the letter, and it wasn't as if there was a real detailed, coherent list of concerns or questions in forms that we could then pursue. It sent a red flag. And that's why to some extent, we took our task to be trying to do what we would like the letter writers to do. Mainly, to be more explicit about sorting out their various concerns. I think we made a sincere effort to do that. We would look at something in the letter like concerns about "playing God" or something like that and say, "Now what's really bothering them? It could be this, or it could be that."[85]

Moreover, the letters from the consultants were also "quite disappointing" because "it still wasn't clear in many cases what was really bothering people and that a lot of work had to be done to try to articulate these concerns [to] make them amenable to discussion."[86] The result was that "any kind of direct input from the religious people just dropped out of the picture; we just went on with it and that's where we started—with their initial letter, but at a certain point, we began to think we had a clear enough idea about what some of the issues were that ought to be addressed that responding to the religious input was not the only concern."[87]

The standard of making the debate amenable to policy discussions, as well as the desire to debunk "misplaced or confused concerns" that could have a "chilling effect on the scientific endeavor," resulted in a wholesale refashioning of the theologians' arguments from the point of view of the bioethics profession. This new approach, accepted by the commission, is apparent in the final draft. While the stated purpose of the report in the previous draft was to "raise, in as precise a fashion as possible, ethical and social issues that ought to be considered along side the medical, scientific, technical and economic issues that already arise in the decision making process," the final report had among its purposes "*clarifying* the concerns *underlying* the simplistic *slogans* that are frequently used" and "identifying the issues of concern *in ways meaningful to public policy consideration*."[88]

The Final Debate

In July 1982 it had been exactly one year since the commissioners had last addressed the topic of HGE and two years since they had begun deliberations.[89] The final debate in the commission involved two outside consultants who had read the draft of the final report. The first was W. French Anderson, chief of the laboratory of molecular hematology at the National Institutes of Health, one of the most famous researchers on HGE at that time, and one of the most influential writers in the debate. The other panelist was Nicholas Wade from the *New York Times,* who had previously written a book somewhat critical of the scientific community in the Asilomar debate. Anderson largely agreed with the report as it stood, but Wade had a different opinion.

Although he had not participated in earlier meetings, Wade recognized the underlying purpose of the report. "Despite acknowledgment of what the churches have asked it to do, [the report] does not in fact address seriously the questions the churches put to it," he began. Rather, he read the report "as an attempt to explain away, to de-emphasize problems, to soothe everyone down, not to get the public all heated up."[90] He claimed that the report had become "captive" to the "vested interest" of, and was thus acting as a "megaphone" for, the scientific establishment, trumpeting their views "almost without any modifications."[91] The motive he identified for the scientists was that they "had a bad experience in the public debate" over biohazards after the Asilomar conference, and wanted to avoid more controversy.

The first response to Wade was from Anderson who, after jokingly acknowledging that he himself was part of the scientific establishment, justified calming the public by stating that the members of the scientific community "were all hampered enormously" by Asilomar. "Work was slowed down . . . and we spent enormous amounts of money" building containment facilities. He concluded that "the scientific community and the medical community are very concerned that that sort of a waste doesn't happen again." Anderson then acknowledged that the draft "does attempt to soothe the issue, but of course I guess my bias—and I guess as part of the scientific community most of us feel this way—the sorts of things which have gotten the public concerned are things which aren't reality anyway: the changing of human nature, the building of robots, and so on and so forth."[92] A commissioner who had recently joined the commission later replied to Wade that perhaps the "larger role" of the commission was "in seeing that the proper information gets to the public, which would be reassuring, which would develop an environment in which the scientists can work with the necessary serenity and concentration that they need."[93]

Wade's criticisms forced the commissioners to respond with justifications for why a government commission should calm people by debunking arguments that caused concern among the public. Jonsen defended "a certain government interest in dispelling misinformation" by claiming that "we have important things to achieve in health and agriculture and all that, and if there is a dominant misunderstanding of the culture we are going to be continuously running up against road blocks."[94] That is, the universal end used by bioethicists—beneficence—justifies dispelling misinformation.

Yet the misinformation that leads the people to put up "road blocks" goes beyond misinformation about "facts," such as the inability of scientists to create plant-people, to "misinformed" forms of argumentation. For example, to illustrate a point Capron invented a story of visiting a "tribe of people who have had very little contact with the Western world, and who were suffering from a disease for which Western medicine had a cure." If these people respond that interfering with the disease "is playing God," they should be given "reassurance that the gaining of new knowledge is not a bad thing, that is, if that is what is meant by Godlike powers, that is indeed what makes humans special. . . . But once one has given proper deference to the basis of public anxiety, I'm puzzled by the notion that it's wrong to provide a reassurance that finding out things in and of themselves, as opposed to the uses to which it's put . . . is a bad idea."[95]

From Capron's tale we see that the notion of misinformation had traveled a great distance from factual misunderstanding to the substantively rational form of argumentation. Capron claimed it was a legitimate task to correct the misinformed argument that "finding out things in and of themselves" is wrong—but this is the substantively rational link between means and ends. He proposed to replace this misguided argument with the argument that means should be evaluated by how they are used—which is the consequentialist link between means and ends in the formally rational form of argumentation. Despite the debate at this meeting, and the criticisms raised by Wade and others, the report remained as it was.

The Content of *Splicing Life:* Does God Make Formally Rational Arguments?

The structure of the final report reveals the formally rational assumptions that were used.[96] The analysis begins with a statement that "the commission seeks ways to realize the benefits without incurring unacceptable risks" in HGE—implying that arguments about problems with HGE would be evaluated by whether HGE was the most efficient means of achieving ends. The report acknowledges that this formally rational method is complicated in this case because calculations are difficult. The greatest complication is that "while some people focus on particular consequences of various applications of genetic engineering technology, others are concerned about the acceptability of genetic manipulation per se. In this context, balancing risks against benefits makes little sense because actions, not consequences, are at issue."[97] That is, there are persons who reject formal rationality entirely, using substantively rational forms of argumentation instead.

The analysis therefore is divided into two sections. In the first section, "the commission considers theological and secular attitudes toward the technology as such, rather than toward its possible consequences, and attempts to *clarify* the nature of these concerns."[98] While the problems identified by the theologians are found throughout the report, what are labeled as their unique arguments are all placed here, under the "playing God" slogan, which is to be "clarified." The second part of the analysis consists of "an examination of the types of risks at issue."[99] Note that the report must find that the arguments outlined in the first section are unfounded, because if HGE is wrong, regardless of consequences, there is no reason to discuss its consequences.

Debunking and Commensurating Ends: "Playing God" and the Theological Arguments

The first section evaluates a series of translations of the claim, made in the letter from the religious leaders, that HGE would allow people to "play God." All these arguments have the substantively rational link between means and end. However, they assume only one translated end for theologians: namely, "not changing what God has already created."[100] "At its heart," the section begins, "the term [playing God] represents a reaction to the realization that human beings are on the threshold of understanding how the fundamental machinery of life works."[101] That is, "playing God" is an expression of awe: humans appear to have powers previously left to God. This first translation makes "playing God" only an expression, not an argument, and is not evaluated further.

The second translation of the phrase "playing God" is that it stands in for the argument that HGE is an arrogant interference with the nature that God has created. This argument is first debunked with the reminder that the public has already given jurisdiction to scientists for the pursuit of the end of interfering with nature: "in one sense all human activity that produces changes that otherwise would not have occurred interferes with nature. Medical activities [such] as the repair or replacement of a damaged heart are in this sense 'unnatural.'"[102] Moreover, none of the theologians consulted by the commission "suggested that either natural reason or revelation imply that gene splicing technology as such is 'unnatural' in this prescriptive sense." Indeed, "human beings have not merely the right but the duty to employ their God-given powers to harness nature for human benefit."[103]

The final translation is that "playing God" represents the opposition to creating new life forms. Once again, it is pointed out that scientists already have jurisdiction over this issue, without public concern, because "if 'creating new life forms' is simply producing organisms with novel characteristics, then human beings create new life forms frequently and have done so since they first learned to cultivate new characteristics in plants and breed new traits in animals."[104] Therefore, one cannot call this a problem "unless one is willing to condemn the production of tangelos by hybridizing tangerines and grapefruits or the production of mules by the mating of asses with horses."[105]

When translated as an argument that HGE is inconsistent with the end of preserving what God has created (or nature), the "playing God" claim makes no sense. It appears nonsensical because the writers create an analogy of HGE, and genetic engineering more generally, to the manipulation of nature, over which scientists already have jurisdiction, with little controversy. Pursuit of this end is shown to be a ridiculous quest for something accepted long ago by people who breed mules and eat tangelos.

Once all the theological arguments have been translated to this one set of substantively rational links between HGE (as the means) and this one end, and once the legitimacy of this end has been dispensed with, all substantively rational arguments and all discussions of ends are invalidated. The report can then move on to evaluating the consequences, or what are called "serious" concerns. What is considered to be specifically religious has been dispatched. The concerns of the theologians can be evaluated, if their arguments can be recast into a formally rational link between means and ends.

The move to this next stage of analysis occurs at the end of the discussion of substantively rational opposition to the creation of new life forms. The section concludes that although the substantively rational arguments about preserving God's creation do not hold up, there are two aspects of this problem that deserve "serious consideration." They warrant consideration because the underlying arguments are based in formally rational, consequentialist reasoning, and not in the substantively rational reasoning that some acts are "inherently wrong." The first "serious" argument is that since created life forms could reproduce, the possibility of "self-perpetuating mistakes" is of concern, but "the point is not that crossing species lines is inherently wrong, but that it may have *undesirable consequences.*"[106]

The second, the "mixing of human and nonhuman genes" into hybrids, sounds like a substantively rational objection, but is eventually translated into a consequentialist one. Raising the problem of whether a group of beings that are "partly human, partly lower animal" could be invented as a slave class, and appealing to the need for universal ends, any claim that this is "intrinsically wrong (and not merely wrong as a consequence of what is done with the hybrids)" will wait upon the development of societal consensus on two points: "what characteristics are uniquely human" and "does the wrong lie in bestowing some but not all of these characteristics on the new creation or does it stem from depriving the being that might otherwise have arisen from the human genetic material of the opportunity to have a totally human makeup?"[107]

The "serious" objection to human-animal hybrids reflected in the phrase "playing God" is "the concern that human beings lack the God-like knowledge and wisdom required for the exercise of these God-like powers." Evoking the stories of Dr. Frankenstein's monster, the sorcerer's apprentice, and the golem, the report notes the legitimacy of the concern that human beings may lack the scientific knowledge to avoid negative consequences. But, once again, this is not supported by substantively rational reasoning, because "if this is the rational kernel of the admonition against playing God, then the use of gene splicing technology is not claimed to be wrong as such but wrong because of its potential consequences."[108] The conclusion is that more scientific data is needed before the development of human-animal hybrids is undertaken, so that people will know what the consequences are. That is, the research must go on so that society can weigh the risks and benefits.

Consequences Section

With the theological arguments having been translated and debunked, the final section of the analysis discusses possible consequences of HGE technologies. The problems addressed here include medical applications, the evolutionary impact on human beings, changes in notions of parental rights and responsibilities, changes in notions of societal obligations and commitment to equality of opportunity, changes in people's sense of personal identity, change in the meaning of being human, eugenics, the issue of how to decide what research to engage in, and the relationship between business and academic research.

Even after the substantively rational link between means and ends has been dispatched, one might still be able to evaluate the secularly expressed ends used by the theologians, to see if they are maximized by HGE. This would result in at least a somewhat thicker debate than would occur if the debate were limited to the Belmont ends. However, it turns out that only ends that are calculable can be considered. Many concerns cannot be evaluated due to the "several types of uncertainty." First, many of the ends of society are not agreed upon, and worse, people in the future may have different ends than people in the present. Second, the development of the technology may itself change people's ends. Finally, consequences cannot be predicted in the distant future.[109]

Although this is not made explicit in the text, when the end for an argument is one of the four universal commensurable ends of the bioethics profession—autonomy, justice, beneficence, and nonmaleficence—then the ends are portrayed as certain, and it is possible to calculate whether the means maximizes the ends. When the end used in an argument is something else, perhaps one not claimed to be universally held, ends become uncertain, and thus the argument cannot be evaluated. For example, the report points out that one of the consequences of HGE may be that with new technological capabilities to enhance health, there may be calls to use the coercive power of the state to ensure that parents fulfill the obligation to have healthy children. Yet, unlike many of the other consequences of HGE, this one is evaluated because it violates autonomy. "Society has traditionally been reluctant to interfere with reproductive choice, at least in the case of competent adults," the report states. "Further developments in gene surgery or gene therapy may lead to further departures from the principle that a competent adult may always refuse medical procedures in nonemergency situations and from the assumption that parenting and reproduction are largely private and autonomous activities."[110]

The principle of autonomy is then fixed, assumed to be important for the present and the future. Not so for other principles or ends. Consider the discussion of the argument that the "increased ability to act for the well-being of the child" could change notions of parental responsibility. Unlike the fixed end of autonomy, in this case it is suggested that ends in the future could be different than they are today: "the boundaries of this responsibility—and hence people's conception of what it is to be a good parent—may shift rapidly. . . . New technological capabilities may change people's view of what counts as a defect. For example, if what is now regarded as the normal development of important cognitive skills could be significantly augmented by genetic engineering, then today's 'normal' level might be considered deficient tomorrow."[111] In other words, what would be considered eugenics today might be acceptable tomorrow because society's ends may change. Another example is the discussion of the possible consequence that HGE will change people's sense of personal identity. "Here again, uncertainty about possible shifts in some of people's most basic concepts brings with it evaluative and ethical uncertainty because the concepts in question are intimately tied to values and ethical assumptions."[112]

Once again, the commission's use of the principles adopted by bioethicists is implicit in the report. In another document, however—one written to summarize all the commission's activities—the commission claimed that in its work it "appealed to a number of ethical principles in its studies," among which three predominated: "that the well-being of people be promoted, that people's value preferences and choices be respected, and that people be treated equitably."[113] This is a list of the three Belmont principles, using different phrasing, before they were expanded to four in Beauchamp and Childress's textbook. There is no discussion of why these ends, developed in the context of a debate about experimentation on humans, are desirable when debating HGE.

Given that the report lumped all the theological *arguments* into "playing God," it should be no surprise that the chapter concludes with the statement that the commission "could find no ground for concluding that any current or planned forms of genetic engineering, whether using human or nonhuman material, are intrinsically wrong or irreligious per se." Moreover, many of the problems identified by the theologians were stripped of their ends, making it impossible to conclude that they were indeed problems. Therefore, "the commission does not see in the rapid development of gene splicing the 'fundamental danger' to world safety or to human values that concerned the leaders of the three religious organizations." Ironically, the chapter ends with a call for the substantively rational debate about ends that the commission had subtly selected out of the process: "the issue that deserves careful thought is: by what standards, and toward what objectives, should the great new powers of genetic engineering be guided?"[114]

The Power of Translation

One of the reasons why *Splicing Life* has been so influential is that it has been viewed as a neutral examination of the debate. It actually examined the problems identified by the theologians, such as the issue of the meaning of being human—problems that had not been a part of the debate since the days of Muller and Ramsey. The commission's examination of the problems brought forward by all participants in the debate has camouflaged how this report, while appearing to take the theologians' concerns seriously, actually changed the meaning of their concerns by using a formally rational type of argumentation to support arguments that, as formulated by the theologians, had been supported with substantively rational claims. What evidence do we have that the theologians would not have agreed to the reduction of their arguments to the phrase "playing God," its subsequent translation into the end of "noninterference with nature," and the final commensuration of their ends to fit with the bioethicists' ends?

Theological Ends

According to theologian Allen Verhey, although the commission deserves credit for trying to make sense of the phrase "playing God," "the phrase does not so much state a principle as invoke a perspective on the world."[115] It is not really an end at all, but rather a way of referring to the entire form of argumentation used by theologians—the "perspective." Moreover, by translating "playing God" into noninterference with nature, the report turns the God that the religious leaders are supposedly accusing the scientists of playing into the "God of the Gaps." In this view of God, "'playing God' means to encroach on those areas of human life where human beings have been ignorant or powerless."[116] God is therefore located in the "gaps" that scientists cannot explain.[117] Every one of the debunked meanings of "playing God" has this view of God: "fully understanding the machinery of life" is knowing what has been unknown to scientists; the nature that human beings are to not interfere with is only the nature they cannot already explain and manipulate; and creating new life forms is wrong only because "playing God" means that human beings lack the scientific knowledge to do it wisely.

The report itself implicitly acknowledges that the consulting theologians did not agree with the end being attributed to their arguments. The first section points out, quite rightly, that all the theological commentators agreed that for the dominant religious traditions in the United States, scientific exploration into areas such as HGE is within the legitimate powers of human beings and does not usurp the powers of God. The religions "respect and encourage the enhancement of knowledge about nature, as well as responsible use of that knowledge."[118] Even the most conservative of the consultants, Paul Ramsey, urged in another context that human beings should "play God as God plays God."[119]

For the theologians, "playing God" is forwarding God's myriad ends, as imprecise, contested, uncertain, and incalculable as these ends may be. The consulting theologians offered several of these ends that were not brought into the *Splicing Life* report. One end is the nuanced version of relieving suffering, described above, and another is that to further God's ends is to have a "preferential option" for the poorest among us.[120] These ends, however, are not calculable, even in secular terms, and thus are omitted from *Splicing Life*. It is unlikely that the theologians thought that only the calculable ends were important.

Other Arguments That Link Means and Ends

The other part of the "perspective" expressed in "playing God," and in theological arguments more generally, is how to determine whether or not means such as HGE support the ends implied in "playing God" as God plays God. *Splicing Life* implied that besides arguments about noninterference in God's creation, the extent to which HGE forwards other ends would have been evaluated by theologians using formally rational, risk/benefit analysis. This link between means and end is used by some theologians at some times, but the substantively rational link between means and ends is pervasive in theology more generally.

As we have seen, Ramsey held this view with regard to a number of ends, and his input to the commission continued this view. But even more liberal theologians than Ramsey do not make all their decisions by weighing consequences. For example, Karen Lebacqz wrote after the release of *Splicing Life* that its reliance on human knowledge in measuring consequences, and trust in "logic and rationality for answers," left "other modes of insight" out of account:

> Prayer, meditation, dreams, the insight that comes as an intuitive response—all these and many more are recognized as sources of knowledge in other traditions but largely ignored in our own. . . . [The danger] is not simply the possible deleterious consequences of new technologies. Nor is it merely the threats posed to deeply held values. It is our tendency to assume that even as we give lip service to human fallibility and ignorance, we nonetheless can trust both in our limited forms of knowledge and in the paradigm that posits knowledge as the key to action.[121]

Theologians who consulted with the commission, both liberal and conservative, did not base their arguments solely on consequences—on the maximization of ends. Nor did any of the theologians argue for the end of noninterference with God's creation. In sum, the requirement that the ethics of the commission be calculable and concrete in order to make the debate more amenable to policy analysis required that the theological ends and method of connecting means and ends be translated into formally rational terms. The translation to this common ethical language was in no way neutral.

What the Theologians Might Have Said

If the form of argumentation that supported the stated problems of the theologians had been included in the report, what might have been different? In the letter to President Carter the religious leaders pledged their own efforts to examine HGE, and the National Council of Churches (NCC) did set up a commission to study HGE, headed by theologian J. Robert Nelson (who also wrote the Protestant theological input for the commission).[122] The report was released at approximately the same time as *Splicing Life*. The results of this strikingly similar process, which occurred at approximately the same time as the commission, are telling.[123]

There are two pertinent differences. The NCC commission was made up of representatives from member denominations and units within the NCC, who were not experts in science or medicine, although they consulted such experts.[124] That is, there were no scientists or bioethicists in the position of making decisions about the report, only theologians.[125] The second difference is that, unlike the President's Commission, the NCC commission was not ultimately speaking to government officials, but rather to the public and members of their religious traditions. This meant that they were not constrained to use formal rationality.

In a comparison of the President's Commission and NCC reports, C. Keith Boone states that while the report of the President's Commission concludes that "the energetic pursuit of gene splicing does not pose a 'fundamental danger' to world safety or human values," these prospects "preoccupy" the NCC study. Not constrained by the scientists' interest in avoiding "speculations," the NCC report, in comparison to *Splicing Life,* "devotes less attention to the proximate and mundane uses of genetic engineering, and more to the remoter, more dramatic uses." Like their substantively rational forebears, these theologians talk about a huge range of topics at once—HGE, gene patenting, engineering of plants and animals, and more.

Most important, while the President's Commission declined in an early draft to have a section discussing the values or ends that HGE should forward, and used only the ends adopted by bioethicists, most of the NCC document is geared toward developing these ends or values. It begins with a short introduction, followed by a chapter on the facts of genetics (excerpted from *Splicing Life*), and then by a substantive chapter on theology. In the theological chapter, which takes up almost 30 percent of the entire report, the authors discuss the theology that would be relevant to setting the ends in this one area. Human beings as co-creators with God and as beings created in the image of God, redemption, and a host of other Christian theological topics are discussed.

Following this, in a section titled "Toward a Value Stance," the authors derive eight ends that HGE and other types of genetic science should pursue. These ends are stated explicitly, and are even numbered.[126] They are not portrayed as universally applicable for all issues, but they are as vague as the principles used by the bioethics profession (i.e., "justice"). Moreover, they are all written in secular language. The President's Commission might have arrived at the same list of ends through secular reasoning. In short, the religious community created a much more substantively rational document than the *Splicing Life* report. They did not have to placate scientists by not raising fears, and did not write the text in a manner amenable to use by unelected government officials.

The Commission and the Rise of Formal Rationality in the HGE Debate

It was not simply concerns about the separation of church and state that resulted in the translation of the theologians' arguments in the *Splicing Life* report. Kass made very similar, substantively rational arguments, using secular language, and his urging to debate the ends that HGE would forward was also dismissed by the commission. All substantively rational arguments were excluded: religious arguments were simply the most conspicuous.

A great deal of the influence of *Splicing Life* on the entire HGE debate lies in the tension between the two roles the commission selected for its work: making policy and educating the public. As Jonsen himself stated during the process, an educational document would not have been restricted to the calculability requirements that were imposed on the report. However, because of the commissioners' insistence that the report be useful for the creation of policy and procedure by unelected officials, it was thinned out, in spite of its additional use for educational purposes in the broader public and the broader HGE debate.

The religious leaders who wrote the letter to President Carter had described the problems they saw with genetic technologies, but had not asked for an evaluation of their arguments. Since the commission agreed to write the report conditional on no serious problems being identified, they could not take the religious leaders' concerns at face value. Instead, they had to explain why the religious leaders saw problems, if in fact there were none. The only conclusion the commission could draw was that the theologians were using bad arguments or had bad information. By the end of the process, it was clear to the theologians that both these conclusions had been reached.

At the meeting where the final draft was discussed, J. Robert Nelson asked to speak to the commission in the public comment period. I interpret his short speech as emergency boundary maintenance to reserve at least some role for theologians in the HGE debate. "Speaking for the National Council of Churches' interest," he began, "I would say that we have not asked this Commission to resolve our theological questions about human life. Your concern, I believe, is not whether some scientists are so-called playing God, or whether such pagan metaphors as Prometheus or Atlas or the crippled Hephaestus describe our current situation, or whether we have made a Faustian bargain, and so on." These tasks he seemed to reserve for the theologians.

What did Nelson want the commission to do? Not to decide what "playing God" really means, but rather to be concerned with "public safety, for public well-being, for public interest, not only of this nation but also at large in humanity." This, he thought, was "the commission's mandate and also its limitation."[127] He reiterated some of the problems that he and other theologians had raised, such as germline HGE, human-animal hybrids, and the patenting of life forms—problems that, contrary to the commission, he thought were not speculations but rather "practical" problems that could be acted upon now.[128]

While the theologians wanted a debate about regulation, they did not want an educational document purporting to show that the only ends invoked in the HGE debate were autonomy, beneficence, nonmaleficence, and justice. *Splicing Life* not only was used for making policy and procedure, but also, as Capron later suggested, transformed the broader debate and even defined the religious objections to HGE in ways that the theologians themselves would not recognize. For example, two years after the release of the report, it had become accepted that the religious view about HGE was encapsulated into the end of "noninterference with nature." In defending gene therapy, an influential participant in these debates stated that "we should not be surprised that discussions of human gene therapy often excite profound, *even religious,* feelings about nature and human life. . . . The insightful analysis of the President's Commission . . . [has] sufficiently aired the issue to establish that meddling with the human genome is no more unnatural or artificial than the other meddling that goes on in a wide range of accepted human pursuits."[129] The need for theologians in the debate was now questionable. As we shall see in chapter 5, their downfall was rapid from this point on.

Chapter 5

Regaining Lost Jurisdictional Ground and the Triumph of the Bioethics Profession

The President's Commission did succeed in calming the public, as executive director Capron reported. With the theological arguments reduced to the phrase "playing God" and treated as "rhetorical claims," the debate over HGE could now move on to "important concerns about means and consequences."[1] Substantively rational arguments were discredited, and the theologians, who up to this point had made the most use of them, were in rapid decline in the HGE debate, resulting in a further thinning of the debate. Formally rational arguments were becoming more influential, as government advisory commissions grew in prominence and the bioethics profession—ideally suited to this new environment—gained stature.

At the end of the *Splicing Life* report, the President's Commission had suggested several possible mechanisms for continuing government oversight. Some in Congress, such as then Rep. Albert Gore, favored the creation of an independent genetics commission. Continuing the tradition among scientists of using government advisory commissions to avoid the imposition of regulations by democratically accountable legislators, a human gene therapy subcommittee of the Recombinant DNA Advisory Committee was quickly created. According to its chair, in a rare admission of the purposes of these commissions, the subcommittee was established "in part to fill a vacuum and in part to forestall congressional action" after the release of *Splicing Life*.[2] Thus, the RAC took over as the decision-maker regarding the ethics of HGE. This meant that participants in the HGE debate, if they wanted to influence what actually occurred with HGE, had to write in a manner suited to this new commission.

That government advisory commissions thus became the consumer and target of ethical arguments about HGE is central to my narrative. Indeed, as I shall show, the writers who targeted government advisory commissions acknowledged as much in their texts. But did the commissions equally "consume" the arguments that were directed to them? From an examination of the archives of the President's Commission, it is clear that before every meeting the staff sent the commissioners a stack of academic papers to read, papers that they thought would be the most helpful for the upcoming deliberations. The commissioners themselves were selected as experts, which means that they were assumed to be well read in the relevant literature. The breadth of such reading is evident in the case of the RAC, whose chair noted that one of its policy-recommending documents "represent[ed] an attempt to distill 15 years of ethical discussion in published articles and books, at public symposia and in government hearings and reports"—"a national and perhaps international consensus" on somatic human gene therapy.[3]

An examination of the texts from the common and influential participants in the debate over HGE during the period from 1985 to 1991 reveals that bioethicists replaced the theologians as the second best represented group of professionals in the debate, behind the scientists. Scientists themselves continued to disengage from the debate—as I argue, because they had formed an implicit alliance with the bioethicists.[4]

When bioethicists attempted to take jurisdiction over ethical decision making from the physicians and medical research scientists who conducted human experimentation, there was bitter resistance. This is because the ends pursued by bioethicists—autonomy, beneficence, nonmaleficence, and justice—as applied to human experimentation challenged the interests of the physicians and scientists. When bioethicists went on to expand their jurisdiction to areas like human genetic engineering, the scientists involved did not really resist. I believe that this is because the four bioethical principles developed for the issue of human experimentation, when applied to HGE, did not really affect how the scientists involved with HGE proceeded, and at the same time provided them with ethical legitimation. That is, principlism could not easily be used to make ethical arguments against HGE.

Two of the four ends of bioethicists (beneficence and nonmaleficence) have long motivated scientists, who have pursued them in their work. When applied to human experimentation, the end of autonomy dramatically affected this research. Applied to HGE, it simply meant that people had to want to be engineered—which, as we will see below, is no real impediment for either somatic or germline engineering. While forwarding justice would have the potential for ending all HGE research—the same money could be spent on inoculating people against common but deadly diseases in Africa—in reality this version of a justice argument was rarely made.

The weakness of principlism for challenging HGE is exemplified by the primary argument that can be created using the bioethical form of argumentation: germline HGE violates the autonomy of not yet existing people. This seems on its face to be a fairly weak and convoluted argument, only created because of the paucity of materials to work with in the bioethicists' form of argumentation. Thus, there was little in the bioethicists' form of argumentation that would limit the actions of scientists interested in HGE, and thus the scientists did not see bioethics as a threat.

By the mid-1980s scientists and bioethicists were using similar forms of argumentation. I would therefore describe bioethics as holding what Abbott calls an "advisory jurisdiction" with the scientists. Scientists could work unimpeded, setting the direction of their research, but bioethicists maintained the right to partially modify the actions of scientists. Had bioethicists created a substantively rational form of argumentation, in which they debated the ends that scientists should pursue, I suspect that scientists would not have been so sanguine and would not have allowed bioethicists to obtain positions of power on government advisory commissions.

The 1985–91 era experienced the further institutionalization of the jurisdictional gains of bioethicists and scientists as they began to ascribe factlike status to their form of argumentation. Scientists and bioethicists, accepting that the Belmont ends were the only relevant ends in debates about any type of HGE, claimed that these ends justified the means of germline human gene therapy as well as the means of somatic human gene therapy. In so doing they reexpanded their jurisdiction almost back to where it had been before Ramsey and other critics challenged the eugenicists in the late 1960s. The substantively rational authors in this debate, though increasingly sidelined, reacted to the further jurisdictional encroachment of bioethics, using a new claim: that HGE as envisioned by bioethicists and scientists would lead to a renewed eugenics movement.

Institutionalizing the Jurisdictional Gains

Figure 4 shows that the scientists, despite sharing the bioethicists' form of argumentation, had their own distinct debates in this era over the use of the means of germline gene therapy for the alleviation of genetic diseases.[5] Their form of argumentation is suggested by the disproportionate use in their texts of risk/benefit analysis, the mechanism whereby one maximizes the ends of beneficence and nonmaleficence.[6]

The jurisdictional settlement described in chapter 3, whereby scientists

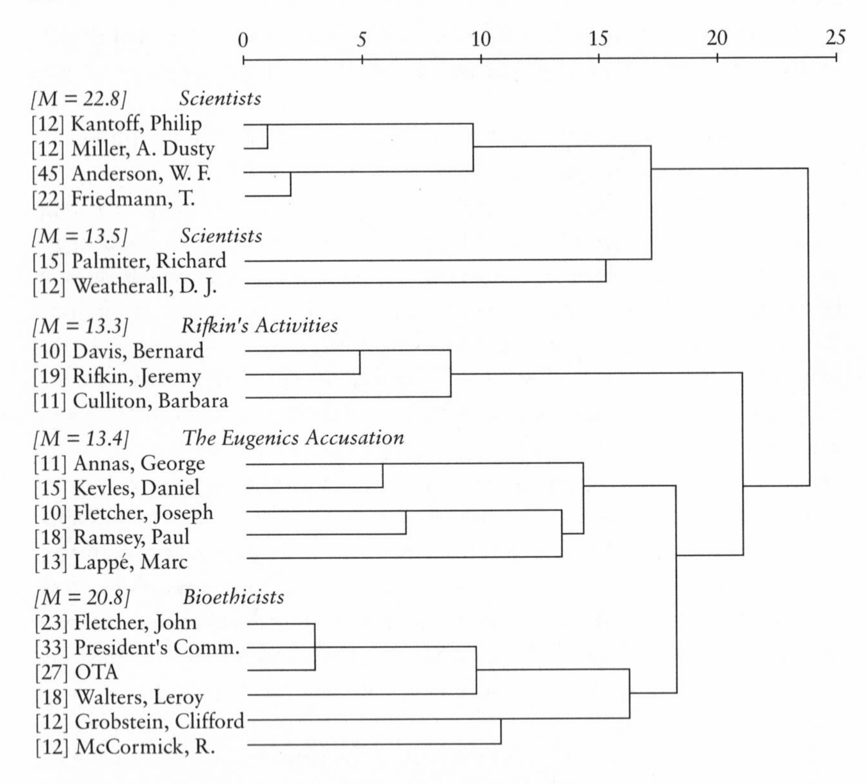

Figure 4 Clustering of most influential authors, 1985–91
NOTE: The number of texts that cite the author is given in brackets. *M* indicates the mean citations per author in the community. The scale at the top is the rescaled distance at which the clusters combine. $N = 79$.

would concern themselves only with the means of somatic gene therapy and not with other types of HGE, is so well institutionalized in this community that most of the texts from influential authors do not even describe alternative views. Review essays written by three influential members of this community are titled "prospects for" *gene therapy,* "progress toward" *gene therapy,* and "the slow road to" *gene therapy.*[7] There is no doubt among these authors that "somatic cell gene therapy" is distinct from other types of HGE because "essentially all observers have stated that they believe that it would be ethical to insert genetic material into a human being for the sole purpose of medically correcting a severe genetic defect in that patient.[8] Similarly, Friedmann states that "few discussions of gene therapy at scientific meetings and in publications still argue its need or potential place in medicine or its ethical acceptability, but rather they emphasize technical questions of efficiency of gene delivery and targeting and selection of suitable disease models."[9]

Moreover, these authors speak of gene therapy as part of one of their secure jurisdictions. "Gene therapy . . . is not a fundamental departure from standard clinical practice," concludes one writer. Its first application would be like that of any other "new procedure in medicine," writes another.[10] As before, gene therapy is metaphorically located in the home jurisdiction by making the case that the means and ends of somatic human gene therapy are the same as those used in a secure jurisdiction. According to one text, "the goal of biomedical research is, and has always been, to alleviate human suffering. Gene therapy is a proper and logical part of that effort."[11]

It was first suggested in the early 1970s that the ethics of somatic human gene therapy would be the same as for any other experiment, and thus by this era it came under the jurisdiction of bioethicists. The form of argumentation used by bioethicists, however, was first connected to somatic human gene therapy in an influential article written in 1980. In that year a scientist at the University of California, Los Angeles, Martin Cline, was found to have attempted human gene therapy without approval from the UCLA institutional review board and, in the opinion of other scientists involved in this issue, prematurely. Writing soon after the revelation of the Cline experiments, Anderson and John Fletcher, who now worked at NIH as assistant for bioethics, wrote an article assuming that somatic human gene therapy is a part of the jurisdiction of scientists and bioethicists. That is, like members of other professions, they defined the problem in a way that made it seem to fit naturally into their home jurisdiction. Raising only the ends from the *Belmont Report,* they presented the ethical problem as Cline's violation of "the fundamental principle" that "it should be determined in advance that the probable benefits outweigh the probable risks."[12]

Citing institutional review board procedures, the *Belmont Report,* and Henry Beecher's exposé of experiments conducted without the consent of the experimental subjects, the authors outlined three criteria to be used for evaluating gene therapy protocols—all designed to calculate whether the benefits outweigh the risks (e.g., whether beneficence and nonmaleficence are maximized). Animal studies should be conducted to determine whether the gene inserted into the target cells will stay in those cells, whether the gene will be expressed in the cell at the appropriate level, and whether it will harm the cell or by extension the animal.[13]

Anderson and Fletcher were keenly aware that government advisory commissions, and not the public, would be the ultimate arbiter of the ethics of HGE. Anderson's 1984 paper, outlining the criteria that should be used to determine when clinical trials of somatic human gene therapy should begin, notes that "the initial protocols designed to carry out gene therapy in patients will probably be evaluated . . . by the Institutional Review Board at the investigator's home institution [and] . . . be approved by the NIH after review by the Recombinant DNA Advisory Committee (RAC)."[14] Anderson was undoubtedly well aware of who needed to be convinced of the ethics of HGE, since his research became the first to be approved by the RAC.[15]

New Policy from the RAC: Institutionalizing Jurisdiction over Somatic Human Gene Therapy

The NIH was and is the primary source of funding for this type of research, and it had not previously approved any research protocols for HGE. The subcommittee of the RAC that had been appointed to forestall congressional action, chaired by bioethicist LeRoy Walters, addressed this issue. Based on the consensus that somatic human gene therapy was just like medicine, and thus came within the scientist/bioethicist jurisdiction, the subcommittee decided that experiments should be permitted to take place if they could surpass the risk/benefit ratio. It began drafting a document titled "Points to Consider in the Design and Submission of Human Somatic Cell Gene Therapy Protocols," to be used by the RAC and NIH for determining whether a particular experiment was ethical.

The subcommittee at this point only institutionalized the jurisdictional settlement that had implicitly been made earlier, according to which the ethics of somatic human gene therapy was the province of medical research scientists and bioethicists, with the remainder of acts that were labeled HGE—such as germline manipulations—remaining outside their jurisdiction. The use of the term *therapy* in the titles of the subcommittee itself and the document it created demonstrates the commitment to the idea articulated by Ramsey, and reiterated in *Splicing Life*, that HGE on the body is "just like" other medical therapies.

Referring to the current federal regulations on institutional review boards—which institutionalized the *Belmont Report* of the bioethicists in human experimentation more generally—the "Points to Consider" document stated that most of the questions researchers have to answer before conducting experiments are those "usually discussed by [institutional review boards] in their review of any proposed research involving human subjects."[16] Reiterating the institutionalized ends, according to subcommittee chair Walters, the most important "areas of concern surrounding human gene therapy" are "anticipated risks and benefits" (beneficence and nonmaleficence), "selection of patients" (justice), "informed consent [and] privacy and confidentiality" (autonomy).[17]

It is interesting to note that the subcommittee also asked two "broader social issues" questions of researchers. These were not of the sort that theologians had been asking, but rather were intended to ensure that "accurate information is made available to the public with respect to such public concerns as may arise" from the study and to ascertain whether or not the researchers "intend to protect under patent or trade secret laws either the products or the procedures developed" in the study.[18] In other words, the only broader concerns to be considered were those of scientists: that the public not become too upset and that the touchy issue of not sharing scientific discoveries with the scientific community be dealt with effectively.

The "Points to Consider" document was accepted by the RAC in September 1985, and became part of government regulation—perhaps the ultimate form of institutionalization.[19] Scientists and bioethicists now had jurisdiction over somatic human gene therapy, and scientists could begin applying to the NIH for funding for these experiments. The first HGE experiment sanctioned by the NIH took place in 1989, and the first attempt to cure a disease directly through human gene therapy was made in 1990. These experiments were briefly blocked by a lawsuit filed by social activist Jeremy Rifkin, but were allowed to resume after a short time.[20]

Instead of stating that germline HGE would not be considered in the future—which would have frozen the jurisdictional boundaries to a great extent—the RAC took the position that "the RAC and its working group will not *at present* entertain proposals for germ line alterations."[21] The means called somatic human gene therapy were now solidly in the jurisdiction of scientists and bioethicists, and with the window left open by the RAC, these professions could quickly move on to try to seize jurisdiction over germline HGE as well.

Jurisdictional Expansion into Germline HGE

Recall that when Anderson spoke about "human gene therapy" in 1972, he was immediately rebuked by Ramsey for his expansive definition of *therapy* (see chapter 3). Ramsey wanted the jurisdiction of scientists to be based on a metaphorical relationship between both the means and the ends of "gene therapy" and those in the jurisdictional home of medical research science, and he was largely successful in achieving this. For example, the President's Commission wrote in *Splicing Life* that "an analogy" with somatic human gene therapy "is organ transplantation, which also involves the incorporation into an individual of cells containing DNA of 'foreign' origin."[22] Somatic human gene therapy was just like any other means conducted on the human body and intended to forward beneficence as an end. For a profession under threat, as scientists were at the time, this was a safe analogy to the most strongly institutionalized part of their jurisdiction. Now that the opponents of the profession had been weakened and their form of argumentation had become increasingly institutionalized, such a safe strategy was no longer necessary.

By the mid-1980s some scientists had begun to question the somatic restriction as they realized that "the need for efficient disease control or the need to prevent damage early in development or in inaccessible cells may eventually justify germ line therapy."[23] Yet an argument for the "therapy" of not yet existing people was not supported by the jurisdictional analogy to treating bodies—as originally pointed out by Ramsey. To gain jurisdiction over germline HGE, scientists would have to sever the metaphorical connection to means conducted in the bodies of their patients. They therefore dropped the metaphor of means and relied instead on a metaphor of ends. They claimed jurisdiction over any means that had the end of forwarding beneficence through the healing of disease—whether in the bodies of existing people or in the germline.

This expansion was facilitated by the decreasing legitimacy of the substantively rational arguments of competitors, as a result of the ever growing influence of governmental advisory commissions and the profession of bioethics. That is, the only relevant ends in the entire HGE debate were increasingly defined as beneficence, nonmaleficence, autonomy, and justice; the success of the scientists' expansion depended on this development.

The common members of the debating communities made up of scientists mostly referred to Anderson for this claim. Although until 1984 he had accepted the boundary of legitimacy to be the somatic-germline distinction in means, beginning in his influential 1985 article he argued that a better boundary is between the ends of "therapy" and "enhancement." Anderson saw "four potential levels of application of genetic engineering" in humans. The first was "somatic cell gene *therapy*," which would result in correcting a genetic defect in the somatic cells of a patient. (This was the existing jurisdictional boundary.) The second level was "germ line gene *therapy*," which would require the insertion of the gene into the reproductive tissue of the patient in such a way that the disorder in his or her offspring would also be corrected. The third level was "enhancement genetic engineering," which would involve the insertion of a gene to try to "enhance" a "non-disease" characteristic, such as height. The fourth level was "eugenic genetic engineering," defined as the attempt to alter or "improve" complex human traits, each of which is coded by a large number of genes (for example, personality, intelligence, character, formation of body organs, and so on).[24]

Although scientists and bioethicists had just institutionalized their jurisdiction over somatic gene therapy, Anderson argued that scientists should have jurisdiction over the second level (germline gene therapy) as well. This is a critical event for our understanding of how the institutionalized form of argumentation affects the substance of arguments in the HGE debate over time. Anderson simply took the ends developed for experiments on the bodies of human beings and applied them by extension to nonsomatic types of HGE by arguing that the new acts also forward beneficence as an end.

In his article Anderson discussed all four classes of HGE, evaluating the science and the ethics of each. For the ethics of somatic human gene therapy, he simply made use of the *Belmont Report*. The second class of acts, germline gene therapy, has as "the critical" ethical question, "should a treatment which produces an inherited change, and could therefore perpetuate in future generations any mistake or unanticipated problems resulting from gene therapy, ever be undertaken?" This question is phrased so that, unlike in the thick debates of previous eras, safety (nonmaleficence) is the only relevant end.[25]

Anderson concluded that "since [germline therapy] is the correction of a genetic defect (albeit in the future), [it] would be ethical and appropriate" if three conditions were met: first, that we had experience with somatic engineering to know whether it would be safe; second, that there be adequate animal studies to determine efficacy and safety; third, that there be "public awareness and approval of the procedure." This final requirement might seem to offer an entry point for other ends into the debate, but since Anderson did not report any possible objections to germline gene therapy except safety, he seems to have meant only that society must agree to the risks involved.[26]

The third class of HGE, enhancement genetic engineering, was ruled unethical, but only for safety reasons: "in short, we know too little about the human body to chance inserting a gene designed for 'improvement' into a normal healthy person."[27] Moreover, if scientists gained that knowledge, Anderson would slide this component of HGE under the "therapeutic" banner, arguing that it would be acceptable to engage in it when "justified on grounds of preventative medicine," such as giving people an extra gene to lower their cholesterol level.

Anderson's fourth and final type of HGE is the most interesting for what it reveals about professional competition. By limiting the use of the term *eugenic* genetic engineering to traits not typically considered to be "diseases" (such as intelligence), he redefined the term from its traditional use. The President's Commission, for example, had stated that "altering the human gene pool by eliminating 'bad' traits is a form of eugenics."[28] Eliminating "bad" traits, however, is what Anderson's germline therapy is intended to do.

Limiting the use of the term *eugenics* in this way shifts the public concern with eugenics onto acts that scientists do not currently want to do. Moreover, distancing germline gene "therapy" from "eugenics" obscures how germline gene therapy as defined by Anderson was one of the dreams of Huxley, Muller, Dobzhansky, and their eugenicist contemporaries—a dream that brought critics such as Ramsey into the debate and raised fears in the public. Anderson thus accepted one of the commonly articulated ends of eugenicists—beneficence—while attempting to distance himself from another, the end of species perfection.

If Anderson's attempt at regaining jurisdiction over the means of germline gene therapy is successful, he will have gained back much of the ground lost by scientists who had to retreat under pressure from critics, all the way back to somatic human gene therapy. The advantage that Anderson had over Huxley, Muller, and Dobzhansky in this struggle was that the ends he used to argue for the expansion had become increasingly institutionalized in the debate. The number of bioethicists in the debate was on the rise, the number of theologians on the decline. By the end of this time period, Ramsey had died, and the new entrants to the HGE debate came to a field where the formally rational type of argumentation used by bioethicists increasingly had the status of "fact."

Bioethicists and Germline HGE

Bioethicists formed a distinct debating community in this era (see fig. 4), and followed Anderson in discussing the ethics of germline human gene therapy, echoing his new jurisdictional strategy.[29] While scientists focused on the Belmont ends of beneficence and nonmaleficence, the bioethicists were concerned with protecting the autonomy of the people whose germlines would be engineered.[30]

Consistent with the narrative I have been building about the bioethics profession, the data show that the common members of this community used the formally rational Belmont ends because they viewed the government as the entity that would evaluate their ethical claims.[31] Examination of the work of the influential authors also reveals that the bureaucratic state is the ultimate consumer of the ethical arguments produced in this community. I offer two pieces of evidence: First, four of the six influential authors in this community are ethicists for the government or chairs of government advisory commissions, or are not individual authors but government entities.[32] Second, an examination of the texts from these influential authors reveals that—especially in comparison to the authors in the other communities in this era—they are writing to influence the government advisory commission that will control the act of HGE. Indeed, of the most influential texts from authors who are not themselves government entities, three out of four begin their first paragraph by stating that a government commission will soon be determining the ethical criteria for HGE, and go on to offer advice on how the decisions should be made.[33] For example, Grobstein suggests that "an oversight body might consider drafting a first-round set of principles" for HGE, which he outlines.[34]

Given that the influential authors in this community tend to have government advisory commissions as their ultimate target audience, their form of argumentation is highly formally rational. The ends in their arguments are by and large the universal, commensurable ends first expressed in the *Belmont Report*.[35] The calculable quality of these arguments tends to result in thin descriptions of the problems of HGE. For example, while in previous eras texts linked abortion, in-vitro fertilization, HGE, and other issues into one deeper problem, such as human control over "natural" processes, the literature in this community discusses only HGE.[36] Linking these problems would greatly complicate the calculations of whether HGE violates the stated ends.

Similarly, the authors in this community have a very limited time frame in describing the problem. Unlike substantively rational authors, who tend to believe that an act is right or wrong regardless of its consequences, the authors in this community need to have the information about the effects of HGE and the ends of the people in the future before they make their evaluation. For example, specifically contrasting his time frame to that of substantively rational authors, one influential member of this community states that "a good working rule in the ethical consideration of technology is to 'refrain from moral judgement on unverifiable possibilities—as notational cases rooted neither in the reality of experience nor a specific context.'"[37]

Taking Back the Germline Jurisdiction with the Scientists

As the ends articulated in the *Belmont Report* became reified, they were increasingly seen by bioethicists as the only form of argumentation for use in analysis for all problems, no matter how distant from the context of human experimentation for which they were created. The jurisdictional expansion from somatic to germline HGE—consistent with the attempts of scientists, discussed above—occurred not through a debate about ends to pursue, but rather through appeals to the increasingly institutionalized ends used by bioethicists.

The writings of bioethicist John Fletcher, the most influential member of this community, show how institutionalized ends are used to expand jurisdiction. Fletcher, at the time a bioethicist at NIH, and W. French Anderson, the leading HGE researcher at NIH, make strikingly similar arguments about expanding their jurisdiction to germline HGE.[38]

Fletcher first reviews somatic gene therapy, and repeats the now institutionalized view that the only ends that should be considered are those in the *Belmont Report*. In his subsequent argument regarding the ethics of germline therapy, he makes the standard professional move of "reduction," where a profession's form of argumentation is claimed to govern another jurisdiction as well.[39] Specifically, he claims that "the reasoning that favors somatic gene therapy could be used to support" germline research.[40]

The bioethics profession portrays itself as a "neutral" enterprise, forwarding universally held and commensurable ends. If there are authors who disagree with a bioethical argument, they cannot be portrayed as disagreeing with the bioethicists' ends, since these ends are argued to be universal. This approach does not always succeed, of course, but in Fletcher's case, he attempts to increase the validity of his claims by arguing that Kass and Ramsey would actually agree with his analysis. As the President's Commission had done, he achieves this through translating their views. The translation is signaled in a subtle way: "Even though these views [of Kass and Ramsey] are premised on theological, philosophical, and ethical beliefs *that many do not hold*, the objections contain a *core* that should and can be addressed."[41] That is, the "ethical beliefs" or ends that are not universal and commensurable will not be further discussed, and the parts of Ramsey's and Kass's arguments that conform with the bioethicists' form of argumentation will be treated as representative of all their arguments. Needless to say, if Ramsey's and Kass's arguments are limited to those used by bioethicists, the two will eventually be described as agreeing with bioethicists that an expansion in jurisdiction is warranted after all.

Fletcher summarizes the arguments of Ramsey and Kass as consisting of three parts. The first is that scientists must show that the therapy "will benefit the subjects and their offspring with no demonstrable harm." Nonmaleficence and beneficence are indeed among the ends used by both Kass and Ramsey.

The second argument is that scientists must show that germline HGE "can be confined to medically necessary goals."[42] This gives the impression that Kass and Ramsey would support the engineering of what Fletcher would call "medically necessary goals." However, this is a great reduction in what they had argued, leaving out their arguments using ends that oppose the Belmont ends. If one examines the pages in Kass's text that Fletcher himself cites to support this claim, Kass actually argues that "the most serious danger from the widespread use of these techniques will stem not from desires to breed a super race, but rather from the growing campaign to prevent the birth of all defective children in the name of population control, 'quality of life,' and the supposed 'right of every child to be born with a sound physical and mental constitution, based on a sound genotype.'"[43] Clearly, Kass is opposed to "medically necessary goals" for germline engineering as well. How could Fletcher have ignored this? He ignores it precisely because it is not part of the rational "core" of Kass's work that is consistent with the institutionalized form of argumentation of bioethicists.

According to Fletcher, the third part of Ramsey's and Kass's arguments concerns "harms that can only be described as metaphysical." The "metaphysical" problems these authors have with HGE will be satisfied if it can be shown that HGE "will not threaten the ethical systems most treasured by society."[44] Fletcher's discussion of the "ethical systems most treasured by society" is the best example of how bioethicists' translation of substantively rational arguments gives them an advantage. Fletcher reduces all that Ramsey and Kass had to say about what is treasured by society that could be lost with HGE to one principle that "exemplifies" these problems—the means of "informed consent," which forwards autonomy as an end, as outlined in the *Belmont Report.*

Thus, for Fletcher, the "core" metaphysical problem identified by Kass and Ramsey is that the descendants of germline-engineered people who do not yet exist have not given their informed consent to be experimented upon. While Ramsey and Kass both made this claim in other contexts, autonomy is clearly not their primary end, and certainly Ramsey, at least, had many more metaphysical concerns. However, by portraying autonomy as the core problem identified by Kass and Ramsey, Fletcher is able to find an answer to it in the bioethics literature. Since parents can give consent for therapeutic experimentation on their children under current bioethics practice, this principle can be extended to parents giving permission to experiment with engineering their children and their children's children and so on through germline engineering. Fletcher thus concludes that Ramsey's and Kass's concerns are satisfied: "the metaphysical challenge" to germline gene therapy "can be met with the argument that human values and ethical systems, exemplified by the consent principle, would not be drastically changed. The benefits of genetic therapy could be gradually and safely presented to society within the ethical systems we know."[45]

Thus it is argued that germline HGE should be brought under the jurisdiction of medical research scientists and bioethicists who use the *Belmont Report.* People might think that Ramsey and Kass were arguing for a thicker debate over the myriad ends that should be applied to germline HGE. But when the ends for which they argue are commensurated into the bioethicists' ends, they are made to appear to agree that a thin debate is more appropriate, and that the Belmont principles are all that are needed in future discussions about germline HGE.

A few years after this text, Fletcher expanded upon his arguments and, following Anderson, argued against the somatic/germline distinction by claiming the factlike status of the Belmont principles. Like Anderson, he redefined *eugenic,* limiting the term to "biological measures employed to improve characteristics in persons who can be generally viewed as normal." Similarly, he held that instead of drawing "a line between somatic cell therapy and germline research, a better line is between curative or preventive gene therapy and eugenic uses of germ cell alteration." He concluded that the placement of this "better line" is supported by "two principles, beneficence and justice [which] will count heavily in the definitions and tests of best interests that will be required to distinguish between tolerable and intolerable genetic experiments beyond somatic cell therapy."[46]

Reaction to the New Jurisdictional Challenge: The Eugenics Accusation

Not all the common authors in this debate had accepted the growing institutionalization of the bioethicists' form of argumentation. There remains a final community of authors, distinct from the others, that remains substantively rational. These writers refer to earlier, substantively rational debates, such as those between Ramsey and Joseph Fletcher. In contrast to the scientists and bioethicists, they are less likely to be interested in the current debate over the risks and benefits of the imminent human gene therapy experiments.[47] They discuss several issues they see as related, such as abortion and artificial insemination, suggesting that HGE is simply one part of a deeper problem about ends.[48] While we might suspect that the common members of this community would be theologians, in actuality during this era the theologians are without a distinct community. Rather, this community consists of a smattering of substantively rational authors from several professions.[49]

While an earlier generation of scientists did not shrink from the use of the term *eugenics,* in this era it had become an epithet. The way that Anderson and John Fletcher had redefined the term to avoid the connection with what they were advocating suggests as much. What really distinguishes this community from the others is that its authors began to more explicitly link the proposed means of HGE of this era with eugenics.[50] To do so, they referred to the authors in the earlier debate: Ramsey, Joseph Fletcher, and Marc Lappé. Ramsey's and Fletcher's views on the subject have been well chronicled. Lappé wrote about the "fallacies of 'genetic control,'" arguing against the threat of genetic load that had animated the scientists of the 1950s through the 1970s. He concluded that scientists continue to try to eugenically control the human genetic destiny due to a "deep-seated aversion most Western scientists (and philosophers) feel towards the chance events that appear to govern genetic systems."[51]

Lappé used ends that are clearly out of fashion in the other communities in this era. Which ends does "genetic control" violate? The first of these is beneficence or perhaps justice, for these techniques "obfuscate the need for solving current problems which do not need novel technical solutions, such as general health care." The second is that genetic control of future generations poses "the threat of dehumanization." Sounding much like Kass in his input to the President's Commission, and using Jacques Ellul's term roughly equivalent to formal rationality ("technique"), he writes that "when technique enters into every area of life, including the human, it ceases to be external to man and becomes his very substance."[52]

In addition to referring to this older debate, common members in this community also engage some of the more recent scholarship on the history of the eugenics movement. When discussing Ramsey, Fletcher, and Lappé, they also discuss the work of historian Daniel Kevles, who wrote what is now the canonical history of the eugenics movement in England and the United States from the nineteenth century to the present day.[53] One of the prominent themes in Kevles's book is the continuity of eugenic thought, beginning with the outright racist and classist proponents, such as Francis Galton, through the Nazis, Muller and the reform eugenicists, genetic screening advocates, and onward to gene therapy. A second, unavoidable theme in the book is that while many scientists and members of the public have considered many human traits to be genetically determined, the evaluation of a trait as "good" or "bad" changes over time. This suggests that good and bad traits are not objective but contingent on cultural context. For example, as recently as the early 1960s, eugenicists had argued that "feeblemindedness" and "sloth" were genetic conditions to be eradicated from the human genome. Any claims by recent scientists that the "disease/enhancement" line was unproblematic enough to begin permanent changes in the germline would seem questionable after reading Kevles's book.

This is why Kevles, whose most influential text was published in the same year as Anderson's and Fletcher's, is not located in the bioethicists' or scientists' community. Since the scientists and the bioethicists who selected Anderson and Fletcher as their influential authors reject the eugenics analogy—with both Anderson and Fletcher using the term in a much narrower sense than it had historically had—Kevles's broader use of the term contradicts the point they are trying to make. For the community currently under consideration, the history of the eugenics movement is critical evidence of the continuity of bioethicists and scientists with the Mullers and Huxleys of a previous generation.

Similarly, another influential member of this community, lawyer George Annas, evokes the specter of eugenics when discussing the relationship between the Human Genome Project and HGE.[54] The Human Genome Project could lead down the slippery slope as increasing knowledge creates demand for genetic intervention. Screening, for example, with an expanded knowledge of the genome, "need not be required, people can be made to *want* it, even to insist on it as their right."[55] Rejecting the view that scientists pragmatically use, that disease is objective, Annas notes that the "normal" genome "will be invented, not discovered." He further points out that "the Nazi atrocities grew out of the combination of a public health ethic that saw the abnormal as disposable, and a tyrannical dictatorship that was able to give the physicians and public health authorities unlimited authority to put their program into bestial practice." The eugenics movement and Nazi historical lessons point to the need for *not* acknowledging scientific jurisdiction over the ethics of this practice because "ethics is generally taken seriously by physicians and scientists only when it either fosters their agenda or does not interfere with it. If it cautions a slower pace or a more deliberate consideration of science's darker side, it is dismissed as 'fearful of the future,' anti-intellectual, or simply uninformed."[56]

While this community has apparently put its hopes on its argument about eugenics, it is still at the margins of the debate.[57] Nor is it likely to become relevant soon, with government advisory commissions increasingly deciding what is ethical. Annas acknowledges as much, noting that the projects called for by the NIH's Ethical, Social and Legal Implications of Human Genome Research program are biased against those who ask fundamental questions. "The brief announcement [of the program] makes it clear that such projects are to be about the 'immense potential benefit to mankind' of the [human genome] project, and focus on 'the best way to ensure that the information is used in the most beneficial and responsible manner.' Those with less optimism apparently need not apply."[58]

The First Post–President's Commission Debate

Following on the success of the President's Commission, in this period bioethicists and scientists institutionalized the jurisdictional claim over somatic human gene therapy that they had earned in previous years. Citing a consensus on the ethics of somatic human gene therapy, a new government advisory commission set the form of argumentation used by the bioethics profession into government regulation, inviting research protocols. By the end of this period, the first somatic gene therapy research was under way.

It was not only the jurisdictional arrangement that somatic HGE should be governed by the bioethicists' form of argumentation that had become institutionalized; rather, the form of argumentation itself came to be reified as normative for all ethical questions. It became increasingly clear to authors in the scientists' and bioethicists' communities that these same ends could justify the means of germline HGE as well. To achieve this, however, they had to change the jurisdictional metaphor. If somatic gene therapy was under the jurisdiction of medical research scientists and bioethicists because the means were analogous to the means in the home jurisdiction of treating diseases in bodies—and because the means forwarded beneficence as an end—then germline gene therapy could not be part of this jurisdiction because there were no bodies to treat. They resolved this problem by changing the jurisdictional metaphor: medical research scientists forward beneficence as an end, through any means. With this claim in hand, scientists and bioethicists began to strive for jurisdiction over the ethics of germline human gene therapy as well.

A resurgence of discussions linking HGE with eugenics came in reaction to this move. Primarily found in a distinct community of substantively rational authors, these discussions questioned the legitimacy of the claim that the means of germline gene therapy forwarded only beneficence ("therapy") and not species perfection ("enhancement") as well, thus calling into question the expansion in jurisdiction. Pointing out that disease is not objective, these authors seemed to be calling for a thicker, more substantive debate, now that the ability to actually conduct germline HGE was drawing near. This thicker debate, however, would not come to pass.

Chapter 6

"Reproduction" as the New Jurisdictional Metaphor: Autonomy and the Internal Threat to the Bioethics/Science Jurisdiction

In 1990 W. French Anderson and his colleagues had conducted the first somatic HGE experiment approved by the Human Gene Therapy Subcommittee of the Recombinant DNA Advisory Committee. Having set its ethical standards in the original "Points to Consider" document, which was revised in 1989, the RAC was now busy calculating only whether submitted protocols met those formally rational standards. Was the risk low enough? Was the potential benefit high enough? Did the research subjects have enough information to make an autonomous decision? Between 1990 and June 1995 the RAC approved 105 more somatic experiments, involving a total of 597 research subjects. The NIH was spending approximately $200 million a year on HGE research, while private funding was estimated to exceed that amount.[1]

Examination of the texts from the common authors in the period 1992–95 reveals a decisive turn away from discussions of public policy, government regulations, and government advisory commissions.[2] Moreover, the texts of the most influential authors place less emphasis on upcoming decisions from these commissions. That is, while the RAC retained decision-making authority, authors tended not to explicitly appeal to it. There are two reasons for this. First, the somatic human gene therapy debate was effectively over, as the RAC was routinely approving research protocols, without changing their ethical framework. As for any study that has to be approved by an institutional review board, this approval process had been subsumed into the everyday exercise of the bioethicist/scientist jurisdiction in which the ethics of an experiment were calculated from the set ends. Second, the RAC had made it clear that it would not be making decisions about germline HGE—yet.[3]

In 1991, however, referring to the texts from the authors in the previous era who argued for expanded jurisdiction, the chair of the Human Gene Therapy Subcommittee, LeRoy Walters, called for "a detailed *public* discussion of the ethical issues surrounding germline genetic intervention in humans." The precedent for the discussion he had in mind was the "public discussion of somatic cell gene therapy that occurred in the years 1969–1988." By the time the first experiments were put forward in 1988 and 1990, "the central ethical questions in somatic cell gene transfer and therapy were well understood by researchers, by politicians, by the press, and by the general public." He stated, "It is, in my view, not too early to intensify and broaden the discussion of germ-line genetic intervention."[4]

In an ideal world, this was the time when the debate should have become thicker, as people were not constrained to use universal ends. This did not happen. Instead the debate remained fixed on the formally rational ends used by the bioethics profession, as a result of the institutionalization of these ends and the continued rise of the profession. Instead of a discussion of ends, what we see in this debate is a further institutionalization of the bioethicists' ends.

The growing institutionalization of the bioethicists' form of argumentation is suggested by the continued growth in the number of bioethicists in this debate. When scientists were being challenged by theologians for jurisdiction in the 1960s, bioethicists and theologians had equal numbers of influential authors, but among the common authors there were many more theologians than bioethicists. By the mid-1980s this had changed, and bioethics was second only to science in producing influential authors. By the time period considered in this chapter, bioethics had the greatest number of influential authors, followed by science, philosophy, law, and finally, theology, which had only one. Trends among common authors are similar (see fig. 1).

The data presented in this chapter reveal that the institutionalization of the bioethicists' form of argumentation has become strong enough that they no longer really argue against theologians and other substantively rational authors, but simply label them as irrational—so far outside the new consensus about legitimate arguments that their views should not be seriously considered.

Still, all is not well for the bioethics jurisdiction in this period. I have repeatedly described the four ends used by bioethicists as "commensurable," by which I mean that all other ends can be translated to one of these four (though with some loss of meaning). For example, I showed in chapter 4 how many of the ends advocated by theologians were translated into beneficence or nonmaleficence, with a lot lost in the translation. However, as theorists doing foundational bioethics have long acknowledged, the four principles are not commensurable with each other. The clearest indication of this problem is that there is no established way to adjudicate conflicts when the pursuit of one end seems to contradict the pursuit of another.[5] For instance, if it appeared that germline HGE would promote autonomy and reduce justice, what claims should then be made about its ethics? Which end should prevail, and why? Bioethicist Robert Veatch proposes several possibilities. First, the ends could be ranked by importance. Second, a system of balancing or weighing the ends could be developed, creating a truly internally commensurable system. Third, one end could be considered critical, and the rest, derivative.[6] While theorists in the bioethics profession work on perfecting their form of argumentation—essentially making it more amenable to calculation and thus more formally rational—one of the above approaches is being adopted in practice.

I have described public bioethical debate for practical purposes as a closed system in which members of various professions debate each other. In reality, of course, the environment in which this system is embedded has been changing due to forces far beyond its control. One major change has been the growing support of autonomy in public life in general, and in other bioethical debates in particular.

While bioethicists and others continued to discuss the four ends in arenas such as human experimentation and HGE, the end of autonomy was forwarded by many other groups in society. Most notably, the amount of discourse produced by social movements promoting autonomy in decisions about abortion certainly outweighed the sum total of bioethicists' arguments. Autonomy over decisions about one's own body was the only end pursued by bioethicists that had constitutional status, following the *Griswold v. Connecticut* and *Roe v. Wade* decisions of the U.S. Supreme Court.[7] Although these rulings initially applied only to birth control and abortion, they were later interpreted more broadly. In essence, one of the ends of bioethicists had been institutionalized as applicable to related topics, but through the efforts of others. The incommensurability among the four ends used by bioethicists was being resolved through the slow but steady rise of autonomy as the only important end to pursue in arguments about HGE.

In this chapter I show this process at work. The HGE debate has always had a community discussing genetic screening and genetic counseling as means whereby the genes of descendants could be manipulated. Once the debate turned from somatic gene therapy to germline gene therapy, authors began to think through the exact means whereby one would engineer a person's descendants. Realizing that germline HGE would occur on eggs, sperm, or embryos, and would use means similar to those used for genetic screening and other reproductive technologies such as IVF, HGE debaters reconnected to the screening and counseling debate. In doing so, they found that this debate used a different jurisdictional metaphor of means—that IVF and the like are "reproduction"—and that this means should forward only the end of autonomy. If one argued that the means of germline HGE was "just like" IVF, the genetic screening of embryos, or abortion, then autonomy would apply. In consequence, the debate would be over because autonomy is constitutionally required.

As I will discuss in chapter 7, this development has the possibility of ending the jurisdiction of bioethics and science over HGE at some point in the future. If the decision about whether to genetically engineer their descendants becomes the autonomous decision of couples, then there will be no public policy on the ethics of the HGE decision itself, only on its funding by the government—akin to the way abortion, in-vitro fertilization, and amniocentesis are currently treated. Returning the debate to the public in this way might in the future allow a return of a more substantive debate, such as that surrounding abortion, because individual decision making is particularistic.[8] But I shall argue that even if this were to happen, theologians are in such a marginalized position that they are unlikely to be able to take advantage of the opportunity. This marginalization is exemplified in the data presented in this chapter that show social activist Jeremy Rifkin to be the most visible proponent of the substantively rational arguments. No other theologian has risen to the status of Ramsey.

Using the Institutionalized Arguments

Not only are bioethicists in this period the most influential professionals in the entire debate, for the first time their particular debating community is by far the most influential of all the communities (fig. 5).[9] Moreover, scientists in this period do not have their own community; rather, the most influential scientists are part of the bioethicists' community—a sign of the strengthening links between the professions.[10]

Although the influential authors among the bioethicists retain their personal connections to government policy making in this period, the common authors among the bioethicists are not discussing public policy or advisory commissions more than anyone else.[11] These bioethicists have taken up the effort to make the means described as "germline gene therapy" part of the jurisdiction of bioethicists and scientists.[12] In spite of the shift away from explicitly referencing government policy making, the common authors continue to use the formally rational Belmont ends.[13]

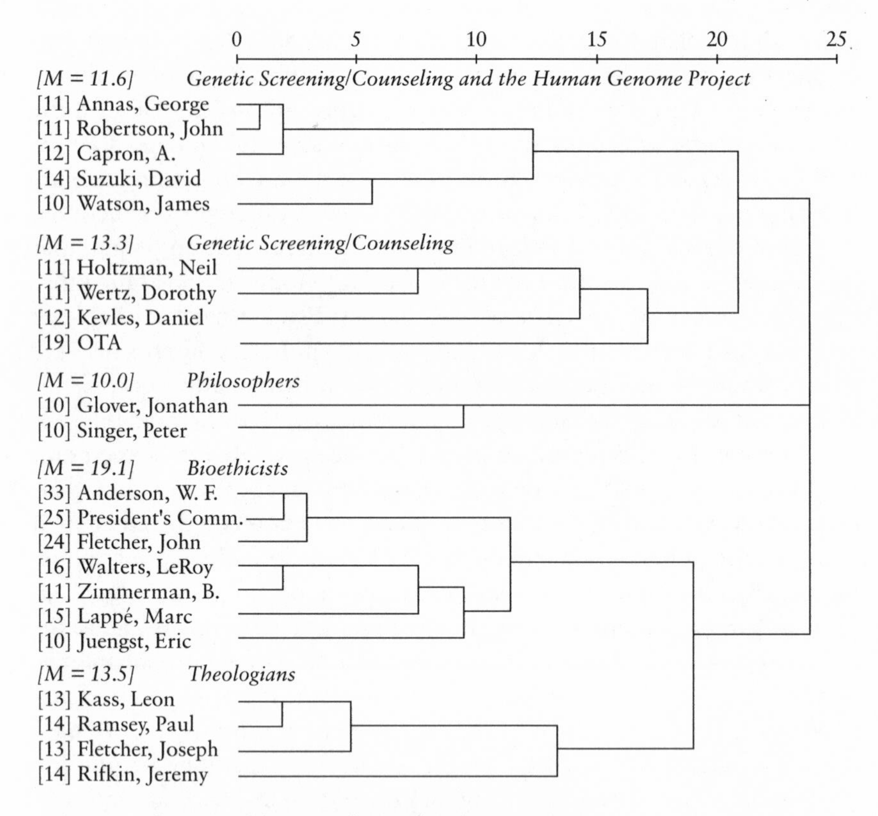

Figure 5 Clustering of most influential authors, 1992–95
NOTE: The number of texts that cite the author is given in brackets. *M* indicates the mean citations per author in the community. The scale at the top is the rescaled distance at which the clusters combine. $N = 68$.

Examination of the texts from the influential authors in this community reveals that somatic human gene therapy is considered without question to be a medical issue—that is, totally under the jurisdiction of bioethicists and medical research scientists. Bioethicists are also using the strength of their jurisdiction to bring germline gene therapy under their jurisdiction by further discrediting influential authors from other communities who dissent from this view.

The manner in which these authors argue against their opponents demonstrates the strength of their jurisdiction and how they have thinned the debate. Bioethicists in this era have institutionalized their form of argumentation so well that they no longer need to argue in a serious way against their competitors; instead, they can simply point out that their opponents are outside the newly taken-for-granted debate. For example, bioethicist Eric Juengst of the NIH, in his introduction to a special journal issue on the ethics of germline gene therapy, makes an analogy using Alfred North Whitehead's three stages of inquiry: romantic, precision, and generalizing. For Juengst, the debate during the 1960s and 1970s on germline HGE, as conducted by such substantively rational authors as Paul Ramsey and Joseph Fletcher, was "primitive," "inspir[ing] anxiety among pioneers" in the debate.[14] While other topics discussed during the 1960s and 1970s, such as organ transplantation and psychosurgery, have "been assimilated quite productively into bioethics' evolution toward clinical ethics and health policy, the subject of human germ-line engineering resists *civilization*."[15] In my terms, "assimilation" means that organ transplantation and psychosurgery have come under the formally rational jurisdiction of bioethics, while germline HGE has not.

In Juengst's view, germline HGE remains "uncivilized," and outside the bioethicists' jurisdiction, due to the influence of the writings of authors such as Ramsey and Fletcher, where the topic did not receive "the kind of *dispassionate* and *systematic* treatment that marks civilized bioethical inquiries."[16] Clearly, there is no need to argue against Ramsey and his contemporaries on their own terms, because of their "passion" and lack of "systematic" thinking. Revealing once again that the form of argumentation of the bioethics profession was created to be useful for unelected government policy makers, Juengst points out that "it is difficult to translate this literature into practical policies for scientific research beyond the cautionary moratoria it has already inspired." Therefore, he wants "to attempt to begin that translation, by moving the discussion of human germ-line modification into the second of Whitehead's three stages of inquiry: the stage of Precision."[17]

Other influential members of this community simply gesture to the new consensus they have erected to dismiss those who remain outside it. LeRoy Walters refers to the "taboo mentality" surrounding germline HGE that is being lifted due to the efforts of "several *respected* ethicists, including John Fletcher . . . and Eric Juengst . . . , and clinicians, including W. French Anderson . . . [who] have ventured to examine germline intervention in a relatively *neutral and objective manner*."[18] The implication is that those responsible for the taboo are not "neutral and objective," like the other influential members of this community he mentions, and are therefore not part of the reasonable debate.[19]

Finally, John Fletcher, writing with W. French Anderson, states "a working rule for ethical debate on new technologies": that authors should "refrain from moral judgement on unverifiable possibilities."[20] This working rule "has perhaps been most frequently violated in the debate about germ-line gene therapy," for "no other topic in bioethical debate stirred such basic emotions or yielded stronger language." Fletcher and Anderson then provide quotations from Kass, Ramsey, and Joseph Fletcher as examples of this "emotion" and "strong language," and violation of the working rule. The work and the form of argumentation of such "emotional" authors do not need to be considered on their own terms. Fletcher and Anderson make clear that without emotion, the new jurisdictional metaphor they suggested in the previous era will be accepted: that scientists forward beneficence through healing disease in bodies or in the species. That is, in the stage of "precision," "morally relevant differences between somatic cell gene therapy and germ-line therapy appear to be less significant than the difference between both of these and enhancement of human traits having little to do with disease."[21] What I have previously shown to be the result of a simple extension of a jurisdictional metaphor has been confirmed as correct due to its correspondence with Whitehead's stage of "precision."

The Rise of Autonomy as the Preeminent End

In almost every period of the HGE debate there has been a separate community that linked its discussions of HGE with discussions of the genetic screening of individuals or their embryos and fetuses for the purposes of determining the genetic traits of children. In the mid-1990s the simplest way for individuals to ensure that their descendants would not have diseases for which they were a carrier would have been either to adopt children or to genetically screen fetuses, aborting those that had the disease. Compared to HGE, these are fairly low-technology solutions that achieve the same ends. Indeed, the genetic screening of fetuses is so common that it has become an institutionalized part of obstetrics practice to ask pregnant women over a certain age or with certain family histories if they would like to have their fetuses screened.

Who should decide if a particular genetic condition warranted an abortion? In the 1960s it was generally not the pregnant woman but rather a panel of physicians who ultimately decided what was ethically proper. The innovation of the pro-choice movement, codified in the 1973 *Roe v. Wade* decision, was that the pregnant woman should decide when it was ethically proper to have an abortion. More specifically, women would decide which genetic traits would be expressed in their children. Genetic screening has therefore been inextricably tied to the abortion debate in the United States.

Although there has been a long struggle by pro-life groups to bring alternative ends into this process, autonomy has been codified by constitutional interpretation by the Supreme Court, thus becoming the only end that is pursued in collective discussions about any aspect of reproduction—whether contraception, abortion, in-vitro fertilization, or other reproductive technologies. If the means of germline HGE is described as the same as the means of IVF, using the metaphor of reproduction, autonomy will become the only end pursued in that area as well.

What would cause the rise of this metaphor in the HGE debate? The metaphor of "therapy" of future bodies was fine for the era in which an ethicist did not really have any sense of how germline HGE would occur. But by the mid-1990s scientists had conducted germline engineering in frogs, and had some sense of how it would be done in humans. As with somatic HGE, the challenge is how to gain access to the cells to be modified. Put quite simply, it was clear that either a couple's germ cells (sperm or egg) or a very early embryo would be engineered, and the easiest way to do this was outside of their bodies. All the techniques for the extraction of reproductive cells, their external fertilization, and eventual implantation in a woman's uterus were already common practice in IVF clinics. Moreover, these clinics were already applying genetic knowledge about these cells (such as whether they were male or female) to decide which to implant in a woman. In the same way that the *Splicing Life* report in the early 1980s argued that the means of *somatic* gene therapy were "just like" organ transplantation, authors could now make the case that these new techniques for conducting *germline* HGE were "just like" what already was occurring in IVF clinics. Another development in IVF clinics, and reproductive medicine more generally, was the absolute primacy of patient autonomy.

In the 1992–95 era two highly related debating communities are discussing the increased precision of new genetic screening techniques.[22] What is interesting about these communities for the larger discussion is that some of these authors who discuss both germline HGE and genetic screening begin to promote the end of autonomy for the means of germline HGE.

Lawyers and Autonomy

Lawyers have been present in fairly equal, low numbers through every period examined in the HGE debate. Only in the 1992–95 era do they have their own distinctive debate, which generally concerns how the increased knowledge about genetic effects produced by the Human Genome Project will affect HGE and genetic screening.[23]

Law professor John Robertson argues for the need for gene therapy of embryos—which would result in germline HGE. His argument is intertwined with the current ethics of other acts undertaken on embryos. He begins by showing that the current ethics of genetic screening of embryos applies to the changing of embryos as well. "If a prenatal diagnosis shows that the fetus has a serious single-gene defect and the mother opposes abortion, prenatal genetic treatment may be essential to prevent the damage that will occur when a gene necessary for normal development is missing." In such a case, "the parents would be free to apply the prenatal genetic therapy to the fetus in order to protect the child that they are bringing into the world."[24]

The reproductive autonomy of parents even trumps the ends of beneficence and nonmaleficence toward the children who would eventually emerge from this process. While some ethicists have argued that "rather than risking damage to future children, affected embryos should be discarded, and only genetically healthy ones should be transferred to the uterus," Robertson sees that this might be against the parents' conscience. "Persons who believe that embryos deserve respect and should be transferred to a uterus rather than merely being allowed to die would find gene therapy an acceptable and possibly an obligatory option." Or, in a similar instance, perhaps not enough eggs were taken from the woman, and after in-vitro fertilization all have the genetic trait that the parents are trying to avoid. "If parents desire to preserve embryos and bring them into the world in a healthy state, gene therapy on the embryo when there are reasonable grounds for trying it, when the parents have been fully informed, and when institutional review requirements have been met is an ethically acceptable option that general principles of civil and criminal law would not prohibit."[25]

In stating what the public policy should be, Robertson first assumes that HGE is reproduction, and then states that autonomy as the only end to pursue in reproduction has been institutionalized through law. The prohibition of germline HGE "would violate a fundamental right," he says, because "gene therapy on the embryo is closely tied to procreative choice. . . . The U.S. Constitution, it may cogently be argued, gives the parent the right to provide his or her children and their descendants with a healthy genome. . . . Properly understood . . . the right to procreate includes a right to practice negative eugenics—to deselect harmful characteristics from future generations."[26]

Analytic Philosophy and Autonomy

Recall that the profession of bioethics was defined by distancing its form of argumentation from both theology and analytic philosophy. More precisely, bioethics distanced itself from the substantively rational form of argumentation of theology, but retained the formally rational type of argumentation of analytic philosophy, while rejecting philosophy's interest in abstract, speculative, and impractical questions. For the first time in the debate, in the 1992–95 period we see emerging a separate community, composed of analytic philosophers, who largely use formally rational arguments, but with a concern for what bioethicists would call abstract, speculative questions.[27] This suggests that philosophy and bioethics are indeed separate professions, and that those analytic philosophers who did not become bioethicists retained their interest in the less practical questions.

The philosophers in this debate produce two types of texts. First, although many mainstream analytic philosophers tended to disdain the impurities required for applied ethical arguments in bioethical debate, analytic philosophers had become the theorists of the forms of argument used in the bioethics profession. Engaged in what I have been calling foundational bioethics, an influential member of this community, Peter Singer, addresses these sorts of questions. To be specific, he is an advocate of a particular form of consequentialist argument called utilitarianism, which is formally rational as I have defined it.[28]

The second type of text written by philosophers in this debate examines more abstract and speculative issues. An example is Jonathan Glover's book, *What Sort of People Should There Be?* Glover's work is similar to Singer's in that it spends a good amount of time on what a legitimate argument is, and also argues for a formally rational type of argumentation. Like Singer, Glover rejects a priori ethics and wants to maximize means to given ends.[29] His most influential book is a wide-ranging discussion of what an earlier generation called "designing our descendants." Glover is in favor of doing so, arguing that what Anderson called "eugenic genetic engineering" is ethically acceptable after all.

Facing the same problem that confounded scientists of an earlier generation, Glover sees that the primary ethical impediment to designing human descendants is the selection of the ends to pursue. While authors such as Ramsey, Huxley, and Muller had debated in order to convince others to accept their ends, Glover thinks that this debate will not be possible in our newly pluralistic public culture. We can get around this problem, he argues, by forwarding only the end of autonomy. What he envisions is a "genetic supermarket," which would meet "the individual specifications (within certain moral limits) of prospective parents." The idea of "letting parents choose their children's characteristics" is then "an improvement on decisions being taken by some centralized body."[30] The ends that have been so controversial then become a matter of the private conscience of each set of parents.

Bioethicists: Moving toward an Exclusive End of Autonomy?

It is not only members of the communities of lawyers and philosophers who begin to use autonomy as the predominant end, but bioethicists as well. For example, Burke Zimmerman's article on the ethics of germline human gene therapy argues that in addition to beneficence the end of "parental autonomy should permit parents to choose to use [germline gene therapy] to ensure a normal child." For Zimmerman, as for other authors in the debate, this focus on autonomy comes from the tendency to equate HGE with reproduction. After outlining how germline HGE would occur through manipulating sperm, eggs, or pre-implantation embryos, Zimmerman concludes that "just as parents with reproductive disabilities are free to choose in vitro fertilization procedures, they should also be accorded the right to subject their viable embryos to screening and selection, or to direct genetic intervention in order to guarantee the health of their children."[31]

An earlier article, written by Gregory Fowler, Eric Juengst, and Zimmerman, states that "discussions of 'human genetic engineering' proceed as if the idea of germ-line genetic intervention in humans is so revolutionary that we have no moral resources for assessing it: it could only represent 'playing God,' because there are no acceptable human games enough like it to suggest the rules relevant to it." However, "this attitude neglects one *very natural* and very relevant source of guidance: the professional ethos that already governs clinical practice in medical genetics."[32] The ethos of clinical practice in medical genetics is forwarding autonomy. Therefore, whether to engage in germline HGE would be determined by the self-defined concerns of the patients: "the primary question that clinical geneticists should ask . . . is simply whether techniques for germ-line intervention will effectively improve the ability to respond to the reproductive health concerns and complaints of their patients."[33]

Registering Concern about Autonomy

While some bioethicists seem to be promoting the trend toward favoring autonomy, others recognize the dangers of this approach. In an empirical study of the ethical decisions of medical geneticists, professor of public health Dorothy Wertz and bioethicist John Fletcher discuss the problem of screening fetuses for the purpose of aborting fetuses of the wrong sex—mostly females. In a survey they conducted, 62 percent of U.S. geneticists said that they would either perform prenatal diagnosis to determine the sex of a fetus for parents who explicitly wanted to have a sex-selection abortion, or they would refer them to someone who would perform the diagnosis. This compares to the findings of a 1972–73 study, in which only 1 percent of M.D.s/Ph.D.s in genetics said they would approve the use of amniocentesis to determine the sex of the fetus to satisfy parental curiosity.[34] The reasons the geneticists gave for conducting sex screening were most often phrased "in terms of respect for patients' autonomy and rights of choice." Moreover, many of the respondents to their survey "regarded sex choice as a logical extension of parents' rights to control the number, timing, spacing, and quality of their offspring."[35]

Wertz and Fletcher give several ethical arguments against sex-selection abortion, the most important of which is that "it undermines the major moral reason that justifies prenatal diagnosis and selective abortion—the prevention of serious and un-treatable genetic disease. Gender is not a disease. Prenatal diagnosis for a nonmedical reason makes a mockery of medical ethics."[36] That is, with autonomy as the only end, beneficence and nonmaleficence become unimportant. More important, without beneficence and nonmaleficence the "enhancement" versus "disease" distinction that Fletcher and others have made in the HGE debate will be of little consequence. As the authors warn, sex selection may be "a precedent for direct genetic 'tinkering' with human characteristics having little or nothing to do with disease."[37]

However, bioethicists cannot argue outright against autonomy as an end, because it is used in their other jurisdictions. Wertz and Fletcher struggle with a solution to balance the bioethicists' ends without letting autonomy predominate. The authors note that "the logical solution" to the problem would be to "simply withhold information about sex, rather than withholding prenatal diagnosis" from patients.[38] This, however, would return jurisdiction over medical decision making to physicians, putting "control into the hands of doctors, not patients, and set[ting] a precedent for a resurgence of medical paternalism."[39] Resisting this paternalism was the basis for bioethicists' capture of jurisdiction over medical decision making in the 1970s and 1980s.

For doctors explicitly practicing sex selection, Wertz and Fletcher suggest that the ethical codes of medical societies be recast so as to ostracize and punish transgressors. Of course, this stops only those who are open about what they are doing. Wertz and Fletcher call instead for a "middle ground" policy in which information about fetal sex is available upon request but is not regularly revealed to the patients. Although "this will not prevent sex selection if someone is determined to do it, it does minimize opportunities for abuse," presumably from patients who would think of a sex-selection motive after having the sex of the fetus reported to them. However, it seems to me that this plan will not stem the practice if people simply learn to ask for their results, in the same way that women by and large have learned that they can ask to have amniocentesis done to check their fetus for Down syndrome. It is clear that outside the bioethicists' jurisdiction over medical decision making by physicians, the pursuit of the end of autonomy is incompatible with their other ends, and may threaten their jurisdiction over means such as HGE.

Coming Full Circle: Back to Hermann Muller

All these developments took place in the context of debates over germline gene therapy. The last type of HGE that had remained outside the bioethicists' and scientists' jurisdiction was what Anderson and Fletcher had called eugenic germline engineering: the use of HGE to forward the improvement of the species as an end. Scientists had abandoned the pursuit of this end in the 1960s and 1970s under pressure from competitors. "Improving the species" as an end could not be supported in an environment where government advisory commissions were the ultimate decision-makers, because it would have given government representatives too much discretion in deciding what constitutes "improvement." The seeming success of the pro-choice slogan "who will decide?" suggests the power of the argument that government representatives cannot make legitimate decisions about means described as "reproductive."

The government could, however, claim that it was ethical to improve the species, as long as each citizen was able to decide autonomously what constituted improvement. In such a case it would be following the rule of allowing liberty to the people—a safe position for a government advisory commission. If autonomy became the end pursued in HGE, all the arguments made by opponents of enhancement HGE would be moot, and the science could proceed if there were enough people willing to pay for it. (This is the case with IVF research today.) Species perfectionism could be pursued again, although not out of a sense of societal obligation, as Muller thought reasonable, but rather out of a desire for "better" children.[40] Autonomy could bring about a voluntary, positive eugenics.[41]

In his 1991 text Zimmerman goes most of the way back to Muller's position. Summarizing arguments against germline HGE, he says that "it is feared that 'enhancement' modification may be ordered by parents, exercising their right of parental autonomy, who wish to guarantee that their children will be significantly above average." Concluding that individuals are already stratified by genetic endowment from their parents, and that society already allows differential "access of opportunity to other factors affecting socioeconomic status," such as education and health services, he asks, "Is there really anything wrong with" germline HGE? In an argument that could have come directly from the pen of Muller or Huxley, he continues: "What about the positive side, of increasing the number of talented people. Wouldn't society be better off in the long run?"[42]

Similarly, once philosopher Jonathan Glover advocates autonomy to avoid the setting of ends, he goes on to discuss changing human nature through germline enhancement. In this dream of species improvement he reaches back to the vision of Hermann Muller, and quotes extensively and approvingly from his 1935 book. In the first, "preparatory phase of history," humankind was the "helpless creature of its environment, and natural selection gradually ground it into human shape." In the second stage of history, according to Muller, humankind reached "out at the immediate environment, shaking, shaping and grinding to suit the form, the requirements, the wishes and the whims of man." In Glover's vision, in humankind's "long third phase," humankind "will reach down into the secret places of the great universe of its own nature, and by aid of its ever growing intelligence and cooperation, shape itself into an increasingly sublime creation—a being beside which the mythical divinities of the past will seem more and more ridiculous, and which setting its own marvelous inner powers against the brute Goliath of the suns and the planets, challenges them to contest."[43]

For many years no influential member of this debate had quoted Muller approvingly: why would his views seem legitimate once again? With the HGE debate increasingly focused on the means of germline HGE to forward beneficence ("therapy") and beginning to discuss its use to forward autonomy ("enhancement"), Muller's views seem acceptable if expressed in formally rational terms. Perhaps the theological profession that opposed Muller during his lifetime will rise again to oppose the recapturing by the scientists and bioethicists of jurisdiction over all means described as HGE. This seems unlikely, however, as the evidence shows that the theological profession's ability to compete in this debate is in jeopardy.

Travails of a Marginalized Form of Argumentation: The Rise of Jeremy Rifkin

In the 1992–95 period there is once again a distinct community of theologians.[44] The most influential authors are still Ramsey, Joseph Fletcher, and Kass—as well as social activist Jeremy Rifkin. As in previous eras, the data show that the common authors are engaging in a debate that links HGE with other issues, all part of a larger, deeper problem.[45] Although it might be considered a good sign for theology that it has its own community of debate again, optimism for the profession's competition for jurisdiction over HGE would be unwarranted. It is indicative of the marginalization of the theological community that its influential authors wrote their texts in the late 1960s and early 1970s. By the early 1990s Joseph Fletcher and Paul Ramsey had both died. While the common authors in this community are disproportionately theologians, of the influential authors, only Ramsey was a theologian at the time the texts were written. It is indicative of the crisis in theology that no one has filled the shoes of the leaders of the previous generation.

What do the younger theologians see in the twenty-year-old work of Ramsey? Reflecting on Ramsey's work after his death, and referring to his need to use ends worded in secular rather than theological terms, D. Stephen Long describes "a sad state of affairs in Paul Ramsey's work." He was "a particularist forced to use the leveling, generalized putatively universal language of modernity in hopes that his particular tradition might find some room in the modern era." Unfortunately, "his work is much too particular to Christianity to be useful as a common, universal politics of speech, and it is much too universal to be useful as a politics of speech for the creation of an alternative Christian community. Yet Ramsey's work does offer an alternative to the dominant ideology undergirding much of ethics, and his work waits for the creation of those institutional practices that will give it life."[46] The one younger, influential author in this community who might perhaps "give life" to Ramsey's ethics is Rifkin. The characteristics of Rifkin's work, however, suggest how marginalized the theologians have become.

If substantively rational forms of argumentation were allowed to be a part of the debate in the forums that represent the public interest, such as government advisory commissions, it seems unlikely that Rifkin would have become an influential member of this community or of the HGE debate in general. The success Rifkin achieved in making his arguments directly to the public suggests that public concerns about HGE were not satisfied by the debate in the advisory commissions. His hope seems to be that the public, acting through its democratically accountable representatives, will control the aspects of HGE he sees as wrong. What results from Rifkin having the public as his audience is a secular, substantively rational form of argumentation that does not originate in any theological conviction.

Rifkin has a long history of trying to appeal to the public to bypass what could be construed as undemocratic authority. In the 1960s Rifkin was a New Left antiwar activist. He later founded a nonprofit organization, the People's Bicentennial Commission, to protest the commercialization of the 1976 bicentennial celebrations, and called for the redistribution of wealth in the country. In 1977, his organization having been renamed the Foundation on Economic Trends, Rifkin entered the HGE debate in a way that would characterize his future participation and partially explain why he has been able to communicate his arguments to the public.

At a 1977 National Academy of Sciences forum on recombinant DNA research, in the middle of the Asilomar controversy, Rifkin and some fellow activists marched down the aisle waving placards and chanting, "We will not be cloned."[47] Fearing disruption of the conference, officials allotted Rifkin time for a statement. While Rifkin spoke, some of his fellow protesters stood in front of the stage wearing stocking face masks, intended to make them look like human mutants, while others lifted a banner bearing a quotation from Adolf Hitler: "We will create the perfect race."[48]

Beginning in September 1983 Rifkin's organization began a strategy of suing the NIH and others to prevent the initiation of experiments that had been approved by the RAC. Between 1983 and 1987 Rifkin's group filed twelve lawsuits.[49] In 1989 he also sued the NIH to block the first HGE experiment on humans that had been approved by the Human Gene Therapy Subcommittee of the RAC.[50]

Scientists and bioethicists in the HGE debate have extremely negative—almost visceral—reactions to Rifkin. On first glance, the reasons seem obvious. The assumption of both professions is that calm, unemotional, logical discussions should occur within agreed-upon procedural frameworks. Rifkin's main method of getting his message out is not intellectual argument, but lawsuits, protests, press releases, and various "guerrilla theater" events.[51] While all other participants in the HGE debate use academic types of arguments, Rifkin's texts are meant to persuade readers, with lessened regard for scholarly convention. As Rifkin says in the introduction to one of his books, he sides "with the opponents of genetic engineering, and this book is intended to reflect that point of view."[52] Stephen Jay Gould, applying academic standards to his review of Rifkin's 1983 book *Algeny,* states that he regards the book as "a cleverly constructed tract of anti-intellectual propaganda masquerading as scholarship. Among books promoted as serious intellectual statements by important thinkers, I don't think I have read a shoddier work."[53] Like sermons, Rifkin's books aim to persuade, not only through the logic of the argument but through oratorical skills and emotion. Rifkin is writing not so much for a scientist like Gould—or for government advisory commissions—as for the broader public. He writes well publicized and widely reviewed book club–type books.[54]

These tactics have allowed Rifkin's arguments to reach the public. In the estimation of Krimsky, on genetics policy Rifkin "has had more impact on the media than any single group or individual in the United States."[55] Even John Fletcher, generally an opponent of Rifkin's views, believes that Rifkin has "to some extent" influenced public opinion in the United States.[56]

It is not only Rifkin's tactics, his opposition to what scientists and bioethicists want to do, or his refusal to follow accepted intellectual standards that makes him so universally opposed by scientists and bioethicists. Rather, it is his attempt to shift the audience of ethical arguments from government advisory commissions back to the public, which had been the audience before the rise of commissions. Worse, from the perspective of those professions that have thrived in the formally rational environment, the positive response Rifkin has received from the public undermines the primary source of legitimacy of government advisory commissions, and of the entire profession of bioethics as well.

Consider the President's Commission, which used the purportedly universal bioethicists' ends to ensure that its conclusions were in the public interest, instead of simply being in the interests of scientists. If these ends are truly the universal ends of the people, and the conclusions of *Splicing Life* logically flow from these ends, then the public should agree with the report. This includes theologians, who were also portrayed as agreeing with the report's conclusions.

The appearance of consensus was diminished shortly after the release of *Splicing Life* when Rifkin orchestrated a "theological letter" about germline HGE, sent directly to Congress and the press, which called for a ban on germline HGE, in opposition to the conclusion of *Splicing Life*. The letter was signed by the leaders of "virtually every major church group in the United States," from Jerry Falwell, president of the Moral Majority, to the president of the liberal United Church of Christ. Also among the signers were the heads of many of the member denominations in the National Council of Churches (NCC) and J. Robert Nelson, who had written the NCC theological document for the commission.[57]

Rifkin received even more media attention than the commission (including the front page of the *New York Times*), and the apparent contradiction was clear. Reflecting the previously held perception that theologians had agreed with the conclusions of *Splicing Life*, it was reported in *Nature* that "the statement appeared to take a much harder line than that espoused by religious leaders and theologians who testified before the President's Commission." Similarly, the article noted that at the congressional hearings on the report held by then Rep. Albert Gore after the report was released, there was "strong support for the commission's approach of increased scrutiny together with no a priori bans on human genetic engineering." Yet "several clergymen who testified before Gore's subcommittee also signed [Rifkin's] resolution."[58] The tension was clear: *Splicing Life* had misstated the theological views, or the theologians upon whom the commission had relied did not represent the mainstream of their traditions, or those who signed Rifkin's letter were outside the theological mainstream. Since the signers of Rifkin's statement were the titular heads of denominations, the last conclusion would not hold. Readers likely concluded that one of the first two explanations was correct, and both undermined the legitimacy of the President's Commission as representing the public view.

The reaction of the established voices in the debate was to imply that the report of the President's Commission did represent a public consensus, so the religious leaders must have been misled by Rifkin. This was supported by labeling the statement as so incorrect as to be undeserving of serious consideration. For example, the executive director of the by then defunct President's Commission called the resolution "knuckle headed," although the signers included a few consultants to the commission and also a Nobel laureate in biology.[59] The thesis that the signers of the statement had been misled was apparently so widely held that a writer for *Science* magazine decided to call them to find out whether they really agreed with what they had signed. He found that "indeed, most of the signatories . . . said that they do want germline cells declared off limits to all genetic engineering, including efforts to correct genetic diseases."[60]

The writer for *Science* did find some who had signed the statement to spur discussion, and this finding was then used to demonstrate that the signers lacked knowledge of the consensus that had been reached in their own fields. The executive director of the President's Commission was quoted as saying that "sometimes when an urgent problem is being ignored, it may be justified to yell 'fire' just to get attention. But a false cry of fire is not needed in this case . . . the subject has actually been discussed intensively for more than a decade."[61]

Rifkin may not follow scholarly conventions in his writing, but he is a very good political activist, and he knows where lie the tensions that can be exploited. In this case he is exploiting a tension in U.S. political culture between what political scientists call the "delegate" and the "trustee" versions of representation.[62] If a representative simply reflects the citizens' desires, he or she is a delegate. If the representative is elected because the citizens trust him or her to lead them wisely, the representative is considered to be a trustee. The tension is that bioethicists and government advisory commissions act as trustees—attempting to lead the public to improve their reasoning and values—while legitimating themselves as delegates.[63]

If it becomes a question whether government advisory commissions actually represent the ends held by the people, then it also becomes a question whether the form of argumentation used by the profession of bioethics contains the universal ends of the people. If not, then the form of argumentation used by bioethicists is not actually a "philosophy of the people," but rather the philosophy of particular interest groups.[64]

Rifkin wisely recognizes that while Americans may trust their officials to lead in areas such as foreign policy, which the citizens feel they do not have enough information to evaluate, when it comes to issues of morality and ethics, the citizens feel themselves qualified to judge, and want delegates. Rifkin exploits this tension quite well, and this, I believe, is why he causes such an apoplectic reaction among scientists and bioethicists.

The Secular, Substantively Rational Claims of Rifkin

Rifkin in his ethical writings speaks directly to the public, bypassing the government commissions and the debate now dominated by bioethicists and scientists. Although he has no clear link to any religious group, he articulates the thick concerns of the theologians who have not turned to the form of argumentation used by bioethicists. Indeed, although Ramsey might not have welcomed the association, in the words of John Fletcher some of Rifkin's work is "a secular voice echoing Ramsey's theology."[65] Theologians active in this debate by and large do not agree with Rifkin's particular arguments, but he is a member of their debating community nonetheless—because he shares their form of argumentation and because his arguments are so public.

Rifkin definitely sees his own work as allied with the concerns of theologians, and tries to appeal to this community in his writings.[66] Rifkin's 1977 book *Who Should Play God?* was written with pastor Ted Howard and is basically a retelling of the pre- and post-Asilomar debate in more sensational terms.[67] Like Ramsey's texts, Rifkin's narrative has as its protagonists such figures as Kass, Callahan, C. S. Lewis, and the French theologian Jacques Ellul. His antagonists are Joseph Fletcher, Joshua Lederberg, Bentley Glass, Bernard Davis, Hermann Muller, and Julian Huxley.

Like Ramsey, who was opposed to the "messianic positivism" and "unspoken premises" of scientists such as Hermann Muller, Rifkin is opposed to what he calls more simply "the scientific approach" to looking at reality.[68] Indeed, for Rifkin as for Ramsey the problem is not HGE per se, but rather the scientific worldview that underlies it. Ramsey connected all sorts of problems, such as cloning, abortion, and the like, to HGE because they all reflected the same underlying problem, and Rifkin does the same. For example, his 1977 text includes discussions of artificial insemination, in-vitro fertilization, artificial wombs, cloning, and genetic screening.[69]

Recall that one of Ramsey's arguments, echoed by other theologians, is that humans should be wary of the idea that they are in total control of their environment. The more recent theological version of this debate creates a spectrum in which, at one end, human beings and their environment are thought of as "created" by God; while at the other end of the spectrum, human beings control their environment, and in the middle, they are "co-creators" with God.[70] The scientists discussed here would generally stand at the "control" end of the spectrum, arguing that human beings are the sole creators. Most of the theologians in the debate would fall in the middle. Rifkin echoes a secular version of this debate, replacing God with "nature" and "the cosmos," and tends toward the "we are created" end of the spectrum, holding that human beings cannot manipulate fundamental parts of nature.

> We behave as if the world were made for us, never stopping for a moment to consider the possibility that we might have been made for the world. If the world is ours for the making, then of course we are more than justified in believing that the reality we are organizing is in accord with whatever lies beyond it. But what if such were not the case? What if the opposite were true and we suddenly discovered that we were made for the world? How different our course would be. Nature's relationship to us would no longer be as important as our relationship to nature. Instead of forcing the cosmos to conform with our behavior, we would have to refashion our behavior to conform with the cosmos.[71]

This is a secular translation of theology, though not a very subtle one. As Alexander Capron has wisely pointed out, Rifkin conflates God with nature, instead of worrying about the relationship between God and nature, which has marked hundreds of years of Christian and Jewish theology.[72] In other words, Rifkin describes God in a way that critics would call a "God of the Gaps," where God lies in what human beings cannot control or understand.[73] However, like the theologians reviewed in the President's Commission case who believed that to "play God" meant to act on our obligations to God, Rifkin makes a secular echo, with nature and the cosmos standing in as secular gods: "Either we are not in need of anything outside of ourselves or we are. If we're not in need of anything outside ourselves, then we can do pretty much as we please. If, on the other hand, we are in need of things outside ourselves, then we are indebted and must take into account more than our own desires when we act. . . . Everything about us has been borrowed. We have been lent by nature. It has given over to us parts of itself, thus precluding their use for an infinite number of other things. The cosmos owes nothing to us. We owe everything to the cosmos."[74] That is, the world is not simply matter for human manipulation (as occurs in HGE), but rather human beings have obligations to nature, to the "cosmos." Of course, Rifkin's argument faces the same challenge as the theological version of this claim: that it is hard to know what God or nature wants human beings to do.

Like the theologians he resembles, Rifkin uses thick arguments that are generally substantively rational. Like other substantively rational authors in this debate, he sometimes uses consequentialist reasoning, but he also makes many arguments that means such as HGE are a priori wrong because they are inconsistent with a system of ends, regardless of the consequences. Moreover, unlike the formal arguments that assume a limited number of ends, his involve defining and debating myriad ends he puts under discussions of human obligations to nature and the cosmos. Assuming substantive rationality tends also to lead to thicker descriptions of, and solutions to, the problem of HGE. In substantively rational arguments, it is one of a range of problems discussed, all of which are indicative of the deeper problem of human beings not obeying nature. Unlike the bioethicists, Rifkin does not focus on the individual but rather discusses what society should do.

Perhaps the feature of Rifkin's arguments that has most angered his critics is his unwillingness to limit his time frame of analysis to the immediate future—a common feature of substantively rational arguments. For example, a good portion of his argument uses the slippery slope metaphor: perhaps somatic HGE of Tay-Sachs disease is not so bad, but from there it is only a slide down that slippery slope to the point where—perhaps a hundred years in the future—people will be engineering skin color and intelligence. Finally, like other substantively rational authors, Rifkin uses arguments that include the solution of banning some technologies, such as germline HGE.

Examination of how authors refer to Rifkin's work reveals that scientists and bioethicists refer to him as an example of the irrational public that must be calmed. John Fletcher, for example, devotes a good portion of one of his most influential articles to debunking Rifkin, concluding that his arguments "appear to be misplaced and largely emotional criticisms."[75]

Theologians in this debate do not necessarily agree with Rifkin's analysis—for example, his conflation of God and nature—but they must address his work because he is the most visible and influential person who is talking about the thicker concerns of importance to them. Consider the liberal Protestant theologian Ronald Cole-Turner, one of the common members of the HGE debate in this era. He is interested in the thickest of questions that tend not to concern bioethicists and scientists: the "theological significance" of HGE and its implications "for the evolution of life on earth." In tune with the substantive debate over ends more than twenty years earlier, before deciding what to do about HGE, Cole-Turner wants to discuss "the purpose of humanity itself within the scheme of creation."[76] He disagrees with most or all of what Rifkin has to say—but Rifkin appears to be the only visible author asking similarly thick questions and discussing concepts like "sacredness." Cole-Turner criticizes Rifkin for thinking of DNA as a "distinctive metaphysical category, off limits to technical manipulation." Rifkin's sacralizing of DNA "is to reduce God to the level of restriction enzymes, viruses, and sexual reproduction." In my terms, Cole-Turner is criticizing Rifkin's "God of the Gaps" theology.[77] For theologians, Rifkin is someone to debate with, someone who is asking the same questions—someone who, at least for Cole-Turner, serves as a foil for arguing that human beings should be cocreators with God.[78] However, it is ironic and indicative of the increasingly marginalized place of theology in this debate, that the most visible proponent of the thicker, substantively rational form of argumentation is using a translated form of theology that theologians do not agree with.

As the data examined for this book end in 1995, the theologians seem to be destined for the margins of the HGE debate. However, despite the bioethicists' and scientists' institutionalization of their jurisdiction, and of their formally rational arguments, there are signs of trouble there as well. The recent upsurge in the use of the end of autonomy by a diverse group of authors suggests that jurisdiction over the ethics of germline HGE may revert directly to the people. The probability of this development, and its implication for the future of the HGE debate, must be weighed.

Chapter 7

Conclusion: The Future of Public Bioethics and the HGE Debate

> The dehumanizing deeds, if they are done, will not be done by obviously evil men and women. They will be the work of those who are moved by the thirst for knowledge, fame and power—but also by beneficent concern for future suffering.
>
> **—Gilbert Meilaender (1990)**

In this book I have demonstrated that the HGE debate has become thinner over time or, more technically, that it has become more formally rational and less substantively rational. Ends are increasingly assumed instead of debated, and the ends that are used are portrayed as universally held by all citizens in the country and are thus quite thin. The increasingly assumed ends are also commensurable in that authors believe that diverse ends can reasonably be translated into a few commensurable ones.

The requirement that these formally rational arguments be calculable has had an effect on the typical descriptions of HGE and solutions offered. While previous arguments included discussion of different technological means, such as abortion, IVF, and HGE, the means of HGE is increasingly considered separately from other problems. Similarly, because it is more difficult to calculate either the effects on, or the ends of, humanity, HGE is increasingly considered to affect primarily the individual. This same concern for calculability also restricts arguments to technologies that are imminent, and for which the effects are well known. Finally, since the substantively rational method of determining the ethics of a means by whether it is consistent with given ends has declined, and decisions are now made by assessing whether the means of the technology maximize the ends, authors are less likely to advocate permanent bans. After all, the debate now goes, people's ends in the future may change, or we may learn more about the effects of the means, so we must have the technology available in the future.

The Expanding Democracy Explanation

Most bioethicists, I suspect, would say that this change grew out of an increasing commitment to democratic values. What I am calling formal rationality they would call the overlapping consensus or the common ethical language that respect for the pluralistic values in our society demands of us. From this viewpoint, the debate became more formally rational because we as a society decided to use only ends that were universally held. We stopped debating ends because in the mid-1970s the *Belmont Report* uncovered the ends that were shared by all the people and that were applicable to all issues having to do with human beings and science and medicine. Thus, the rise of the bioethics profession, which uses formal rationality, is seen as natural—not the work of interested individuals, but simply part of the flowering of democracy in America.

The evolution of the HGE debate as described in the previous chapters suggests that this explanation of the bioethics profession is deficient in several ways. First, it implies that all collective decision making in pluralistic societies changed from a "pluralistic consensus," interest-group model to an "overlapping consensus" model in the 1960s or 1970s.[1] But this is false. In liberal democracies legislative debate—the pluralist model—is still where we flesh out the competing ends. Congress has debated and established multiple ends that American society should pursue—not necessarily universally held or commensurable ends—since the 1960s or 1970s when democracy was said to have expanded. For example, elected officials have decided that the preservation of ecological diversity is an end to be pursued, resulting in the Endangered Species Act of 1973. They have even implicitly decided that abortion is to be discouraged, through the passage of the Hyde Amendment (1976), as a consequence of some loosely defined end of preserving all forms of life that are genetically human. Not all issues have changed—only some have been "civilized" by the bioethical form of argumentation—and the debate over those that have changed has purposely been moved to government advisory commissions.

Second, this explanation is wrong because the order of events it presents is the reverse of what actually took place. First came government advisory commissions, which needed a formally rational type of argumentation, and the overlapping consensus model, which is a close relative of formal rationality. The real force was not a public demanding democracy, but rather scientists pushing for government advisory commissions, which led to commissions educating the public using commensurable ends.

Finally, this explanation is deficient because it does not address the role of jurisdictional competition in spreading formal rationality. As bioethicists institutionalized their jurisdiction over human experimentation, obtained by appealing to the new audience of government advisory commissions, they and their formally rational type of argumentation became the legitimate way to describe the ethics of human experimentation in all contexts. When another issue, such as HGE, appeared, there was no need to decide what the public's "overlapping" ends really were. Rather, since the profession of bioethics and its form of argumentation already existed, they were able to slide right into this new role. Formal rationality spread because this profession had already gained jurisdiction over related areas.

The Macro-historical Process Explanation

The second common explanation for the rise of formal rationality, among both sociologists and members of other professions, is that it results from some inevitable process outside human control—be it "modernization" or the progression of human thought. As I outlined in chapter 1, some bioethicists imply such a view with theories of inevitable "stages" of debate that end with the most formally rational stage.[2] Moreover, the entire Weberian tradition, including the Habermasian strand, also implies this inevitability. In the words of a Weber interpreter, Weber believed that a formally rational social life "is no mere possibility, but the inescapable fate of the modern world."[3]

We can envision weak and strong versions of this perspective. In the strong version, implicit in accounts by bioethicists and philosophers, actors in these debates recognize the new constraints of modernity and act accordingly. In other words, there were no interests exercised, the debate simply shifted as people's consciousness collectively changed. This version is repudiated by the history of the HGE debate, which shows that change in the debate did not simply happen, but rather was contingent upon one group moving the location of the debate from the public and its representatives to unelected government advisory commissions.

The weak version of this explanation assumes struggle and a zig-zag path toward a slowly evolving formal rationality. Thus, the argument would go, the general conditions are present for the rise of formal rationality, and debates will become formally rational at different stages as actors who benefit from those general conditions eventually win out. There is, of course, no way to conclusively evaluate claims like this, where negative findings can be explained by saying, "It just hasn't happened yet."

However, some additional leverage can be obtained by comparing the HGE debate to the debate over abortion.

The modern debate about abortion appeared as an issue at approximately the same time as HGE, yet it has resisted assimilation to the "civilized" debate of bioethicists.[4] That is, the arguments about abortion have retained debates about ends, and methods of linking means and ends, that are not considered legitimate in the bioethicists' form of argumentation. The bioethicists' encounter with debates about embryos, fetuses, and abortion has, for this very reason, been a difficult one. In 1979 the Ethics Advisory Board (EAB) of the U.S. Department of Health, Education and Welfare, which followed the National Commission, produced a report on research on human embryos. The report suggested that while the embryo is entitled to profound respect, this respect is not the same as that offered to persons. The report recommended that embryo research should demonstrate its safety and efficacy, that embryos should be sustained in vitro for only fourteen days, that gametes should be obtained only through informed consent, and that embryo transfer should be conducted only with married couples.[5] The report was ignored by the Carter administration, and the EAB was shut down in 1980 after only two years of operation. Since department regulations required the EAB to approve any research involving embryos, yet there was no longer an EAB, no research with human embryos was funded by the federal government between 1981 and 1992. Explicit moratoria on such experiments were written into various congressional bills, and the successor body to the President's Commission was prevented for two years from doing any work because of congressional debates over abortion.[6]

The consideration in the abortion debate of ends that are not considered legitimate by bioethicists has caused them great frustration. For example, John Fletcher and W. French Anderson, writing about germline HGE research, discuss the various "obstacles to research which originate in the federal sector," which contradict the institutionalized Belmont ends they are pursuing. They note that scientists funded by the federal government are forbidden to study the genetics of human embryos, limiting knowledge on issues such as cancer and genetic disease. "This obstacle violates every ethical principle of biomedical research," they state. "Are pre-embryos more valuable or more to be protected in research than children with cancer?"[7] Clearly, the constituents of the lawmakers who have placed these constraints on federal policy have an additional end that outweighs the beneficence that Fletcher and Anderson refer to: what is somewhat imprecisely called "the sanctity of human life" by the pro-life movement.

However we may personally evaluate this particular dispute, what is more important for the theme of this book is how the abortion debate has remained both somewhat substantively rational and independent from the bioethicists' form of argumentation. This suggests that there is nothing inevitable about the rise of the four ends used by bioethicists specifically, or the rise of formal rationality more generally.

Other features of the abortion debate also support the argument in this book. The reason why the abortion debate has remained at least somewhat substantively rational, while the HGE debate has become formally rational, is that the abortion debate has retained the public as the ultimate decision-maker and the HGE debate has not. Politicians on state and federal levels determine whether abortion is going to occur—whether through votes or the appointment of justices. These politicians are influenced by the public and various interest groups, from the Right to Life Committee to Planned Parenthood. In sum, the rise of formal rationality was not natural or inevitable, but as I have shown, required moving the issue away from the public and toward less democratically accountable arenas of decision making. Thus, the thinning of the HGE debate might have been prevented if actors at various points had been able to keep decision making in the hands of the public.

For example, in the mid-1970s, amid public concern about biohazards that ultimately resulted in the Asilomar controversy, the public could have been organized into competing social movements, pro and con, to have a more substantive debate about the hazards of genetically engineered microorganisms and genetic engineering more generally. Yet, as others have shown, the scientists were particularly adept at diverting the controversy through temporary self-imposed moratoria and the use of government advisory commissions.[8] In the mid-1980s the extraordinarily diverse group of religious leaders opposed to germline HGE, organized by Jeremy Rifkin, could have formed a permanent, religiously based social movement organization, just as a similarly constituted group of pro-choice religious leaders in 1973 created a coalition to defend abortion rights.[9] Instead, the group opposing germline HGE seemed to last only as long as its press conference.

In short, in the HGE debate, as in the abortion debate, what was needed was for somebody to start a social movement. But the emotion and sense of outrage that could have fueled such a movement were placated with assurances that government advisory commissions were handling the situation and would recommend policy to account for the concerns of the public. The government advisory commissions reported back to the public using formally rational arguments that had the effect of calming the public and diffusing any social movement activity. Contrary to the "onward march of rationality" thesis, public bioethical debates do not inevitably become more formal, but rather such change is contingent on the debaters themselves.

A New Explanation for Change

The data on the HGE debate suggest that both the "expanding democracy" explanation and the "onward march of rationality" explanation are deficient. What these explanations lack is attention to the human actors who actually write all those texts that collectively make up the HGE debate.[10] Texts do not simply arise from their social-structural context, but are intentionally produced by people with their own social contexts and interests. The theoretical framework used in this book takes a first step toward accounting for this by assuming that writers have a simple interest in promoting their vision of what is correct. To continue to promote their views in the debate, authors need legitimacy in order to warrant their continued salary from a university, think tank, or scientific laboratory. More critically, they need to be read by the people who actually make the decisions about HGE. An author's or a community's influence on a debate also depends on the beliefs of those who make the decisions. If it is the public that decides which arguments are the best, one group of authors will be legitimate. If government advisory commissions are making the decisions, another group of authors will be legitimate.

Legitimacy is not simply an individual matter. One of the findings of this research is that the legitimacy of authors is structured by their profession. It is professions that are collectively competing for legitimacy—for what I have called jurisdiction. Thus, a change in the dominant form of argumentation in the HGE debate is explained by showing how a particular profession gained jurisdiction over the work of promulgating the ethics of HGE, and why that profession adopted the form of rationality that it uses. It turns out that change in who provides legitimacy to arguments—from the public to the bureaucratic state—is a primary explanatory factor of both who gains jurisdiction and what form of rationality the profession uses. Given this new perspective, we can now summarize why the HGE debate changed from substantively to formally rational, and also explain the ethical consensus that has formed in the debate over what types of HGE should be allowed.

The HGE debate of the 1950s and the 1960s was dominated by reform eugenicist scientists, such as Hermann Muller, Julian Huxley, and Theodosius Dobzhansky. To them, HGE was synonymous with what was later called germline HGE. This debate was substantive and thick—and largely about ends. The ends pursued were not only what later came to be called beneficence (the concern over the "genetic load" of the species) but also ends that are best described as theological. Muller and his contemporaries saw in germline HGE the ability to forward the end of giving humans a new meaning and purpose in life after Darwinism had discredited theological meanings and purposes.

By and large, the reform eugenicists wanted the population to be more like themselves: intelligent, genetically fit, and free of tendencies toward "criminality" or "sloth." However, a younger generation of scientists coming to prominence in the early 1960s began to worry that the ability to engineer the future of the human species was imminent, and called for public debate on the topic, beyond the debate among eugenicists. At the same historical moment, scientific jurisdiction over public affairs was being questioned more broadly in society by the New Left and environmental movements, among others.

With the public beginning to question whether jurisdiction in the HGE debate should be retained by the scientists, the theological profession had an opportunity to counter scientists such as Muller, who were seen as infringing on the theological jurisdiction. Theologians entered the HGE debate with a very substantive form of argumentation. Ends were debated, not assumed, because proponents hoped to convert readers to accept their ends. Some authors advocated a link between means and ends such that means must be consistent with a system of ultimate ends, not maximizing ends. This was the debate at its thickest.

As I showed in chapter 3, the scientific community became fearful that this new democratic control of science would threaten its home jurisdiction of basic research. The uninformed public, acting through its congressional representatives, might curtail the funding for science—a process that, from the scientists' perspective, had already begun. The scientists struggled to take ethical decision making out of the hands of the popularly elected officials and into government advisory commissions, where debate could be controlled. Their success in this effort changed the HGE debate forever.

The new arbiter of the ethics of HGE was the government advisory commission, not the public, and what was considered to be a legitimate argument changed accordingly. Government advisory commissions prefer formally rational arguments, for reasons outlined in chapter 1, and a new profession arose that could take advantage of the changed environment. The profession of bioethics created a formally rational type of argumentation that became the accepted way to make claims in the HGE debate. The primary competitor to bioethics, theology, could not take advantage of this new environment because a shift to formal rationality would threaten its own core jurisdictions having to do with transcendent belief, which cannot be expressed in formally rational terms. Thus began the decline of theology in the HGE debate.

To mitigate the threat to their core jurisdiction, scientists also defensively redefined the HGE technology they were developing, and connected these means to their home jurisdiction using a more defensible metaphor. They no longer claimed to forward the end of providing meaning, nor did they want to develop means to change the genes of future generations. Rather, they wanted to develop the means of somatic human gene therapy—the manipulation of the body to forward the end of beneficence. These means and ends were at the unquestioned core of the jurisdiction of medical research scientists.

As I have shown in the case of the President's Commission, government advisory commissions did much to institutionalize the formally rational type of argumentation and to delegitimate authors using substantively rational forms of argumentation—most notably theologians. By the mid-1980s the profession of bioethics had grown in influence in the HGE debate, and government advisory commissions had become the final arbiter of ethical claims. Bioethicists and their formally rational type of argumentation were well suited to survive in this environment, and both prospered.

With the four ends of the bioethics profession increasingly institutionalized, scientists and bioethicists took advantage of the opportunity to move the boundaries of their jurisdiction back toward where they had been in the 1950s and 1960s. The end of beneficence justified the focus on somatic gene therapy, but there was nothing in the ends that suggested why beneficence had to be limited to means that affected the bodies of already existing people. Beneficence could justify "therapy" of future generations, because ends that would have contradicted this idea were no longer legitimate—like the idea expressed by theologians that human beings should be wary of assuming that they control everything. Thus, scientists dropped the metaphor linking the means of HGE with means in their home jurisdiction, and simply made the metaphorical link of ends. Scientists developed any means necessary to forward beneficence, they argued. By the late 1990s, as the bioethicists' form of argumentation became increasingly institutionalized, autonomy became increasingly dominant as the only end to pursue.

In practical terms, the shift to formal rationality in the HGE debate has meant that the professionals who write to and for the public have severely constrained their debate at a time when the technology is actually becoming possible and decisions are becoming more critical.

Formal Rationality, Public Bioethical Debates, and Public Discourse

Beyond the particularity of the HGE case, and even the particularity of public bioethical debate, this study fits into a more general intellectual debate. Recognition of the importance of studying the assumptions behind public discourse has grown in many social science disciplines over the past twenty years. For example, the social problems literature is almost defined by the insight that social problems are not objective, but exist only when a group has spent the resources necessary to label a problem a problem.[11] Similarly, recent "social constructivist" approaches to social movements emphasize that people do not revolt due to objective conditions, but only when they have a language that defines their collective situation as unjust.[12] Finally, many studies have demonstrated the societal effect of the dominance of certain forms of discourse over others. *Habits of the Heart,* for example, shows how a discourse of individualism affects people's actions in society, and other studies explain the sources of discourses of compassion, narcissism, and the self.[13]

In congruence with these studies of public discourse, this book has been concerned to explain the growth and effect of formal rationality in public discourse. The most influential studies in this particular area have been the highly theoretical work of German social theorist Jürgen Habermas. He sees formal rationality as the province of both the state and the market, and substantive rationality as dominant in the life-world—roughly the different communities in a pluralistic society where debate about ends or values occurs. The primary problem of modern society is that due to the "colonization" of the life-world by the state and the market, substantively rational discourse is replaced by the formal, leaving in jeopardy people's ability to conceive of proper ends.

Nothing in this book contradicts this narrative. However, since Habermas is not conducting empirical research and is operating at such a high degree of abstraction, much is left to explain. For example, why some debates would resist this colonization is unclear, and given that the details are not a part of his explanation, the process also appears inevitable.

There have been few empirical studies of this process.[14] Sociologist Wendy Espeland is one scholar who has studied the social sources of forms of rationality. In her examination of a debate over the building of a dam, she shows how different groups endorsed different forms of rationality, with the decision of which rationality to use being a political accomplishment of one group over another. Moreover, she shows that the rise of formal rationality was not inevitable, and that the attempt to impose formal rationality on a group that used substantive rationality provoked resistance.[15] My study complements Espeland's, focusing more on interprofessional struggle and the state, and covering a much greater time range. From these few studies it is too early to provide any generalizations about this process in public life more generally, but if Habermas and other, more theoretically inclined writers are correct about its importance, much greater attention to this topic by scholars is needed.

Recent Developments in HGE Technology and the HGE Debate

As one would expect from the rapidity with which the technology of HGE changes, and the pace of change in the debate over HGE as well, much has happened since 1995, the last year of quantitative data that was available when my data collection ended. In general, the trend described in the previous chapters has continued. The debate has come full circle, with discussions of whether the disease/enhancement distinction is worth retaining, and whether germline research should be approved by the RAC. With the exception of the formally rational type of argumentation underlying the arguments, Hermann Muller would have been able to take part in today's debate.

Questioning the Disease/Enhancement Distinction

Since the early 1990s the science of HGE has become much more capable. The RAC has been asked to approve protocols for the genetic engineering of diseases that are less life threatening, such as rheumatoid arthritis.[16] Thus, the conditions to be treated by "therapy" have started the slide away from conditions universally viewed as bad. For example, what about HGE that provides a person with increased immunity from cancer? Is that treating disease, or enhancing the individual? Is this the first step toward the brave new world envisioned by the reform eugenicists?

In 1997 the RAC called a special meeting to discuss the ethics of the disease/enhancement distinction, introduced by Anderson and Fletcher in the mid-1980s and subsequently institutionalized in the debate. That the RAC called the meeting and selected the panelists tells us something about the existence of a debate independent of the state in the public sphere. The RAC would later be setting policy, and it was also shaping the public opinion that it would purportedly follow in setting that policy—throwing into question the very existence of an independent public sphere in the HGE debate. But for now, let us consider the technology the scientists at that meeting were envisioning and their prediction of what the ethical consensus would be. Scientist Theodore Friedmann opened the conference with this stunning statement:

> Will technology make [enhancement] feasible? Probably. Are there appropriate therapeutic and *cosmetic* goals? Probably. Are there applicable ethical or policy norms? Yes, probably, but they are going to be imperfect for some time and we should agree that there will be no universal agreement on what those norms are. Are there new or severe ethical problems? I would say yes but I will leave it to the

> much more enlightened discussion of the ethical issues. Finally, is it going to be socially acceptable? Yes, I think so, to some extent and, therefore, I would conclude that previous pronouncements on the unacceptable nature of any enhancement therapy have probably been facile and probably incorrect.[17]

Later Friedmann stated that "in the next year or two we will begin to see therapeutic results and . . . the techniques, I think, for achieving those therapeutic results will likely be similar to or identical to those that could be employed in enhancement manipulation."[18] He mentioned a few types of enhancements that were under discussion: "physique, modification of the aging process, changes in memory, and then from there, intellectual targets, personality targets, behavioral and cognitive issues, and from there even modification of the genetic influences that affect our moral decision making."[19]

Here was a prediction from one of the top scientists in this field that enhancement is imminent, that therapeutic and cosmetic goals are appropriate and will be socially acceptable. There was no consensus at this conference, but the discussion shows how rapidly the ethical consensus is changing. One of the last barriers to scientific experimentation was being reassessed only twelve years after the first argument for the disease/enhancement distinction and only a few years after that distinction had begun to be widely accepted. With this barrier breached, then all acts of HGE—even enhancements of intelligence or strength, be it in the soma or the germline—would fall under the jurisdiction of bioethicists and scientists. The debate is not over yet, however, and as of early 2001 the RAC has not agreed to any protocols for obvious enhancements.

The Recommended Germline Debate

Erik Parens has noted that when the technological capacity was still lacking, there was agreement that germline HGE should not be done. "But inevitably, once you have the technical capacity, everything changes."[20] I do not mean to imply any technological determinism here. Rather, when a technique could not yet be done, scientists were not pressing to include it in their jurisdiction. But once a scientist wanted to use a technology, and the case could be made that it would forward one of the four ends, then scientists and bioethicists would begin to question the previously established jurisdictional boundary.

Although scientists increasingly want to conduct germline HGE experiments, the RAC has yet to approve any. The scientists' arguments for this have been classically formally rational: the means of germline HGE would forward the end of beneficence more than the means of somatic HGE. Using arguments that maximize beneficence, the scientists would experience little opposition from within the bioethics/science jurisdiction. (The end itself would not be questioned, because people wanting to debate ends had been previously marginalized in the debate.)

Why would germline HGE more effectively forward beneficence than somatic HGE? To oversimplify, scientists have not yet gotten somatic HGE to work very well.[21] Of the many problems, the most obvious has been that it is difficult to change enough cells in a fully grown body to change bodily function. For example, to use somatic HGE to relieve ADA (adenosine deaminase) deficiency, one must change the genes in the bone marrow stem cells. The first recipients of engineered genes in Anderson's 1990 experiment had to remain on other drugs to treat their medical problems because not enough cells had been changed to produce the missing enzyme.

The growing consensus among scientists in the late 1990s was that germline HGE would be much easier to accomplish than somatic HGE.[22] Only the one fertilized egg cell would need to be changed, and as it multiplied, the changes would be expressed in every cell of the person's body, including the reproductive cells, and would be passed on to the next generation.

The suggested technology for engaging in germline HGE has also taken a creative turn. The previous assumption had been that germline HGE would be carried out like somatic HGE. The DNA would be placed in a virus that incorporates itself into a cell, so the virus could be used to "deliver" the desired genes to the cell. At a conference held in 1998, however, a new strategy was revealed: Scientists were attempting to create artificial "extra" chromosomes, containing whatever genes one wanted to put on them, which would be inserted into the fertilized egg. Humans created in this manner would have 47, not 46 chromosomes in every cell of their body. Artificial chromosomes have been patented, and companies have been formed to exploit this technology.[23]

Recall that in 1991 the chair of the RAC called for an extended discussion by the public about the ethics of germline HGE. A conference, billed as the "first significant public forum anywhere" focused exclusively on germline HGE, was held at the University of California, Los Angeles, in March 1998. Although it was billed as a public forum, and was covered by reporters for the *New York Times, Newsweek, Washington Post,* National Public Radio, and *Nature,* the meeting is more accurately described as a conference where some members of the scientific community worked out what they saw as the ethics of germline HGE. Organized by scientists, the panel consisted of nine scientists, a bioethicist, and an expert on public policy. The bioethicist was John Fletcher, who had previously argued for bringing germline HGE into the scientific/bioethical jurisdiction.

There were no critics of germline HGE on the panels, nor was there much discussion of ends outside of the institutionalized Belmont ends. The primary debate was not over ends, but over whether the ends of nonmaleficence and beneficence would be maximized by germline HGE; formally rational maximization of means to ends was assumed. The primary reason that germline HGE should not occur yet, according to most participants, was that not enough research had been conducted to demonstrate its safety.

At this conference, enhancement was not considered to be much of a problem. James Watson, former director of the Human Genome Project and codiscoverer of the structure of DNA, stated at one point, "If you could cure a very serious disease, stupidity, that would be a great thing for the people who otherwise would be born seriously disadvantaged."[24] According to *Nature*, the "lone voice" advocating a sharp line between enhancement and disease was W. French Anderson, who had created the distinction over a decade previously.[25] Excerpts from the conference proceedings reveal that he was opposed to enhancements only for safety reasons, suggesting that once this hurdle was cleared, there would be no legitimate reasons in this community not to engage in enhancement.[26]

Beyond beneficence and nonmaleficence, the end of autonomy was taken very seriously by conference participants, who paid special attention to the bioethicists' argument that a person who does not yet exist has not given informed consent to have his or her genome changed. The scientists have taken this argument so seriously that they are designing their technology to satisfy it. For example, one of the conference organizers suggested that "the designer gene for, say, patience could be paired with an on-off switch," and "the child would have to take a drug to activate the patience gene. Free to accept or reject the drug, he retains informed consent over his genetic endowment."[27] Another ethical concern to be circumvented with technology involved beneficence and nonmaleficence—specifically, the possibility that something would go wrong when the gene was passed to future generations. Geneticists are working around this by creating genes that self-destruct in cells that become eggs or sperm so that, while the entire future body of the embryo would be transformed, true germline HGE would not occur. The process, already tried with animals, involves designing a string of genes flanked by the molecular version of scissors. These scissors are activated by an enzyme made in cells that become eggs or sperm. Once activated, the genetic scissors cut out the introduced gene so that it is not passed on to future generations. An article on the conference did point out that "there is no easy technological fix for another ethical worry"—that germline HGE would violate the final Belmont principle of justice by allowing the rich to engineer their children to give them advantages over the poor.[28]

Beyond the scientific creativity, these technological developments demonstrate how firmly institutionalized the ends of the bioethicists have become: they are to be inscribed into the design of the technology itself. Moreover, news reports on the conference presented these ends as the only ones open for debate. Now that germline HGE is apparently imminent, we have a debate that is constrained to disagreements over whether the technology would maximize a set of institutionalized ends. "Safe enough" or "not safe enough" will be the primary debate the public sees, while debates based on other ends, or about such ends, are forgotten.

Finally, the scientists at the conference expressed the usual fear that the public will restrict their scientific freedom. In a series of policy recommendations arising from the conference, social control over the technology was directed toward the commissions in the federal bureaucracy where formal rationality would, according to my argument, forward the scientists' interest, and away from more democratic forums. The recommendations called for the RAC to begin consideration of germline proposals, the FDA to regulate germline research, the NIH's Human Genome Program to start investigating the ethics of germline HGE, and the new federal bioethics commission to begin discussing patenting human genes. Arguments were made against social control by more democratically representative bodies, which would allow the public's ends to be debated: "No state or federal legislation to regulate germline gene therapy should be passed at this time."[29] Watson, a longtime defender of scientific freedom, said after the conference that "scientists should proceed unhindered towards germline engineering."[30]

If this was truly the first public forum on germline HGE, does this one conference, limited to the scientists' perspective, satisfy the discussion in the public realm called for by the previous chair of the RAC? Apparently so for some advocates, because immediately following the meeting the chair of the RAC was quoted as saying that the RAC's effort to rewrite its guidelines, last updated in 1990, would "be an opportunity for the committee to revisit" the germline HGE issue.[31] At the RAC meeting immediately following the UCLA conference, one of the scientists on the RAC also took note of the conference and raised the issue of whether the RAC should begin discussing germline HGE.[32] Another RAC member concluded that "the RAC is the ideal forum for having a societal debate of the germ-line gene transfer issue," and the RAC subsequently endorsed a recommendation to invite experts to discuss germline HGE. However, since it was pointed out by another RAC member that the "public reaction" to germline HGE "will be quite different from the opinion of some scientists," the RAC agreed that the discussions should not be viewed as an endorsement of germline HGE and that the intent of the tentative discussions it was about to undertake should be "clearly articulated to the public, perhaps as a statement published in the Federal Register."[33] That is, the RAC wanted to avoid the public outcry that in the past has resulted in calls for more democratic control over the ethical decision making about HGE.

Human Cloning

In 1997 it was announced that researchers in Scotland had cloned an adult sheep, creating a lamb named Dolly. The debate that ensued offers an opportunity to further demonstrate and clarify my argument. I have shown how the public bioethical debate among the professionals has become more formally rational. However, despite the efforts of many advocates of formal rationality, it is unclear whether the public has taken up this form of argumentation, or whether it retains a store of substantively rational arguments. Put in a language common to academics today, does the public resist the discourse of the elites? I have also provided evidence that when decision making lies more directly with the public—as it did before the 1980s in the HGE debate and as it does in the abortion debate—the debate has remained substantive. The cloning debate suggests that the public does indeed retain a store of substantively rational arguments, although they are not very well articulated, probably due to the lack of elites to put them in coherent form. Further, scientists, bioethicists, and government advisory commissions do try to commensurate the public's ends and translate their arguments to make them compatible with formal rationality.

In the case of Dolly, scientists were able to take the genetic material from a cell of an adult and place it into an egg whose nucleus had been removed. They then implanted the zygote into an adult for gestation. Applied to humans this would be a particularly precise form of germline HGE, because one could determine with certainty the germline of one's descendants by examining the genes, or simply the traits, of the person chosen for cloning.

The news of Dolly caused a public furor. The litany of arguments as to why we should be outraged about this newfound ability to clone revealed that the public was debating many more ends than the four used by the bioethics profession. Moreover, the public seemed willing to conclude that cloning should never occur because it was inconsistent with these ends (and not because it might be calculated to not maximize ends). Reacting to public outrage, President Clinton forbade federal funding of human cloning research and requested that an already existing government advisory commission, the National Bioethics Advisory Commission (NBAC), report on the ethics of human cloning in ninety days.

On first examination, the furor surrounding Dolly the sheep would seem to contradict the findings in this book. At the hearings on human cloning commissioners heard from a few substantively rational theologians as well as from secular substantively rational authors, such as Leon Kass. The NBAC also commissioned a paper from a scholar on religious perspectives on cloning. Its final report had an entire section on religious perspectives on cloning, and repeated verbatim many arguments that did not fit within the formally rational type of argumentation now institutionalized in the field. For example, the final report referenced debates about the ends of scientific exploration, a debate that has not been encouraged by scientists or bioethicists. Referring to this particular passage, Erik Parens later wrote, "Can anyone name another government sponsored report in which a phrase like 'the moral ambiguity of our own cleverness' appears?"[34]

If my argument is correct, how could a government advisory commission have engaged in such a substantively rational debate, involving myriad ends beyond the usual four? How could theologians and others have made claims about acts needing to be consistent with ends, not simply to maximize ends? The answer is that, for this issue, and unlike most other decisions by government advisory commissions, the ultimate arbiter of the ethical decision was the public. There was no doubt from the beginning that a democratically elected official would be held accountable for the decision on this issue—be it the president or the Congress. Before the NBAC had even begun to deliberate, in reaction to the public outrage members of Congress had introduced bills to ban human cloning. A bank of news cameras was filming the proceedings as the NBAC reached its final judgment.[35] The NBAC had to publicly air the public's concerns.

The need for the NBAC to let the public know that it was legitimately handling the situation allowed the theologians back into the bioethical debate in a very visible way. But were the theologians invited because their arguments might contribute to the NBAC's understanding of human cloning? Obviously some commissioners thought that the theologians provided insight. However, I agree with the author of the religious perspectives background document for the NBAC, who later stated that "the contributions of the religious perspectives were deemed politically important and ethically insignificant."[36] To put a slightly sharper point on it, theologians were no longer seen as serious contributors to the debate itself, but rather served as spokespersons for communities within a pluralistic public that was closely watching the cloning debate.

Let us consider the reasons for inviting theological input. The first reason, it seems clear, was that the president of the United States had defined the problem theologically: "any discovery that touches upon human creation is not simply a matter of scientific inquiry, it is a matter of morality and spirituality as well."[37] With the president telling the NBAC that he thought cloning had religious implications, it would be surprising if the NBAC did not try to obtain religious input.[38]

Commissioner James Childress later offered five reasons why the NBAC had invited theological input. The first was that religious communities "shape the moral positions taken by many U.S. citizens on new technological developments." Second, religious traditions often make secular arguments. Third, the NBAC wanted to know whether there was a consensus among religious groups about cloning. Fourth, it wanted to start a national debate about cloning, and, implicitly, religious communities would frame the debate. Finally, and most tellingly, noting that effective public policy requires assent from the governed, the NBAC wanted to gauge the "nature, extent, and depth of opposition to those policies" by religious groups.[39] In sum, while the third reason given implies that theologians might have been able to offer insight to the process, the other four reasons suggest that theologians represent the views of interest groups that must be brought into any compromise position.

The public had also defined the cloning problem even before the NBAC had addressed it. Had the NBAC limited the debate to the bioethical status quo, it would have appeared to the public that the commission was dodging the real issues. The public was extremely negative about cloning, and for reasons that scientists found to be ridiculous. Popular magazines speculated about creating legions of Michael Jordan and Hitler clones or creating copies of people that, being genetically the same, would somehow overcome the nurture part of the nature/nurture debate and act just like the original. Some even wondered whether it might be possible to create genetically identical bodies as spare-parts storehouses.

The public was repulsed, although it could only express this repulsion through concerns about cloning Hitler. Because the public framed the topic with repugnance, and the public was paying attention, repugnance would have to be addressed. Thus, other substantively rational authors besides theologians were brought in by the NBAC. Kass was invited to testify, and as one would expect, he did not conform to the type of argumentation laid out in the *Belmont Report*.

While bioethics had evolved to require calculable arguments about ethics, Kass argued that "in crucial cases repugnance is often the emotional bearer of deep wisdom beyond reason's power fully to articulate it." Making decisions on the basis of "deep wisdom" that could not be articulated was not a notion that would resonate well with bioethicists, who by and large reject the epistemological standard of intuitionism. A serious debate about ends had not really occurred since the setting of the bioethicists' ends in the *Belmont Report* over twenty years before, and Kass seemed to want a debate about what the ends of science itself should be: "you can strike a blow for the human control of the technological project for wisdom, for prudence, and for human dignity. The prospect of human cloning so repulsive to contemplate, in fact, provides the occasion as well as the urgent necessity of deciding whether we shall be slaves of unregulated progress and ultimately its artifacts or whether we shall remain free human beings to guide our technique towards the enhancement of human dignity." The conclusion of his testimony was a long quotation from the late Paul Ramsey on the necessity of a substantively rational link between means and ends that could result in permanent bans: "A man of frivolous conscience announces that there are ethical quandaries ahead that we must urgently consider before the future catches up with us. By this he often means that we need to devise a new ethics that will provide the rationalization for doing in the future what men are bound to do because of the new actions and interventions science will have made possible. In contrast . . . a man of serious conscience means to say in raising urgent ethical questions that there may be some things that men should never do. The good things that men do can be made complete only by the things they refuse to do."[40]

Kass's and others' arguments give the human cloning report a substantively rational tone. The presence of substantively rational arguments, as well as input from substantively rational authors, however, does not diminish the force of my general argument because this input was the result of the public being the ultimate decision-maker in the process. It is important to note that the NBAC report is not as thick as it first appears, and that when it came to actually reaching ethical conclusions about what to do about cloning, the broader, substantive debate receded into the background. The commissioners recommended a temporary moratorium on creating a child through cloning, but the only argument they could agree upon to support this policy was that cloning is not safe. That is, it violates the end of nonmaleficence. The commission also recommended that any legislative moratorium expire after three to five years, so that we could decide—if we so chose—to engage in human cloning. Moreover, the commission did not endorse any substantively rational arguments or ends outside of those listed in the *Belmont Report.* Rather, the thicker arguments provided by people who testified were essentially reported without any attempt to evaluate them.

One commissioner has suggested that because the president gave the

NBAC only ninety days to reach a conclusion on the issue, it would have only enough time to "describe and not to resolve the ethical arguments."[41] Nonmaleficence was, in this view, a stop-gap for the real substantive debate the commission did not have time for. What would the final report have been like if the NBAC had had the two and a half years that the President's Commission had had to work on *Splicing Life,* and not just three months? It seems quite possible that given the time to work through the arguments of someone like Kass, the commission might have translated his substantively rational arguments to formally rational arguments, as the President's Commission had done.

The power of the bioethicists' form of argumentation is suggested by Childress's description of the process. Recalling being questioned as to whether the NBAC process had used the principles of bioethics, he argued that the principles had been used, and that the commission had calculated ("weighed") to see if the ends had been collectively optimized:

> I responded that NBAC's concern for safety reflects the principle of nonmaleficence and that NBAC at this time could not identify benefits of human cloning that outweigh the risks to children (a consideration of beneficence) or claims of autonomy in reproduction or in scientific inquiry strong enough to outweigh the risks to children. In addition, concerns about respect for persons, including their dignity as well as their autonomy, surfaced in discussions about objectifying and commodifying children. I would argue that these principles, and others, were transparent in NBAC's deliberations. . . . At the very least, the commission's consensus reflects its views about the respective weights of three prominent moral principles—nonmaleficence, beneficence, and respect for autonomy—in the context of recommending public policies regarding human cloning.[42]

Although Childress noted that some people testifying before the NBAC argued for the case that cloning was wrong regardless of consequences, the commission clearly did not take that view, using instead a consequentialist view. Moreover, the use of the four thin principles meant that a good number of nonreducible arguments—those that could not be commensurated to one of the four ends—did not have to be evaluated. In his words, "consensus on the safety argument meant, however, that NBAC did not have to try to resolve the larger debate about the probable impact of acts and especially of widespread practices of human cloning on social values that are not reducible to harms or wrongs to individuals."[43]

As noted by critics, not having this debate on "social values that are not reducible to harms or wrongs to individuals" means that once cloning can be demonstrated to be safe, it will be considered ethical, as the NBAC's perspective becomes institutionalized. Given the constraints of a government advisory commission, the commissioners did an exemplary job. The real question is why such debates are taking place in commissions that operate under these constraints.

To date, despite the recommendation by the NBAC that Congress pass a temporary moratorium, no bill has been passed. In July 1998 it was reported that scientists had cloned mice, which are more similar to humans in this regard than sheep are. "If you can clone from mice, humans should be easier," said one prominent scientist.[44]

The Future of Professional Jurisdictions

It is clear that the conclusions in the debate over HGE will largely depend upon which profession retains jurisdiction. While predicting the future is dangerous, some generalizations can be made about the future prospects for the primary professions that have been competing for jurisdiction over HGE.

The Future of the Bioethicists' Jurisdiction

While the bioethics profession has been strengthening its jurisdiction over ethical decision making in public bioethical debates, it has grown even stronger in its other jurisdiction of decision making in hospitals (clinical bioethics). The bioethicists have also increasingly taken on the trappings of a classic profession, with a professional association of 1,500 members; 200 centers, departments, and programs; and an academic degree (a master's in bioethics from the University of Pennsylvania and a Ph.D. in bioethics from various other universities).[45] There is even a debate about licensing: who is qualified to be a clinical bioethicist and to offer ethical judgments in hospitals?[46] Licensing, and other internal and external controls over who is a legitimate member of the profession, is the hallmark of an increasingly successful profession.

The organization that accredits hospitals has required since 1998 that every hospital have a mechanism for resolving ethical problems that could arise. While institutions where human research is conducted have been required to have such a committee for decades, extending this requirement to nonresearch hospitals expands the scope of the bioethicists' jurisdiction tremendously. Preliminary reports suggest that the primary mechanisms are for hospitals to have access to an "ethics consultant," who would most likely be trained in the profession of bioethics, or to establish an ethics committee, whose members would likewise require some training. The bioethicists' growing strength in their home jurisdiction suggests an increasing ability to compete in adjacent jurisdictions.

Although bioethicists do not yet have complete dominance over public bioethics, as they do in the clinical setting, their jurisdiction in these debates continues to strengthen. Put simply, there are more and more bioethicists taking part in these debates, and fewer members of other professions. Moreover, there are more government advisory commissions spread throughout the executive branch, and other bureaucratic institutions, such as drug companies, have begun to create similar commissions.

However, bioethics as a profession is going through a period of self-reflection as it passes the twentieth anniversary of the *Belmont Report.* Several texts have questioned whether the Belmont ends are the proper ends, and a 1999 conference, held under the auspices of the National Bioethics Advisory Commission, also called those ends into question.[47] Some competitors to principlism have been suggested, such as casuistry and virtue theory.[48] There is also a burgeoning subfield, the sociology of bioethics, which tends to show that people do not make ethical decisions as bioethicists think they do or want them to do.[49] It is too early to tell whether this self-reflection will threaten the jurisdiction of bioethics or change the form of argumentation in the profession in any fundamental way.

In my estimation, it seems unlikely that the bioethics profession will move away from a formally rational type of argumentation, because its core jurisdiction of clinical bioethics now depends upon it.[50] Bioethics professionals and their form of argumentation are now a part of the bureaucratic structure of the scientific research enterprise and patient care. The institutions involved are ultimately responsible to the bureaucratic state: hospitals lacking legitimate ethics mechanisms will be denied accreditation and thus Medicaid funds, while research institutions may be denied government research funding or may have such funding removed. All these decisions are made in the deepest recesses of the bureaucratic state, far from democratic accountability. Given this structure, a social movement of some type would have to convince those who oversee these decision-making entities, such as the Office for the Protection of Research Risks at the NIH and the Joint Commission for the Accreditation of Health Care Organizations, that formal rationality is not required. Given the bias toward formal rationality in the bureaucratic parts of the state, this would be quite a challenge. If principlism is retained by the bioethicists who practice clinical bioethics, it will most likely be retained by those who participate in public bioethical debate, because to have two forms of argumentation would weaken the profession, opening it to attack.[51]

While principlism fits the environment of bioethics professionals quite well, its internal logic is threatened from within the bioethics profession itself. As noted in chapter 6, many participants in the HGE debate have begun to elevate autonomy as the sole end to be forwarded. If autonomy becomes the sole end to be pursued in HGE, then the type of ethical debate examined in this book would end, and the profession of bioethics would have to retool itself or fade into irrelevance. While advocates of formal rationality tend to describe the problem of HGE as one facing individuals, not society, they nonetheless consider the individual's ethics in relation to the ethics of the society. If autonomy becomes the only end, then ethics become individual, and what any individual wants to do about HGE is of no concern of anyone else. Most notably, it will be of no further concern to bioethicists, whose form of argumentation is premised on what can be legitimated as being in the collective interest.

The Future of the Scientists' Jurisdiction

While bioethics seems to be facing some degree of challenge, it is hard to see anything but further opportunities for scientists. Their success has been nothing short of stunning. At every turn they have succeeded in moving the debate away from the public and the public's representatives into government advisory commissions where representatives of the profession are disproportionately represented and where the form of argumentation furthers their interests.

Looking back over the first thirty years of the HGE debate, we see that the alliance with the profession of bioethics has also forwarded scientists' interests quite well. In the area of HGE, there seems not to have been a single moment when a mainstream scientist wanted to conduct an experiment and bioethicists said no. Individual scientists have been slowed down, but only out of concern for safety (nonmaleficence), which is consistent with the interests and beliefs of the scientific community. This is due not to some moral failure on the part of individual bioethicists, but rather to the form of argumentation they use, and that form of argumentation has been encouraged by the environment of government advisory commissions. There is almost no way within their form of argumentation to challenge the use of any means if it can be shown to forward beneficence and nonmaleficence. There is also no way to set alternative ends that could be used to challenge the desires of scientists.

The inability of public bioethical debate to constrain scientists' actions has been noted by others as well. Consider the view of Daniel Callahan, who arguably founded the debates in public bioethics in the late 1960s, and shortly thereafter became a critic of what I call its increasingly formally rational type of argumentation. Looking back, almost thirty years after he began work in the field, he argues that public bioethics "has not taken on, for careful examination, the implicit models of human life and welfare and the human future that lie behind the biomedical research enterprise." Worse, from his perspective, "if it has not been utterly captured by [the scientific] enterprise, it has mainly stood on the sidelines, wagging its finger now and then."[52] Of course, it could be argued that everything scientists have suggested has been in the public's interest, so there was nothing to disagree with. However, as this book has shown, scientists have attempted to keep decision making regarding HGE away from the public, which suggests that the public might have disagreed, given the opportunity.

The Future of the Theologians' Challenge for Jurisdiction

It seems clear from the evidence presented in this book that the real problems for the theologians began when the decision making on HGE and other public bioethical questions shifted away from the public and toward government advisory commissions. Unless there is another shift back toward the public, theologians have little chance of gaining jurisdiction. Their form of argumentation is simply not compatible with the formal rationality required in such contexts, if they are to retain their primary jurisdictions. Even if they engage in secular translation of theological ends, their ends are not calculable and thus do not fare well in debates.

This problem was identified many years ago. In 1978 Gustafson was already calling for theologians who participate in debates such as the HGE debate to pay more attention to what he calls "the home folks who might care more about what they have to say."[53] Herein lies the choice for the public voice of theology, and it is a choice that theologians have faced many times before. Theology can change the society either by trying to speak to (or for) everyone, or by speaking to its primary constituency in the hope that this group will grow as people join the faith. In Christian terms, they could lead a church or a sect.

Trying to "speak for" everyone, as bioethicists do with their universal ends, is hopeless and a violation of the separation of church and state. The theologians did pretty well trying to "speak to" the public, who could form their own opinions and contact their elected officials, while avoiding speaking only to their narrow group (sectarianism). Speaking to everyone does not violate the separation of church and state, but is actually protected by it. This seems to be what the theologians have always had in mind. Ramsey, for example, wrote that he did not believe that it was "possible to do Christian ethics for those who do not share Christian convictions," but that he would still write for the public. "Readers who are not Christian are invited to read as if overhearing an ongoing conversation from which they may learn something." This education, or, more properly, evangelism, was part of what he called a "culture-forming impulse" of theology.[54]

If there is room for theologians in public bioethical debates, it is in focusing the ends of communities that may be interested in their arguments or, at theology's most inclusive moments, as organizers of cross-community deliberation about ends. Theology's future in the primary debate about HGE, focused on what government advisory commissions should recommend, is bleak.

Lessons for Future Public Bioethical Debates

I have taken the normative position that a substantively rational debate about what a society should do is critical for any democratic country. In this view, the citizens would listen to the professionals debating HGE and would communicate among themselves, and groups would develop the ends that they wished to forward, bringing these demands to the public's elected officials. This is akin to what Habermas thinks should happen in the life-world; following the Weberian strands in Habermas's thinking, if we cannot have substantive debate, we become slaves to the means that we ourselves have produced. This does not mean that we should not also have formally rational debates. Michael Walzer, for example, believes that "minimal morality" (what I call thin debate) is "very important . . . but it can't substitute for or replace the defense of thickly conceived values."[55] The Habermasian tradition also acknowledges the need for both life-world and system, assuming they keep to their respective places.

I recognize that there is a very large debate among political theorists about the Habermasian tradition, and that there are many theorists who argue that setting a few commensurable and universal ends is actually more in keeping with the liberal democratic tradition. That is, some political philosophers would argue that having only a formally rational debate is preferable. In this version of political liberalism, "the state may not dictate any particular vision of the good life."[56] That is, it cannot set any ends at all that an individual does not hold, for to do so would violate the individuals' conscience. The advantage of what theorists call an "overlapping consensus" is that, since the overlapping ends are thought to be universally held, they are not dictating any "particular version of the good life." Nobody's conscience is violated from being on the losing end of a political debate over an issue such as abortion or HGE.

One problem with this formally rational debate is that any ends that truly overlap all groups in society will be extremely thin. Bruce Jennings asks whether "we want to say that in order to achieve a morally authoritative kind of consensus we want participants to deliberatively bracket their differences and search for that common ground they share, no matter how narrow or minimalist it might be?"[57] These thin ends will also probably not be usable to ask any of the really important questions: "it is but a thin friendship that remains harmonious only so long as the friends never touch on anything that really matters to them."[58] This concern rings true in the HGE debate. As I have noted, the deeper concerns of theologians and others could not be translated to these minimalist, common ends, and this made their questions unexaminable, leaving important questions about HGE unanswered.

Another trouble with the formally rational type of argumentation that presupposes universal overlapping ends is how these ends are discovered. Returning to the Habermasian tradition, this is articulated in the concern about the scientization of politics and the rise of a technocracy, where elites will determine what the interests of society actually are.

In calling for deliberation over ends in society, political theorists Amy Gutmann and Dennis Thompson identify the primary problem with elites determining what the ends of society are without consulting the citizens directly:

> Some of us (and perhaps all of us sometimes) believe that we already know what constitutes the best resolution of a moral conflict without deliberating with our fellow citizens. Assuming that we know the right resolution before we hear from others who will also be affected by our decisions is not only arrogant but also unjustified in light of the complexity of issues and interests at stake. If we refuse to give deliberation a chance, we forsake not only the possibility of arriving at a genuine moral compromise but we also give up the most defensible ground for maintaining an uncompromising position: that we have tested our views against those of others.[59]

A recent advocacy advertisement in the *New York Times*, criticizing the utopian dreams of scientists, sums up well the problem with elites setting ends. After summarizing the future possibilities in HGE—making "designer babies" and the like—the advertisement asks: "So, where do *you* stand on this? Has anyone ever asked?"[60] This is a critical challenge in a democratic country. In the final analysis, I feel that the case for relying upon formally rational argumentation in public bioethical debates is normatively inferior to also having a substantive debate about ends.

How then do we reinvigorate a substantive debate among the citizens? The conclusion most readers of this book might reach would be to shut down government advisory commissions and to exile professionals from these debates. That is, of course, too simple. There are legitimate reasons why the debate has evolved in this way, along with the less legitimate reasons outlined in previous chapters. To conclude, I will outline some of the pros and cons of the ways that public bioethical debates could be structured, and offer some (perhaps utopian) suggestions for reform.

The Public and Its Elected Representatives or Government Advisory Commissions

A large part of the explanation for the decline of substantive rationality in the HGE debate is the shifting of the location of ethical decision making from the public and its representatives to government advisory commissions. This led to the dominance of formal rationality—a negative occurrence, according to my normative position. Before deciding to have some form of direct democracy—a national referendum or some similar mechanism for making public bioethical decisions—we have to acknowledge the disadvantages of letting the public decide on its own, acting through its elected representatives.

The main disadvantage of public decision making is also as close to an official justification for government advisory commissions as exists: the public lacks the knowledge to make effective decisions. For example, a very small percentage of the public really understands how genetic inheritance works. This is evidenced by the cloning debate, where many people seemed to think that if you cloned Adolf Hitler, the clone would be able, by taking advantage of his genetic composition, to seize control of a country in the midst of a depression and try to take over the world. This tension between democracy and reluctance to fully entrust the average citizen with the power of democracy is of course embedded in the American political process: the United States is, after all, a representative democracy and not a direct democracy. The Senate, for example, was intended to be at some distance from the passions and excesses of the general public.

It has also been noted that reliance on the public to organize its opinion into associations and to appeal to its representatives—what political theorists call the pluralist model—is not quite democratic in practice, because people have differential access to the public sphere and their elected officials. The Protestant theologians of the 1960s, for example, were primarily mainline Protestant, not evangelicals, and the part of the public that would be most amenable to their message was a much more powerful group in this era than that of evangelicals. This disproportionate power is exemplified by the disproportionate number of members of Congress who have been mainline Protestants.[61]

I have chronicled in this book what I consider to be the negative effect of government advisory commissions. But before abolishing them, we must consider their advantages. First, and most obviously, if the public or its representatives are unwilling to engage in debate and to come to conclusions, at least someone, somewhere, is talking and making decisions. Second, as I mentioned above, since these commissions do not directly involve the public, which lacks expert knowledge, better decisions are made if these experts have the public's interests in mind. Third, from the Weberian tradition we can recover a final advantage: While Weber bemoaned formal rationality, exemplified in capitalism, for its corrosive effects on society, he simultaneously acknowledged that it allowed for an increase in efficiency. Similarly, there is a certain efficiency in government advisory commissions that create bureaucratic mechanisms for the resolution of ethical problems.

Using Professionals to Speak to and for the Public

The primary advantage of using professionals to structure debates in public bioethics is precisely that they are professionals: they have specialized knowledge that the public lacks. For example, compared to a very low level of understanding in the public, all the writers in this debate understand the distinction between germline and somatic HGE. However, this study has also shown that professionals engage in a form of systematic paternalism toward the public they purportedly represent. Time after time it has been demonstrated that it does not matter what the public wants at a given time; the various professionals feel they know better what is *actually* in the public's interest. Recall how scientists seemed to know, for example, that it was in the public's interest to have its fears calmed in the *Splicing Life* report. On a more subtle level, bioethicists claim to know what the public's ends are with regard to HGE, even though the public has never been asked. Even the professionals who speak more to the public than for it, such as the theologians, are not innocent here either. Ramsey for one seemed to assume that at least a majority of the public shared some sort of generic mainline Protestantism—an assumption that was probably false.

Some of this belief among professionals that they know what the public would want if the public had the knowledge and time to examine the issue stems from their expert knowledge. However, part of this unwillingness to follow the public comes from a concern for professional interests. Scientists feared public decision making on HGE because they feared for their jurisdiction and funding in other areas. Similarly, bioethicists are trained in their form of argumentation, which was originally created for questions of human experimentation, so to maintain the coherence of their overall jurisdictional claims, they end up making the same arguments in areas such as HGE. Professions have interests, and these interests can distort their conclusions about what the public would want.

The public needs professionals in these debates for their expertise, but it would be better to avoid the biases that come with using them. Similarly, there are advantages and disadvantages of using government advisory commissions. Here I give some tentative suggestions for how to have the best of both worlds.

Recommendations

We need thick public debates about ends, as well as thin debates about how to advance these ends once they have been agreed upon by the public.

We also need to use the advantages that accrue from professionals and government advisory commissions, while avoiding the drawbacks.

To continue with the language of this book, we could have "ends commissions" and "means to ends commissions" (though surely we would use less pedantic titles in practice). For example, an ends commission could be presented with a field such as human genetic science and asked to identify the ends that should be pursued. The group could even start with the four ends of the bioethicists, and then deliberate as to whether these are the four ends that should be forwarded, or whether some different ones should be substituted. The commission might decide that some ends are linked to the possible means in a substantively rational manner, so that only if the means was consistent with the end would it be approved. It might see other means as linked in a formally rational manner, so that the means would be approved if it maximized the end or a series of ends. It is quite possible that the commission might finally select ends that would suggest that the means of germline HGE should never be used. Likewise, the public might decide that autonomy was the preeminent end to be forwarded, and in a substantively rational, not a formally rational, manner.

How would such an ends commission work? Although this question lies beyond my expertise, some examples are instructive. In 1990 the state of Oregon created a commission to determine which medical procedures would be covered under Medicaid. The commission conducted "an elaborate process of consultation," including a survey of state residents to see how they "would rate the quality of life resulting from the various conditions and treatments." It also held forty-seven meetings throughout the state, involving more than a thousand people, "to identify community values in health care."[62] This process had many flaws, but it seems to be a starting point. It seems to me at least to be better than starting with a set of assumed values that a commission of professionals could apply. The rapidly growing number of scholars studying "deliberative democracy" suggests that workable mechanisms may be created in the not so distant future.[63]

The evolution of the HGE debate does suggest some cautions. For instance, we should be very cautious about having government advisory commissions—or any organization distant from democratic legitimation—orchestrate a deliberation over ends. The people conducting the deliberation would have to have legitimacy as representatives of the public. Even worse, as demonstrated by the case of the President's Commission, is the situation when these commissions see themselves as both trying to change the ends that society holds and creating policy that advances those ends. The RAC has also recently embraced this dual task in addition to its policy-recommending role: it considers one of its "most important roles" to be "public education" as well as being a "forum for airing a wide range of public concerns" about HGE.[64] A recent panel of the American Association for the Advancement of Science suggests a similar mechanism.[65] The problem with this is exemplified by the President's Commission. The form of argumentation required to suggest policy and procedure is incompatible with the search for the true ends held by society, and can distort this search. Establishing a commission to deal both with ends and with means toward ends should be discouraged.

Professionals claiming expertise on the issue at hand should be on the means to ends commission but not the ends commission. I do not see the justification for assuming that a diversity of professions represents the diversity of moral opinion in the United States. Surely, if the purpose is to represent the different religious and philosophical traditions that should be setting the ends for our scientific means, government advisory commissions should not require their potential members to possess a Ph.D. If anything, that seems to guarantee that only religious and philosophical traditions consistent with the Enlightenment philosophies that dominate American universities will be involved in the discussion. Is including a Jew, a Catholic, an evangelical Protestant, a liberal Protestant, a feminist, an atheist, a communitarian, and a libertarian on a government panel any more arbitrary than assuming that a biologist, a physician, a sociologist, and a bioethicist can somehow represent the values of the public? Perhaps such a panel would be hopelessly deadlocked. However, supporters of the current system need to articulate how a consensus derived from the exclusion of some groups in our pluralistic society is not in and of itself a violation of the liberal democratic values that these debates purportedly foster.

Others have suggested something similar. In reaction to how scientists use these commissions to forward their interests, it has been suggested that scientists not be allowed to serve on them. George Annas, writing in the *New England Journal of Medicine,* has suggested a regulatory agency to oversee research in HGE, human embryos, and other technologies. This agency, he argues, must be composed almost exclusively of nonresearchers and nonphysicians "so it can reflect public values, not parochial concerns."[66]

A good compromise between the risk of scientists, or other professionals, advancing their professional interests over public interests, and the risk of having uninformed people make decisions on technical matters is to have the ends commissions composed of nonprofessionals, and the means to ends commissions made up of professionals, such as scientists. A lay-dominated ends commission could decide, for instance, that nonmaleficence is the only end to forward in germline HGE, and the professionally dominated means to ends commission could determine whether some specific act of germline HGE would indeed be safe. I should add that the form of argumentation of the bioethics profession makes it perfectly suited for participating in these means to ends commissions. Examining the writings of bioethicists reveals that they are quite expert at determining whether a very particular act is consistent with or maximizes some set ends. To take a classic question: should a physician inform a patient that he or she has Alzheimer's disease (thus advancing the end of autonomy) if such information will also harm that patient by negatively affecting his or her coping mechanisms (violating the end of nonmaleficence)?[67] The profession of bioethics seems to be the best at resolving these questions. If an ends commission decided that its ends to forward in genetic research were beneficence, nonmaleficence, and maintenance of the current specificity of genetic change possible in the reproductive act, I have no doubt that bioethicists could determine which, if any, forms of HGE advanced these ends.

Of course, it is important to point out that we already have the equivalent of ends commissions, called legislatures. Elected officials set ends within the constraints of not contradicting the ends institutionalized into the Constitution. If there has been a thin debate, it has been because elected officials have been more than willing to pass controversial issues on to government advisory commissions. And, as always, it is up to the public to hold legislators accountable for their lack of action.

Perhaps the suggested structure of these ends and means to ends commissions is naive, or politically impossible. My recommendations are not so much a blueprint as an ideal-type to which we can compare alternatives. Of course, this is all for naught if the more teleological interpretations of Weber are correct: that nobody has ever seriously resisted the iron cage. Frankfurt school theorists like Habermas see the new social movements of Europe as possible defenses of the substantively rational life-worlds from the iron cage.[68] Setting public ends through deliberative democracy practices could have the same effect.

Coda

Do we want to engineer the human species? Do we want to make all human beings genetically resistant to disease? Do we want to genetically engineer human beings to be more intelligent? These are all questions that were asked in the 1950s and 1960s, and which, after a hiatus of a few decades, are being asked once again. I do not have an answer to these questions. Nor does society have an answer at this time. In the not too distant future, you may read in the newspaper that the RAC is debating whether to allow a germline genetic enhancement experiment that would improve the resulting baby and that baby's descendants. That newspaper article will, in the name of balance, interview experts from opposing sides and the television show *Nightline* will have them on for a live debate. Thus, the parameters of debate will be defined for us. If the rise of formal rationality in this debate continues, and the bioethics profession continues to increase its presence, the two experts may be dueling bioethicists, one saying that to allow the experiment will violate the end of nonmaleficence, because it might not be safe, the other saying that to stop the experiment would violate the end of autonomy of parents to engineer their children as they see fit. Since they would be speaking not to a government advisory commission, but to the public, they could have a debate about ends, but they will not. Their formally rational type of argumentation has been institutionalized by their profession.

The ends forwarded by the dueling bioethicists will be implicitly presented as legitimate due to deep processes—perhaps the need for democracy, or even the deep structure of modernization. It will not be apparent that any human agents institutionalized these ends and not others. I hope that from this book it will be apparent that there were once other ends that were part of this debate, ends that are now lost from view, and that the ends we are presented with were not inevitable or natural, but ultimately under our control.

Appendix: Methods and Tables

Obtaining the Population of Texts

To obtain a population representative of the debate about HGE as a whole, I defined the population as all books and academic journal articles published on this topic from 1959 to 1995. A facsimile of the population was created through the use of bibliographic sources. Between 1973 and 1995 a consistent and data-rich bibliographic source, the Bioethicsline database of the National Library of Medicine, was used to represent the population. I downloaded the almost 52,000 items in this database and extracted the 989 items that had as their "primary topics" genetic intervention, gene pool, gene therapy, or germ cells.

For the years before the creation of the Bioethicsline database in 1973, I used numerous and overlapping bibliographic sources about human genetic engineering to create the population. Throughout the 1970s the Institute of Society, Ethics and the Life Sciences (later to become the Hastings Center) published bibliographies of issues of "society, ethics and the life sciences," classified by topic.[1] From these I gathered all the references collected under topics that represented HGE.[2]

Another source was a bibliography, "Social and Psychological Aspects of Applied Human Genetics," which contained exactly 100 items that were scanned into electronic form.[3] Finally, in 1972 the Library of Congress wrote a report titled *Genetic Engineering: Evolution of a Technological Issue,* which contained an appendix of "selected references of interest," resulting in another 100 items.[4] Removal of redundancies from all the sources resulted in 476 items in the pre-1973 population. The 989 Bioethicsline texts were combined with the non-Bioethicsline population, resulting in a universe of 1,465 items published between 1959 and 1995.[5]

There is a legitimate concern that the different methods used for including texts in these various bibliographic sources might bias my results. The most obvious concern is that the method of searching out texts for inclusion is different between the pre-Bioethicsline and the Bioethicsline data. While there is no objective method to determine whether there is bias, I believe that any such bias is minimal because the early bibliographies appear to have the same degree of comprehensiveness as Bioethicsline. Bioethicsline "seeks to be comprehensive for all English-Language materials . . . which discuss ethical aspects of the topics and subtopics" it searches for, including, but not limited to, books, chapters of books, and journal articles.[6] The earlier bibliographic sources seem equally comprehensive, given that there were so few texts produced before the mid-1970s and that the bibliographies include texts from extremely obscure sources.[7]

Perhaps more important is potential bias in the Bioethicsline data because it is the basis for three of the four time periods analyzed. This bibliography was constructed in the following manner. First, the librarians directly monitored ninety-nine journals and newspapers for items that might be included. Thirty-eight very diverse reference tools and databases were also searched for possible items: the Humanities Index, *Library Journal,* Philosopher's Index, Sociological Abstracts, and many more.[8] Books purchased by the Georgetown University library are also scanned for possible inclusion. Not surprisingly, as journals and indexes come and go over time, their use by the Bioethicsline staff changes. The staff monitors twice as many journals now as in the very first year, but the distribution by professional interest (science, theology, and so forth) remains similar except for an increase in the number of journals now exclusively dedicated to bioethics.[9] This reflects the well-documented emergence of this profession, not a change in the sources used by the librarians. The final step in the selection process involves the senior bibliographer, who reviews all possible items for inclusion, using the general rule that the item to be included must pertain to "health care" or "biomedical research," and at least 50 percent of its pages must discuss issues with "ethics or related broadly defined public policy implications." This standard of inclusion has not changed over the years, and therefore provides no source of bias.

The staff of Bioethicsline made changes in the procedures over the years to compensate for the explosion of literature on this and similar topics. Bioethicsline has slowly but steadily increased the number of items included in the database in each year, from 1,185 in 1974 to 3,518 in 1994. Although the effort involved has doubled, in an attempt to keep up with this increased production the senior bibliographer of Bioethicsline pointed out in an interview that the "literature has exploded beyond the extent to which our staff has exploded." The staff have dealt with this in part by reducing the number of news stories included, by including only one for each event, not one from each of the major papers they track, and by reducing the number of "small un-footnoted items," such as newsletters from organizations and letters to the editors of journals. These changes will not confound analyses in this project because items without citations, such as newspapers and letters, were not used.

Citation Collection and Analysis

To identify debating communities in different eras, the citations were collected from a random sample of the defined population. Before describing this process, I will explain how highly cited texts are more influential in a debate than noncited texts, and how authors who tend to cite the same authors are grouped in the same debating community.

What Is a Citation?

Using citations as data obviously relies on a theory of why one author cites another—a subjective account that the outside observer has no access to. Cronin, following many information scientists, notes the "essential subjectivity" of citation practices and outlines reasons why one author cites another: as an acknowledgment of intellectual influence, as a form of reward or income, as a way to settle priority claims, or as a tool of persuasion.[10] In addition, citations can be "affirmative or negational," "organic or perfunctory," "evolutionary or juxtapositional," or "conceptual or operational." More mundane reasons, such as forgetfulness, personal vendettas, audience, size of the paper, and an author's knowledge of the field also are factors. Many of these more mundane concerns about citing behavior involve idiosyncrasies of individuals that should average out in large samples, simply providing more "noise" in the data.

In response to this uncertainty regarding what a citation actually measures, there has evolved a fairly large literature critical of citation studies. Currently the debate in the literature is between those who see the citation as a measure of "intellectual influence" and those who see it as an attempt to "persuade the reader of the importance and validity of their work."[11] This distinction is important for testing many of the hypotheses in the Merton-influenced sociology of science, such as the "Ortega hypothesis," the idea that science "advances" in large part due to the contributions of average scientists.[12] The critique is that if citations are methods of persuasion, famous scientists and their articles are "overcited" compared to their actual intellectual influence—leading to a "Matthew effect" for famous scientists in one formulation or a "Nielsen rating" for scientists in another.[13] In an empirical study of citing practices in the humanities, Brooks found that intention to persuade was by far the most common motivation, leading him to conclude that "authors can be pictured as intellectual partisans of their own opinions, scouring the literature for their justification."[14]

My purposes in this book permit me to avoid the debate between the "persuasion" and "influence" positions because I need only a weak theory of citing behavior. My assumption is that an item is cited either because it was important to the intellectual development of the text or because it was important enough to others in the community that the citation would be persuasive. As texts to learn from, or to be used as persuasion, cited texts are "important"; entrants to the debate must familiarize themselves with these texts before they begin. Moreover, "negational citations," in which an author argues against another writer, also show that the cited author is important to the debate. Authors typically only argue against others who have developed a following and are seen as worth arguing against.

Identifying canonical texts in this manner has allowed researchers to identify communities of authors, often called "specialties," "research areas," or "invisible colleges." Studies, at least in the physical and social sciences, have found that co-citation clusters coincide with actual network links between researchers.[15] The assumption is that co-citation clusters (texts or authors that tend to be cited by the same texts) are groups of researchers "engaged in the joint exploration of related problems."[16]

Obtaining Citation Data

Methodologically, this technique requires some choices that flow from the phenomena in question. It is important to note that the sociologists of science, information science specialists, and information companies that developed these data sets and techniques have focused on the physical sciences where journals are the primary mode of knowledge dissemination.[17] Sociologists, for understandable reasons, have tended to use the citations previously gathered by the Institute for Scientific Information (ISI). Because the ISI collects citations from journal articles, however, reliance on these data has meant that only such citations could be examined. This does not significantly bias studies of physical science, where most knowledge is transmitted via journals, but similar examinations of the social sciences and the humanities using only journal articles are biased toward particular forms of discourse because knowledge in those fields is often disseminated through books rather than journal articles.

Preexisting data therefore could not be used for this project. Working from a printed list of all 1,465 items in the population and a random number table, I collected the citations from a sample of the population. At this point some difficult methodological decisions had to be made. First, an item needed to have cited something to be useful to a citation study, so an item without citations was not selected, and another item was drawn in its place. These were primarily news stories from newspapers, popular magazine articles, and editorials, but also included some journal articles, books, and chapters from books that did not have citations. In practice, if an item did not contain four or more citations, it was not included in the sample, because the primary technology of a citation study is measures of association between citations.[18] I examined 765 randomly selected items from the universe.[19] Of these, 345 fit the parameters for inclusion in the study as described, and I collected their citations, as well as any information about the author.[20]

All the citations from these texts were scanned into a computer and compared, using a program I wrote that gave each unique citation an identifying number.[21] Subsequent coding of citations from the texts checked the unique citation list to see if that citation had already been coded by a previous text and, if it had, to assign the same unique citation number to that text.[22] In all, 16,020 citations were coded from the 345 citing texts (an average of 46 per text). In these 16,020 citations there were a total of 11,349 separate cited items, of which 9,313 were cited only once in the data (not cited by any other citing item). Stated differently, 2,036 items were cited more than once. This proportion of items cited only once is comparable with studies of the social science literature.

From the citation data, I aggregated the cited items by author to derive a separate data set of unique author numbers with citing texts.[23] The result was that in the 16,020 citations from the citing texts, there were 6,614 authors, of whom 2,158 were cited more than once. The authors who were cited more than once were placed into a matrix with 2,158 unique author variables (columns) and 345 citing texts (rows) with a 1 in a cell if that citing case (row) cited that author (column) or a zero if it did not. This data matrix was used for all the subsequent citation analyses.[24]

Identifying Debating Communities

To reveal the communities in these data, clusters of co-cited authors are identified using cluster analysis, a technique long used by sociologists of science. This type of analysis requires that the data be broken into time periods, so I divided the analysis into four periods: 1959–74, 1975–84, 1985–91, and 1992–95. These divisions are primarily made to place similar numbers of citing cases in each time period for statistical purposes. Note that the amount of literature produced in the universe is heavily skewed toward recent years, as the debate on HGE have grown markedly. The boundaries between time periods were adjusted a bit to fit with events external to the system of professions that should have influenced the form of debate in use. The debate surrounding the 1975 Asilomar conference was a pivotal moment in the broader debates about HGE, and is therefore selected as a cutoff point. Nineteen eighty-four was chosen as the second cutoff because it is the year after the release of the report *Splicing Life* by the President's Commission, which preliminary analysis had suggested transformed later debates. There is no clear event in the 1990s that should be thought of as a transitional moment, so the post-1985 data are cut at 1992, which permits a nearly equal split in the data.

For the citation clustering algorithm to be interpretable, the number of influential authors to place in communities must be reduced through the use of a citation threshold. The trade-off here, as in most social science research, is between precision and logistic reasonableness. I set the cutoff at a different level for each time period in order to have roughly equal numbers of influential authors under analysis in each period. The number of influential authors in each period did end up being different because I did not want to differentiate between authors with equivalent citation counts.[25]

Cluster Analysis

Cluster analysis is a method of determining which cases are most similar to each other by comparing their characteristics (variables). For example, in the biological sciences, which pioneered these techniques, animals (cases) could be evaluated according to measures (variables) such as weight, number of appendages, and the presence of wings or feathers. A clustering of the cases would show, based on these variables, that monkeys are more similar to humans than they are to squirrels or woodchucks. In my research, the question was which top-cited authors were most similar to each other based on the texts that cited them.

This seemingly simple process is actually quite complicated. The first complication arises in how the analyst measures similarity between cases, which should depend on the nature of the data. In my case, all the characteristics of the cases (variables) are binary, with a 1 indicating that an author was cited by the citing text and a zero indicating that he or she was not cited. This limits the possible similarity measures to only a few possible association coefficients, the choice of which is driven by the importance my theory attributes to the absence of a match between two cases (that is, if they are both *not* cited by the same text).[26] If the nonmatches are included in the analysis, cases will look similar because they both lack the same features.

In my theory, the citation to an author is indicative that (1) the texts of the author were known to the writer, and (2) the texts of the author were useful to the writer as a descriptor of a position in the debate at that time. Therefore, the theory is that two writers cite the text of the same author if they both know of these texts and if they see these texts as the most communicative description of a position. However, if two writers both *do not* cite an author's texts, this could mean either (1) that the texts are not known to the writer, or (2) that the texts are not useful to the writer as a descriptor of a position in the debate at that time. It would be inappropriate to base the estimation of similarity on measures of nonciting since it is not possible to know whether the writers did not cite the same text because they did not know about it, or because they did not want to cite it.[27]

Several measures are used to calculate similarity for binary variables while excluding analysis of nonmatches, including the "Jaccard," "Dice," and the "Binary Lance and Williams Nonmetric dissimilarity measure."[28] In practice, these differ by the weight they place on matches versus mismatches—resulting in the same configurations of clusters but with different amounts of space between them, allowing for differential ease in interpretation. I created the similarity matrix using the "Lance and Williams" because it can also be used for multidimensional scaling.[29]

The next step in the process is to "cluster" the similarity values so that cases that are the most similar (most often cited by the same text) are in the same cluster. To continue the analogy from biology, assuming our data and our common understanding of the animal world are correct, humans, monkeys and other primates should form a distinct cluster, while whales, dolphins, and porpoises should form another.

Clustering opens another plethora of choices, once again dependent on the underlying theoretical model and character of the data. There are two broad families of methods for combining items—hierarchical and iterative. I used hierarchical methods because iterative methods require prior knowledge of the number of clusters that will be found in the data, which was clearly not the case here. The hierarchical methods work in stages. At the first stage, the two most similar authors are linked to form the first cluster. Authors are combined—or added to previously created clusters—in order from the most similar to the least similar. Eventually the authors will be joined into one large cluster after *N* steps, where *N* equals the number of authors being clustered. There are many choices for how these authors are clustered and where to stop the clustering process between the extremes of "one cluster for each case" and "one cluster for all cases." I used the complete linkage method, which is the standard in citation studies.[30]

There is no objective method of determining how many clusters are present in the data. The best method available is to stop agglomerating clusters when the size of the similarity between joined clusters begins to increase between steps. Unfortunately, a graph of the agglomeration coefficients (not shown) reveals that the coefficients never flatten out. Therefore, I use a modified "jump test" where I cut off the clustering process when the similarity between clusters takes a large jump at a point where there are fewer than ten clusters—the maximum that could be substantively analyzed.[31] The dendrograms for these analyses in each period can be found in the beginnings of chapters 2, 3, 5, and 6.

Information about the Authors in a Community

A critical distinction made in the analysis is between influential authors and common authors in a community. *Influential authors* are those who in each era were sufficiently highly cited by the *common authors* to surpass the threshold to be included in the cluster analysis. Since there are, by design, far fewer influential authors, more in-depth data can be gathered on each one of them. Less rich data are available for the common authors as well.

Determining Profession

The professions of both the influential and common authors were determined using the same method. As a first data source for the profession of the common members of the communities, while collecting the citations for each of the 345 citing texts, I also collected any information in the book or article that pertained to the author. For journal articles this typically consisted only of the author's university and departmental affiliation. For books there was often a page "about the author." Approximately 60 percent of the citing texts had information on departmental and institutional affiliation. Bibliographic resources were used to fill in the remaining cases.[32] For the same logistical reasons as above, and consistent with most citation studies, only the first author was coded.

Of course, the institutional and departmental affiliation of an author are not perfect measures of profession. Another possibility consistent with the professions literature would be to code the training of the author—placing analytic philosophers, M.D.s, population biologists, and theologians in distinct categories. This approach would be more difficult logistically, given that information about an author's training or degree is typically not available from the citing items themselves. Further, it is particularly misleading for investigating the profession of bioethics that emerges in the 1970s. Up until the last few years the profession of bioethics has lacked a uniform credential by which to establish membership in the field. In the early days of the profession, bioethicists were trained as medical doctors, theologians, and philosophers. The important distinction is between those who identified with the bioethics profession and those who did not. Therefore, I coded authors by the profession that they most strongly identified with at the time of writing—their primary academic appointment—regardless of how they had been trained. The professions of influential authors were analogously determined.

The authors were coded into mutually exclusive professional categories.[33] First, "bioethicists" are those authors whose primary identification is with a "bioethics" or "medical ethics" department or center, whether within a medical school or outside it. The "theology" profession comprises persons employed by a seminary, denomination, or religious congregation, or those who teach in a religion department of a college or university.[34] Note that the many persons in "bioethics" departments who have theological degrees are not included because they have, by my definition, left the profession. Thus, for example, John C. Fletcher is defined as a theology professional in the early years of the sample and is later defined as a bioethics professional as he goes from working in a seminary to working in a bioethics department.

"Medical doctors" (including psychiatrists) are those whose primary affiliation is with a hospital or medical school, but not with a medical ethics or bioethics program within a hospital or medical school. The broad "scientist" profession consists of those who identify with any physical or biological science. This includes microbiology, biology, physics, and population genetics. The "philosopher" profession, as distinct from the "bioethicist" profession, consists of those authors whose primary appointment is in a philosophy department. Similarly, "lawyers" are those who teach in law schools or who have jobs in law firms, but not primary appointments in medical ethics programs. Less prevalent professions coded in the data are the following: journalism; medical humanities/medical education; sociology and science studies; political science and public policy; nonphilosophical humanities (e.g., English literature); public health/social medicine; nursing and psychology; and public interest groups/government agencies.

Content of Representative Texts from the Influential Members

The content of the arguments of both the influential and common members of the communities is critical. To determine the content of each of these influential authors' work, I read their most cited text during each time period, and coded it descriptively and interpretively.[35] The second most cited text for each author was also coded if it had more than 50 percent of the citations that the top-cited text had. In cases of ties, both texts were coded. Lists of these texts for each time period can be found in tables A-2 through A-5.

Content of the Texts of the Common Members

In addition to the profession of each of the common members of a community, a form of content analysis data for each text written after 1973 is available. Each text in the population drawn from the Bioethicsline data has "keywords" assigned to it; a system used by library professionals to describe the content of texts. Keyword data have been used fairly extensively by sociologists of science, who call the technique "qualitative scientometrics" or "co-word analysis."[36] Keyword data offer easily obtained information on the content of large numbers of texts, with the usual trade-off in precision that use of such data entails.

Over the years the list of keywords in the Bioethicsline database has grown to a total of 651 unique terms that can be applied by the bibliographers.[37] In addition, each of the keywords indicates whether the text is "primarily" or "secondarily" about the content. I translated this textual database to numbers, assigning a value of zero if the keyword was not used for an item, 1 if the keyword was of "secondary" importance to the item, and 2 if the keyword was of "primary" importance to the item. Since these data were constructed for a different purpose, I cannot measure the subtle concepts, such as the form of rationality assumed in argumentation, with the degree of precision I would like. Given these limitations, these data did nonetheless offer important insights into the communities.

The coding scheme used by the staff at Bioethicsline is hierarchical. That is, there are keywords that are considered "narrower" topics from other keywords, which are "wider." For example, the keywords "cystic fibrosis" and "hemophilia" are considered narrower keywords of the keyword "genetic disorders." According to the senior bibliographer, the staff has a "narrowest term rule," such that the narrowest term applicable is used for the item and the wider term is reserved for general discussions (for example, discussions of genetic disorders rather than of hemophilia in particular). What remains unclear is the extent to which the wider term is also used with the narrower term. The senior bibliographer implied that the staff has not been consistent in this matter, and empirically many of the items with narrower terms are also labeled with the wider term. This could be, however, because the item discusses the narrower in the context of the wider, such as a discussion of genetic disorders focusing on hemophilia. Therefore, when I use a wider keyword, the narrower keywords are considered to be representative of the wider keyword. So, for example, when compiling keywords for analysis, "genetic disorders" is coded as "2" for all items that have a "2" for the "genetic disorders" keyword *or* the "hemophilia" keyword (or the keywords for eight other genetic disorders). When I am interested in a narrower term, I do not include the wider terms.

I searched for keywords in the keyword thesaurus produced by the Bioethicsline staff that represented three different features of texts.[38] First, I searched for keywords that indicated that the text described a means in addition to HGE (itself recognized primarily through the keywords "gene therapy" and "genetic intervention"). For example, many texts that discussed HGE also discussed the means of "genetic screening" or "abortion."[39] Second, I searched for keywords indicating that the text discussed particular ends. Here the selection was limited to determining which texts discussed the ends first articulated in the *Belmont Report* (see chapter 3). The use of the keyword "risk/benefit analysis" reveals that the author is discussing how to balance beneficence and nonmaleficence. The use of keywords "confidentiality" and "informed consent" indicates that the author is discussing mechanisms to promote the autonomy of experimental subjects. The keyword "autonomy/freedom" is a broader term that indicates autonomy above and beyond the institutionalized mechanisms of autonomy and informed consent.[40]

Finally, I searched for keywords indicating that the text discussed HGE in relation to government oversight. Three keywords were identified that met the specified usage thresholds: "advisory committees," "government regulations," and "public policy." The keyword "advisory committees" indicates the text discusses one of these committees, such as the President's Commission or the Recombinant DNA Advisory Committee. The keyword "government regulations" refers to texts that discuss micro-level regulations promulgated by bureaucratic authority. For example, the texts during the Asilomar controversy that discuss which laboratory procedures and types of ventilation systems would be required by the government are typically coded as being about regulations. "Public policy" is a term that covers discussions of the type "we as a society should do X because X is more ethical than Y." That is, it is rather general and usually does not refer to specific regulations or proposed legislation.

In chapter 1 I distinguished between the bureaucratic authority of the state, which tends to select formally rational arguments, and authority that is more representative of the public, which will more often select for substantive rationality. It is important to note that the keywords "public policy" and "government regulations" are not indicative of government oversight as pursued by democratically accountable representatives of the public. "Regulation" definitely refers to rules promulgated by NIH and other agencies, and "public policy" does not imply the involvement of the public. The "legislation" keyword reflects discussion of actual laws raised by representatives of the public, but could not be included in the HGE data because it was so rarely used (which is itself indicative of the lack of public involvement on the HGE issue after the mid-1970s). Only 2.2 percent of texts that were primarily discussing the means of gene therapy discussed legislation, whereas 23.2 percent of the texts that were primarily about abortion discussed legislation—reflecting the difference between a debate that is largely taking place away from the public and one that is primarily taking place in public.

That these keywords were coded to the texts for purposes other than the research in this book warrants further explanation of how they were created. After an item has been selected for inclusion in the Bioethicsline database by the senior bibliographer, it is given to one of four professional bibliographers, who assigns keywords to the item after "reading" it. The quotation marks around "reading" here point to the issue of the precision in measurement of the keywords, a source of possible error in these data. When assigning keywords, the bibliographers actually do read the entire item if it is anything but a book, and then assign keywords. For books, according to the senior bibliographer, "*read* is too strong a word." They read the introduction and conclusion in the normal sense of the term and then scan the rest. The bibliographers work in pairs, with one checking the work of the other in an informal technique of intercoder reliability. Although this is, according to the senior bibliographer, much more in-depth analysis than that found in most other bibliographic databases, it does suggest that a slight loss of precision is inevitable when using this type of data. However, any measurement error will not be correlated with any of the other variables in the analysis because this same method was used for all texts and has been used since the inception of the database. One possible source of error is that the number of keywords used per item has increased slowly over time, but this has been controlled for by including in statistical models a measure of the number of keywords coded to each text.[41]

Generalizing about Communities

There is no formula to determine the weight given to the data about the influential authors versus the common authors. Rather, all the data are weighed interpretatively to come up with generalizable statements that can be made about each community. However, I only assert that a particular community is dominated by a profession if the common authors are disproportionately from that profession. This raises question of analysis of the quantitative data from each community.

To claim that a community disproportionately comprises one profession, I conduct regression analysis where the dependent variable is the number of influential authors in a community cited by a common author. The independent variable of interest is the profession of the common author. The analysis uses a negative binomial regression model because the dependent variable is a count, although there are models where a Poisson regression was more appropriate and hence was used.[42] Models also controlled for the total number of texts that were cited because different professions have different citation practices.

Models to determine the content of the texts of the common authors are extremely similar. The dependent variable is the same. The independent variable of interest is whether the text that is doing the citing has the keyword coded onto it, and the negative binomial regression models control for the total number of texts that were cited and the total number of keywords coded onto the text, because both of these measures are structured by profession. Models are Poisson, when appropriate. Keywords are coded zero, 1, or 2 (zero if the topic is not found in the text, 1 if it is present but of lesser importance to the text, 2 if the topic is central to the text).

The President's Commission Case Study

The investigation of this case study occurred in two stages. First, the transcripts from the five commission meetings that dealt with the HGE report, five rough drafts of the report, correspondence, internal memoranda, and notes from the meetings were all gathered from the National Archives in College Park, Maryland, and from the National Bioethics Reference Library in Washington, D.C. These were scanned into the computer and descriptively coded using a qualitative data analysis program.

Open-ended interviews with key commissioners and staff were conducted both in person and, less ideally, over the telephone. Interviewees were given the option of being anonymous or nonanonymous with selected moments of being anonymous. Each of the nine interviews lasted between twenty minutes and two and a half hours. Interviewing began in December 1996 and was completed in July 1997. The interviews were transcribed, and descriptively coded with the codes that had remained important after the analysis of the documentary data.

The interviewees, their role in the commission at that time, and the time and location of the interviews are listed below.

Staff Interviews

Allen Buchanan, staff ethicist, 27 December 1996, Mt. Horeb, Wisconsin
Alexander Capron, executive director, 14 July 1997, Bethesda, Maryland
Renie Schapiro, professional staff, 23 December 1996, Middleton, Wisconsin
Jeff Stryker, research assistant, 19 May 1997, by phone
Alan Weisbard, assistant director for law, 26 December 1996, Madison, Wisconsin
Dan Wikler, staff ethicist, 31 December 1996, Madison, Wisconsin

Commissioners

Albert Jonsen, professor of ethics in medicine, University of California, San Francisco, 8 May 1997, by phone

Arno Motulsky, professor of medicine and genetics, University of Washington School of Medicine, 24 July 1997, by phone

Consultant

Norman Fost, director, Medical Ethics Program, University of Wisconsin–Madison, 23 December 1996, Middleton, Wisconsin

Table A-1 Distribution of Authors by Profession

Profession	1959–74	1975–84	1985–91	1992–95
		Common Authors		
Medical doctors	16.9 (20)	12.8 (24)	10.0 (25)	11.6 (19)
Scientists	34.7 (41)	21.3 (40)	17.3 (43)	15.2 (25)
Lawyers	5.1 (6)	8.5 (16)	9.6 (24)	7.3 (12)
Philosophers	2.5 (3)	5.3 (10)	7.2 (18)	14.6 (24)
Bioethicists	1.7 (2)	6.4 (12)	12.9 (32)	14.6 (24)
Theologians	17.8 (21)	18.6 (35)	10.8 (27)	8.5 (14)
Journalists	6.8 (8)	4.8 (9)	2.0 (5)	3.0 (5)
Medical humanities	0	.5 (1)	2.8 (7)	0
Sociology	3.4 (4)	4.3 (8)	2.0 (5)	2.4 (4)
Political science	.8 (1)	2.1 (4)	1.6 (4)	4.3 (7)
Humanities	2.1 (4)	1.2 (3)	2.4 (4)	0
Public health	.8 (1)	.5 (1)	.4 (1)	1.2 (2)
Public interest/ government	5.1 (6)	9.6 (18)	17.7 (44)	11.0 (18)
Nursing	0	1.1 (2)	0	0
Psychology	1.7 (2)	.5 (1)	0	0
Missing data	2.5 (3)	1.6 (3)	4.4 (11)	3.7 (6)
Total	100% (118)	100% (188)	100% (249)	100% (164)
		Influential Authors		
Medical doctors	5.3 (1)	9.5 (2)	10.0 (2)	4.5 (1)
Scientists	63.2 (12)	47.6 (10)	30.0 (6)	13.6 (3)
Lawyers	0	0	5.0 (1)	13.6 (3)
Philosophers	0	0	0	9.1 (2)
Bioethicists	5.3 (1)	14.3 (3)	20.0 (4)	18.2 (4)
Theologians	5.3 (1)	14.3 (3)	10.0 (2)	4.5 (1)
Journalists	15.8 (3)	4.8 (1)	5.0 (1)	0
Medical humanities	0	0	0	4.5 (1)
Sociology	0	4.8 (1)	0	0
Humanities	0	0	5.0 (1)	4.5 (1)
Public health	0	0	0	4.5 (1)
Public interest/ government	5.3 (1)	4.8 (1)	15.0 (3)	22.7 (5)
Total	100% (19)	100% (21)	100% (20)	100% (22)

NOTE: Numbers in parentheses are the number of cases in a category.

Table A-2 Texts of Most Influential Authors, 1959–74

Name [total citations]	Citations	Citation for Text
Muller, Hermann J. [24]	8	1959. "The Guidance of Human Evolution." *Perspectives in Biology and Medicine* 3:1–43.
	7	1965. "Means and Aims in Human Genetic Betterment." Pp. 100–115 in *The Control of Human Heredity and Evolution,* ed. T. M Sonneborn. New York: Macmillan.
Dobzhansky, Theodosius [23]	17	1962. *Mankind Evolving.* New Haven: Yale University Press.
Huxley, Julian [20]	6	1963. "Eugenics in Evolutionary Perspective. *Perspectives in Biology and Medicine* 7:155–87.
	5	1961. "The Humanist Frame." Pp. 13–48 in *The Humanist Frame,* ed. Julian Huxley. London: George Allen and Unwin.
Lederberg, Joshua [17]	5	1966. "Experiment Genetics and Human Evolution." *Bulletin of the Atomic Scientist* 22:4–11.
	5	1963. "Biological Future of Man." Pp. 267–73 in *Man and His Future,* ed. G. Wolstenholme. Boston: Little, Brown.
Ramsey, Paul [17]	9	1970. *Fabricated Man: The Ethics of Genetic Control.* New Haven, Conn.: Yale University Press.
Sonneborn, Tracy [16]	12	1965. *The Control of Human Heredity and Evolution.* New York: Macmillan.
Wolstenholme, Gordon [16]	11	1963. *Man and His Future.* Boston: Little, Brown.
Medawar, Peter [15]	9	1960. *The Future of Man.* New York: Basic Books.
Glass, H. Bentley [15]	4	1965. *Science and Ethical Values.* Chapel Hill: University of North Carolina Press.
	2	1971. "Science: Endless Horizons or Golden Age?" *Science* 171:23–29.
	2	1970. *The Timely and the Timeless.* New York: Basic Books.
	2	1969. "For Full Technological Assessment." *Science* 165:755.
	2	1965. "The Ethical Basis of Science." *Science* 150:1254–61.
Edwards, Robert G. [13]	4	1969. "Early Stages of Fertilization In Vitro of Human Oocytes Matured in Vitro." *Nature* 221:632–35. (with B. D. Bavister and P. C. Steptoe)
	3	1970. "Fertilization and Cleavage In Vitro of Preovulator Human Oocytes." *Nature* 227:1307–9 (with P. C. Steptoe and J. M. Purdy)
	3	1970. "Human Embryos in the Laboratory." *Scientific American* 233, no. 6:44–54. (with Ruth E. Fowler)

(*continued*)

Table A-2 (continued)

Name [total citations]	Citations	Citation for Text
Taylor, Gordon [13]	12	1968. *The Biological Time Bomb.* New York: World Publishing.
Fletcher, Joseph [13]	5	1966. *Situation Ethics: The New Morality.* Philadelphia: Westminster Press.
	5	1971. "Ethical Aspects of Genetic Control." *New England Journal of Medicine* 285:776–83.
Huxley, Aldous [12]	9	1932. *Brave New World.* London: Chatto and Windus.
Rosenfeld, Albert [11]	10	1969. *The Second Genesis: The Coming Control of Life.* Englewood Cliffs, N.J.: Prentice-Hall.
Neel, James V. [11]	3	1954. *Human Heredity.* Chicago: University of Chicago Press. (with William J. Schull)
	2	1967. "Some Genetic Aspects of Therapeutic Abortion." *Perspectives in Biology and Medicine* 11:129–35.
Beecher, Henry K. [11]	6	1966. "Ethics and Clinical Research." *New England Journal of Medicine* 274:1354–60.
	3	1969. "Human Studies: Protection for the Investigator and His Subject Is Necessary." *Science* 164–1256.
	3	1968. "Ethical Problems Created by the Hopelessly Unconscious Patient." *New England Journal of Medicine* 278:1425–30.
Dubos, René [11]	5	1965. *Man Adapting.* New Haven, Conn.: Yale University Press.
Crow, James [10]	5	1961. "Mechanisms and Trends in Human Evolution." Pp. 416–31 in *Evolution and Man's Progress,* ed. H. Hoagland and R. Burhoe. Boston: American Academy of Arts and Sciences.
Kass, Leon R. [10]	4	1972. "New Beginnings in Life." Pp. 15–63 in *The New Genetics and the Future of Man,* ed. M. P. Hamilton. Grand Rapids, Mich.: William B. Eerdmans.
	4	1971. "Babies by Means of In Vitro Fertilization: Unethical Experiments on the Unborn?" *New England Journal of Medicine* 285:1174–79.
	3	1971. "The New Biology: What Price Relieving Man's Estate?" *Science* 174: 779–88.

NOTE: Numbers in brackets are the total citations to an author for all texts.

Table A-3 Texts of Most Influential Authors, 1975–84

Name [total citations]	Citations	Citation for Text
Ramsey, Paul [25]	19	1970. *Fabricated Man: The Ethics of Genetic Control.* New Haven, Conn.: Yale University Press.
Fletcher, Joseph [21]	13	1974. *The Ethics of Genetic Control: Ending Reproductive Roulette.* Garden City, N.Y.: Anchor Books.
	7	1971. "Ethical Aspects of Genetic Control." *New England Journal of Medicine* 285:776–83.
Dobzhansky, Theodosius [18]	8	1962. *Mankind Evolving.* New Haven, Conn.: Yale University Press.
Davis, Bernard [18]	10	1970. "Prospects for Genetic Engineering in Man." *Science* 170:1279.
Lappé, Marc [17]	6	1972. "Moral Obligations and the Fallacies of 'Genetic Control.'" *Theological Studies* 33:411–27.
	4	1972. "Ethical and Social Issues in Screening for Genetic Disease." *New England Journal of Medicine* 286:1129–32. (with J. Gustafson and R. Roblin)
	4	1973. "Allegiances of Human Geneticists: A Preliminary Typology." *Hastings Center Studies* 1:63–78.
Kass, Leon [17]	9	1972. "Making Babies: The New Biology and the 'Old' Morality." *Public Interest* 26:18–56.
	6	1971. "The New Biology: What Price Relieving Man's Estate?" *Science* 174: 779–88.
Wade, Nicholas [16]	4	1977. *The Ultimate Experiment: Man-Made Evolution.* New York: Walker.
	3	1977. "Gene-splicing: At Grass-roots Level a Hundred Flowers Bloom." *Science* 195:558–60.
	3	1976. "Recombinant DNA: A Critic Questions the Right to Free Inquiry." *Science* 194:303–6.
Callahan, Daniel [16]	3	1972. "New Beginnings in Life: A Philosopher's Response." Pp. 90–106 in *The New Genetics and the Future of Man,* ed. M. Hamilton. Grand Rapids, Mich.: William B. Eerdmans.
	3	1973. *The Tyranny of Survival.* New York: Macmillan.
Etzioni, Amitai [15]	10	1973. *Genetic Fix.* New York: Macmillan.
	5	1968. "Sex Control Science and Society." *Science* 161:1107–12.
Muller, Hermann J. [14]	6	1950. "Our Load of Mutations." *American Journal of Human Genetics* 2:111–76

(*continued*)

Table A-3 (continued)

Name [total citations]	Citations	Citation for Text
	4	1959. "The Guidance of Human Evolution." *Perspectives in Biology and Medicine* 3: 1–43.
Friedmann, Theodore [14]	9	1972. "Gene Therapy for Human Genetic Disease?" *Science* 175:949–55. (with Richard Roblin)
	4	1971. "Prenatal Diagnosis of Genetic Disease." *Scientific American* 225 (Nov.): 34–42.
Fletcher, John C. [13]	3	1972. "Moral Problems in Genetic Counseling." *Pastoral Psychology* 23:47–60.
	3	1972. "The Brink: The Parent-Child Bond in the Genetic Revolution." *Theological Studies* 33:457–85.
Sinsheimer, Robert [13]	6	1975. "Troubled Dawn for Genetic Engineering." *New Scientist* 16:148–51.
	4	1977. "An Evolutionary Perspective for Genetic Engineering." *New Scientist* 73: 150–52.
	4	1970. "Genetic Engineering: The Modification of Man." *Impact of Science on Society* 20:279–91.
Anderson, W. French [13]	7	1972. "Genetic Therapy." Pp. 109–24 in *The New Genetics and the Future of Man*, ed. M. Hamilton. Grand Rapids, Mich.: William B. Eerdmans.
	5	1980. "Gene Therapy in Human Beings: When Is It Ethical to Begin?" *New England Journal of Medicine* 303:1293–96. (with John Fletcher)
Watson, James [12]	4	1971. "Moving toward the Clonal Man: Is This What We Want?" *Atlantic Monthly* 227:50–53.
	3	1968. *The Double Helix*. New York: Atheneum.
Lederberg, Joshua [12]	8	1966. "Experimental Genetics and Human Evolution." *Bulletin of the Atomic Scientist* 22:4–11.
	4	1971. "Genetic Engineering or the Amelioration of Genetic Defect." *Bioscience* 20:1307–10.
	4	1972. "Biological Innovation and Genetic Intervention." Pp. 7–27 in *Challenging Biological Problems*, ed. J. A. Behnke. New York: Oxford University Press.
Edwards, Robert G. [12]	4	1970. "Human Embryos in the Laboratory." *Scientific American* 233, no. 6:44–54. (with Ruth E. Fowler)
	4	1971. "Social Values and Research in Human Embryology." *Nature* 231:87–91. (with David J. Sharpe)

Table A-3 (continued)

Name [total citations]	Citations	Citation for Text
Milunsky, Aubrey [12]	4	1973. *The Prenatal Diagnosis of Hereditary Disorders.* Springfield, Ill.: Charles C. Thomas.
	3	1977. *Know Your Genes.* Boston: Houghton Mifflin.
McKusick, Victor [11]	7	1978. *Mendelian Inheritance in Man.* 5th ed. Baltimore: Johns Hopkins University Press.
McCormick, Richard [11]	4	1974. "To Save or Let Die: The Dilemma of Modern Medicine." *Journal of the American Medical Association* 229:172–76.
	2	1969. "Notes on Moral Theology." *Theological Studies* 30:635–92.
Berg, Paul [11]	7	1974. "Potential Biohazards of Recombinant DNA Molecules." *Science* 185:303 (with David Baltimore, Herbert Boyer, Stanley Cohen, Ronald Davis, David Hogness, Daniel Nathans, Richard Roblin, James Watson, Sherman Weissman, Norton Zinder)

NOTE: Numbers in brackets are the total citations to an author for all texts.

Table A-4 Texts of Most Influential Authors, 1985–91

Name [total citations]	Citations	Citation for Text
Anderson, W. French [45]	26	1984. "Prospects for Human Gene Therapy." *Science* 226:401–9.
	18	1980. "Gene Therapy in Human Beings: When Is It Ethical To Begin?" *New England Journal of Medicine* 303:1293–97. (with John Fletcher)
	18	1985. "Human Gene Therapy: Scientific and Ethical Considerations." *Journal of Medicine and Philosophy* 10:275–91.
President's Commission [33]	27	1983. *Splicing Life: The Social and Ethical Issues of Genetic Engineering with Human Beings.* Washington, D.C.: U.S. Government Printing Office.
Office of Technology Assessment [27]	6	1987. "New Developments in Biotechnology, 2: Public Perceptions of Biotechnology." Washington, D.C.: U.S. Government Printing Office.
	5	1983. "The Role of Genetic Testing in the Prevention of Occupational Disease." Washington, D.C.: U.S. Government Printing Office.
Fletcher, John C. [23]	7	1985. "Ethical Issues in and beyond Prospective Clinical Trials of Human Gene Therapy." *Journal of Medicine and Philosophy* 10:293–309.

(*continued*)

Table A-4 (continued)

Name [total citations]	Citations	Citation for Text
	5	1983. "Moral Problems and Ethical Issues in Prospective Human Gene Therapy." *Virginia Law Review* 69:515–46.
	5	1990. "Evolution of Ethical Debate about Human Gene Therapy." *Human Gene Therapy* 1:55–68.
Friedmann, Theodore [22]	14	1989. "Progress toward Human Gene Therapy." *Science* 244:1275–81.
	7	1972. "Gene Therapy for Human Genetic Disease?" *Science* 175:949–55. (with R. Roblin)
	7	1983. *Gene Therapy: Fact and Fiction.* Cold Spring Harbor, N.Y.: Cold Spring Harbor Laboratory.
Rifkin, Jeremy [19]	11	1977. *Who Should Play God?* New York: Dell. (with Ted Howard)
	8	1983. *Algeny.* New York: Viking Press.
Walters, LeRoy [18]	14	1986. "The Ethics of Human Gene Therapy." *Nature* 320:225–27.
Ramsey, Paul [18]	14	1970. *Fabricated Man: The Ethics of Genetic Control.* New Haven, Conn.: Yale University Press.
Palmiter, Richard D. [15]	8	1982. "Dramatic Growth of Mice That Develop from Eggs Micro-injected with Metallothionein-growth Hormone Fusion Genes." *Nature* 300:611–15. (with R. Brinster, R. Hammer, M. Trumbauer, M. Rosenfeld, N. Birnberg, R. Evans)
	6	1985. "Transgenic mice." *Cell* 41:343–45. (with R. Brinster)
Kevles, Daniel [15]	13	1985. *In the Name of Eugenics.* New York: Alfred A. Knopf.
Lappé, Marc [13]	5	1972. "Moral Obligations and the Fallacies of 'Genetic Control.'" *Theological Studies* 33:411–27.
	3	1972. "Ethical and Social Issues in Screening for Genetic Disease." *New England Journal of Medicine* 286:1129–32. (with J. Gustafson, R. Roblin)
	3	1976. "Ethical and Scientific Issues Posed by Human Uses of Molecular Genetics." *Annals of the New York Academy of Science* 265: 1–208. (with R. Morison)
	3	1984. *Broken Code: The Exploitation of DNA.* San Francisco: Sierra Club Books.
Weatherall, D. J. [12]	4	1985. *The New Genetics and Clinical Practice.* 2d ed. Oxford: Oxford University Press.
	3	1988. "The Slow Road to Gene Therapy." *Nature* 331:13–14.

Table A-4 (continued)

Name [total citations]	Citations	Citation for Text
McCormick, Richard [12]	3	1981. *How Brave a New World?* Garden City, N.Y.: Doubleday.
	2	1988. "The Shape of Moral Evasion in Catholicism." *America* 159:183–88.
Grobstein, Clifford [12]	7	1984. "Gene Therapy: Proceed with Caution." *Hastings Center Report* 14:13–17. (with M. Flower)
Miller, A. Dusty [12]	3	1984. "Expression of a Retrovirus Encoding Human HPRT in Mice." *Science* 225:630–32. (with R. Eckner, D. Jolly, T. Friedmann, I. Verma)
	3	1981. "A Transmissible Retrovirus Expressing Human Hypoxanthine-phosphoribosyltransferase (HPRT): Gene Transfer into Cells Obtained from Humans Deficient in HPRT." *Proceedings of the National Academy of Science USA* 80:4709–13. (with D. Jolly, T. Friedmann, I. Verma)
	3	1986. "Redesign of Retrovirus Packaging Cell Lines to Avoid Recombination Leading to Helper Virus Production." *Molecular Cell Biology* 6:2895–2902. (with C. Buttimore)
Kantoff, Philip W. [12]	7	1987. "Expression of Human Adenosine Deaminase in Nonhuman Primates after Retroviral Mediated Gene Transfer." *Journal of Experimental Medicine* 166:219–34. (with A. Gillo, J. McLachlin, C. Bordingnon, M. Eglitis, N. Kerman, R. Moen, D. Kohn, S. Yu, F. Karlsson, J. Ziebel, E. Gilboa, R. Blaese, A. Nienhuis, R. O'Reilly, W. Anderson)
	6	1986. "Correction of Adenosine Deaminase Deficiency in Cultured Human T and B Cells by Retrovirus-Mediated Gene Transfer." *Proceedings of the National Academy of Science USA* 83:6563–67. (with D. Kohn, H. Mitsuya, D. Armentano, M. Sieberg, J. Zwiebel, M. Eglitis, J. McLachlin, J. Hutton, W. Anderson)
Culliton, Barbara [11]	3	1985. "Gene Therapy: Research in Public." *Science* 227:493–96.
	2	1989. "Gene Test Begins." *Science* 244:913.
	2	1989. "Gene Transfer Test: So Far So Good." *Science* 245:1325.
Annas, George J. [11]	3	1989. "Who's Afraid of the Human Genome?" *Hastings Center Report* 19:19–21.

(*continued*)

Table A-4 (continued)

Name [total citations]	Citations	Citation for Text
	2	1990. "Legal and Ethical Implications of Fetal Diagnosis and Gene Therapy." *American Journal of Medical Genetics* 35:215–18. (with Sherman Elias)
Fletcher, Joseph [10]	4	1971. "Ethical Aspects of Genetic Controls." *New England Journal of Medicine* 285: 776–83.
	3	1974. *The Ethics of Genetic Control: Ending Reproductive Roulette.* Garden City, N.Y.: Doubleday.
Davis, Bernard [10]	2	1970. "Prospects for Genetic Intervention in Man." *Science* 170:1279–83.
	2	1983. "The Two Faces of Genetic Engineering in Man." *Science* 219:495.
	2	1987. "Bacterial Domestication: Underlying Assumptions." *Science* 235:1329–35.

NOTE: Numbers in brackets are the total citations to an author for all texts.

Table A-5 Texts of Most Influential Authors, 1992–95

Name [total citations]	Citations	Citation for Text
Anderson, W. French [33]	16	1989. "Human Gene Therapy: Why Draw a Line?" *Journal of Medical Philosophy* 14: 681–93.
	13	1985. "Human Gene Therapy: Scientific and Ethical Considerations." *Journal of Medical Philosophy* 10:275–91.
President's Commission [25]	21	1983. *Splicing Life: The Social and Ethical Issues of Genetic Engineering with Human Beings.* Washington, D.C.: U.S. Government Printing Office.
Fletcher, John C. [24]	9	1992. "Germ-line Germ Therapy: A New Stage of Debate." *Law, Medicine and Health Care* 20:26–39. (with W. French Anderson)
	5	1985. "Ethical Issues in and beyond Prospective Clinical Trials of Human Gene Therapy." *Journal of Medicine and Philosophy* 10:293–309.
	5	1990. "Evolution of Ethical Debate about Human Gene Therapy." *Human Gene Therapy* 1:55–68.
Office of Technology Assessment [19]	7	1987. *New Developments in Biotechnology—Background Paper: Public Perceptions of Biotechnology.* Washington, D.C.: U.S. Government Printing Office.

Table A-5 (continued)

Name [total citations]	Citations	Citation for Text
Walters, LeRoy [16]	9	1991. "Human Gene Therapy: Ethics and Public Policy." *Human Gene Therapy* 2: 115–22.
	7	1986. "The Ethics of Human Gene Therapy." *Nature* 320:225–27.
Lappé, Marc [15]	8	1991. "Ethical Issues in Manipulating the Human Germ Line." *Journal of Medicine and Philosophy* 16:621–39.
Ramsey, Paul [14]	9	1970. *Fabricated Man: The Ethics of Genetic Control.* New Haven, Conn.: Yale University Press.
Rifkin, Jeremy [14]	7	1983. *Algeny.* New York: Viking Press.
	4	1985. *Declaration of a Heretic.* Boston: Routledge and Kegan Paul.
Suzuku, David [14]	12	1989. *Genethics.* Cambridge, Mass.: Harvard University Press. (with P. Knudtson)
Kass, Leon R. [13]	4	1972. "New Beginnings in Life." Pp. 15–63 in *The New Genetics and the Future of Man,* ed. M. P. Hamilton. Grand Rapids, Mich.: William B. Eerdmans.
	3	1985. *Toward a More Natural Science.* New York: Free Press.
Fletcher, Joseph [13]	12	1974. *The Ethics of Genetic Control: Ending Reproductive Roulette.* Garden City, N.Y.: Anchor Books.
Capron, Alexander [12]	5	1985. "Unsplicing the Gordian Knot: Legal and Ethical Issues in the New Genetics." Pp. 23–28 in *Genetics and the Law III,* ed. A. Milunsky and G. J. Annas. New York: Plenum Press.
Kevles, Daniel [12]	12	1985. *In the Name of Eugenics.* New York: Alfred A. Knopf.
Annas, George J. [11]	3	1990. "Mapping the Human Genome and the Meaning of Monster Mythology." *Emory Law Journal* 39:629–64.
	3	1992. *Gene Mapping: Using Law and Ethics as Guidelines.* New York: Oxford University Press. (with Sherman Elias)
	2	1989. "Who's Afraid of the Human Genome?" *Hastings Center Report* 19: 19–21.
Wertz, Dorothy [11]	5	1989. "Fatal Knowledge? Prenatal Diagnosis and Sex Selection." *Hastings Center Report* 19:21–27. (with John Fletcher)
	3	1989. *Ethics and Human Genetics: A Cross-Cultural Perspective.* New York: Springer-Verlag. (with John Fletcher)

(*continued*)

Table A-5 (continued)

Name [total citations]	Citations	Citation for Text
Robertson, John A. [11]	3	1985. "Genetic Alteration of Embryos: The Ethical Issues." Pp. 115–27 in *Genetics and the Law III,* ed. A. Milunsky and G. J. Annas. New York: Plenum Press.
	3	1990. "Procreative Liberty and Human Genetics." *Emory Law Journal* 39:697–719.
	2	1987. "Pregnancy and Prenatal Harm to Offspring: The Case of Mothers with PKU." *Hastings Center Report* 17:23–33. (with J. Schulman)
Zimmerman, Burke [11]	11	1991. "Human Germ-line Therapy: The Case for Its Development and Use." *Journal of Medicine and Philosophy* 16:593–612.
Holtzman, Neil [11]	11	1989. *Proceed with Caution.* Baltimore: Johns Hopkins University Press.
Watson, James [10]	6	1990. "The Human Genome Project: Past, Present and Future." *Science* 248:44–48.
	3	1991. "Origins of the Human Genome Project." *FASEB Journal* 5:8–11. (with R. Cook-Degan)
Singer, Peter [10]	4	1979. *Practical Ethics.* Cambridge: Cambridge University Press.
	3	1990. *Animal Liberation.* 2d ed. New York: New York Review.
Glover, Jonathan [10]	9	1984. *What Sort of People Should There Be?* Harmondsworth: Penguin Books.
Juengst, Eric [10]	5	1991. "Germ-line Gene Therapy: Back to Basics." *Journal of Medicine and Philosophy* 16:587–92.

NOTE: Numbers in brackets are the total citations to an author for all texts. The sum of the top cited individual items may surpass the sum of the total cited items for that author because many citing items cite more than one cited item from that author in the same article.

Notes

Introduction

1. National Bioethics Advisory Commission, *Cloning Human Beings: Report and Recommendations of the National Bioethics Advisory Commission* (Rockville, Md.: NBAC, 1997).

2. The last three words of this definition remove from our vision methods of transforming the genes of descendants that have much longer histories, such as selective mating, and more recent technologies of genetically testing for, and possible termination of, fetuses and embryos with certain genetic conditions.

3. LeRoy Walters and Julie Gage Palmer, *The Ethics of Human Gene Therapy* (New York: Oxford University Press, 1997), 17.

4. Jesse Gelsinger died in September 1999 in a clinical trial at the University of Pennsylvania. See Sheryl Gay Stolberg, "Teenager's Death Is Shaking Up Field of Human Gene-Therapy Experiments," *New York Times,* 27 January 2000, A20. Although people have questioned for years whether somatic gene therapy is effective, in early 2000 it appeared that a success was at hand. See Thomas H. Maugh, "Gene Therapy May Have Cured 3 Infants," *Los Angeles Times,* 28 March 2000, A1.

5. Ola Mae Huntley, "A Mother's Perspective," *Hastings Center Report,* April 1984, 14–15.

6. Gina Kolata, "Scientists Place Jellyfish Genes into Monkeys," *New York Times,* 23 December 1999, A1.

7. Walters and Palmer, *Ethics of Human Gene Therapy.*

8. Claudia Mickelson, "Meeting Overview and Historical Background: Human Gene Transfer Research in the U.S.," paper presented at NIH First Gene Therapy Policy Conference, "Human Gene Transfer: Beyond Life-Threatening Disease," Bethesda, Md., 11 September 1997.

9. Frederic Golden and Michael D. Lemonick, "The Race Is Over," *Time,* 3 July 2000, 18–23.

10. Of course, we always are changing the human species genetically, if through no other technology than keeping people with genetic conditions alive who otherwise would have died before reaching reproductive age. The reaction is, I believe, to the degree of conscious intention and control over human evolution.

11. Walters and Palmer, *Ethics of Human Gene Therapy,* 127.

12. Paul Ramsey, *Fabricated Man: The Ethics of Genetic Control* (New Haven, Conn.: Yale University Press, 1970), 1.

13. On the decline of public intellectuals, see Russell Jacoby, *The Last Intellectuals* (New York: Farrar, Straus and Giroux, 1987).

14. I borrow the "thick/thin" metaphor from Steve Hart, although his

scheme is different from mine. See Stephen Hart, "Cultural Sociology and Social Criticism," *Newsletter of the Sociology of Culture Section of the American Sociological Association* 9, no. 3 (1995). For similar use of the metaphor, see Michael Walzer, *Thick and Thin* (Notre Dame, Ind.: University of Notre Dame Press, 1994).

15. Walzer, *Thick and Thin;* James M. Gustafson, "Moral Discourse about Medicine: A Variety of Forms," *Journal of Medicine and Philosophy* 15 (1990): 127.

16. Jürgen Habermas, *Toward a Rational Society: Student Protest, Science and Politics* (Boston: Beacon Press, 1970), chap. 6; Alan Wolfe, *Whose Keeper? Social Science and Moral Obligation* (Berkeley: University of California Press, 1989); Walzer, *Thick and Thin.*

17. "Overlapping consensus" comes from John Rawls, "The Idea of Overlapping Consensus," *Oxford Journal of Legal Studies* 7, no. 1 (1987): 1–25.

Chapter 1

1. See John Fletcher, "Moral Problems and Ethical Issues in Prospective Human Gene Therapy," *Virginia Law Review* 69 (1983): 542. Paul Ramsey and Leon Kass were the first authors to make the "informed consent of not yet existing people" argument. See Paul Ramsey, *Fabricated Man: The Ethics of Genetic Control* (New Haven, Conn.: Yale University Press, 1970), 133–34; Leon R. Kass, "Babies by Means of In Vitro Fertilization: Unethical Experiments on the Unborn?" *New England Journal of Medicine* 285, no. 21 (1971): 1174–79. The irony of these two authors having invented the idea will become apparent in later chapters.

2. I shall show this in chapter 5. Thomas Murray, reviewing and critiquing the arguments about germline HGE, says that what has "struck most commentators as distinctive and significant" about germline HGE is that "the principle of informed consent that justifies many of the risks of medical experimentation cannot apply in any simple way to the descendants of subjects of gene therapy research. These persons were not yet conceived at the time of the research, and hence cannot be said to consent to risks to their own genome. Second, whatever harm might be caused by gene therapy might be magnified manyfold if it were passed onto future generations rather than dying out with the subjects of the current research." See Thomas H. Murray, "Ethical Issues in Human Genome Research," *Federation of American Societies of Experimental Biology Journal* 5 (1991): 58. My evaluation of the content of the debate generally agrees with Murray's.

3. Robert L. Sinsheimer, "The Prospect for Designed Genetic Change," *American Scientist* 57, no. 1 (1969): 134.

4. Ibid.

5. Ramsey, *Fabricated Man,* 144.

6. Max Weber, *Economy and Society,* 2 vols. (Berkeley: University of California Press, 1968), 85.

7. Rogers Brubaker, *The Limits of Rationality: An Essay on the Social and Moral Thought of Max Weber* (Boston: George Allen and Unwin, 1984), 36.

8. Weber made a distinction between subjective and objectified rationality that is important for the case at hand (Donald N. Levine, *The Flight from Ambiguity* [Chicago: University of Chicago Press, 1985], 152–62). Subjective rationality guides the behavior of individual actors. Subsumed under subjective rational-

ity is Weber's famous distinction between action that is considered *by the actor* to be the calculated maximization of means to end (instrumental rationality) and action considered to be consistent with the ends (value rationality). Objectified rationality occurs where courses of action are evaluated in terms of institutionalized norms, where "rationality is embodied in the social structure and confronts individuals as something external to them" (Brubaker, *Limits of Rationality,* 9). Therefore, formal and substantive rationality are the objectified versions of instrumental and value rationality (Levine, *Flight from Ambiguity,* 161; Wendy Espeland, *The Struggle for Water: Politics, Rationality and Identity in the American Southwest* [Chicago: University of Chicago Press, 1998], 35). Technically, according to Weber's analysis, an individual's argument cannot be objective and thus cannot be either substantively or formally rational, but rather must be categorized as value rational or instrumentally rational. I will use the objective or institutionalized terms when discussing individuals because I am concerned with their arguments *becoming* institutionalized, or with the extent to which their arguments reflect institutionalized arguments.

9. Jürgen Habermas, *The Theory of Communicative Action, Vol. 2: Lifeworld and System: A Critique of Functionalist Reason* (Boston: Beacon Press, 1987); Habermas, *The Structural Transformation of the Public Sphere* (Cambridge, Mass.: MIT Press, 1989).

10. Robert Wuthnow, *Between States and Markets: The Voluntary Sector in Comparative Perspective* (Princeton, N.J.: Princeton University Press, 1991), 298.

11. Brubaker, *Limits of Rationality,* 37.

12. I should acknowledge that, beyond Weber and Habermas, I have found insight into the distinctions that flesh out the broader forms of rationality in the work of several other scholars: Stephen Hart, "Cultural Sociology and Social Criticism," *Newsletter of the Sociology of Culture Section of the American Sociological Association* 9, no. 3 (1995); Steven M. Tipton, *Getting Saved from the Sixties: Moral Meaning in Conversion and Cultural Change* (Berkeley: University of California Press, 1982); Michael Walzer, *Thick and Thin* (Notre Dame, Ind.: University of Notre Dame Press, 1994); Jeffrey Stout, *Ethics after Babel* (Boston: Beacon Press, 1988); Renee C. Fox and Judith P. Swazey, "Medical Morality Is Not Bioethics—Medical Ethics in China and the United States," *Perspectives in Biology and Medicine* 27, no. 3 (1984): 336–60; Renee C. Fox, *The Sociology of Medicine: A Participant Observer's View* (Englewood Cliffs, N.J.: Prentice Hall, 1989); Gilbert Meilaender, *Body, Soul, and Bioethics* (Notre Dame, Ind.: University of Notre Dame Press, 1995); James M. Gustafson, "Moral Discourse about Medicine: A Variety of Forms," *Journal of Medicine and Philosophy* 15 (1990): 125–42; Daniel Callahan, "Bioethics," in *Encyclopedia of Bioethics,* ed. Warren Thomas Reich (New York: Macmillan, 1995), 247–56.

13. Ramsey, *Fabricated Man,* 30 (emphasis in original).

14. LeRoy Walters and Julie Gage Palmer, *The Ethics of Human Gene Therapy* (New York: Oxford University Press, 1997), 81.

15. Ramsey, *Fabricated Man.*

16. Bentley Glass, "Science: Endless Horizons or Golden Age?" *Science* 171 (1971): 29.

17. Max Weber, *From Max Weber: Essays in Sociology,* trans. H. H. Gerth and C. Wright Mills (New York: Oxford University Press, 1946), 147.

18. Brubaker, *Limits of Rationality,* 37.

19. Wendy Nelson Espeland and Mitchell L. Stevens, "Commensuration as a Social Process," *Annual Review of Sociology* 24 (1998): 313–31.

20. Joseph Fletcher, "Ethical Aspects of Genetic Controls," *New England Journal of Medicine* 285, no. 14 (1971): 779.

21. This insight is derived from Gustafson, whose analogous authors "are seldom interested in specific acts except insofar as they signify a larger and deeper evil or danger." See Gustafson, "Moral Discourse about Medicine," 130.

22. John Fletcher and W. French Anderson, "Germ-Line Gene Therapy: A New Stage of the Debate," *Law, Medicine and Health Care* 20 (1992): 26.

23. J. B. S. Haldane, "Biological Possibilities for the Human Species in the Next Ten Thousand Years," in *Man and His Future,* ed. Gordon Wolstenholme (London: J. and A. Churchill Ltd., 1963), 337–61.

24. Bruce Jennings, "Possibilities of Consensus: Toward Democratic Moral Discourse," *Journal of Medicine and Philosophy* 16, no. 4 (1991): 454. For a similar distinction to those of Jennings, see the discussion of constitutional and procedural theories in Amy Gutmann and Dennis Thompson, *Democracy and Disagreement* (Cambridge, Mass.: Harvard University Press, 1996).

25. Jennings, "Possibilities of Consensus," 456; John Rawls, "The Idea of Overlapping Consensus," *Oxford Journal of Legal Studies* 7, no. 1 (1987): 1–25.

26. U.S. Congress, Office of Technology Assessment, *Biomedical Ethics in U.S. Public Policy—Background Paper* (Washington, D.C.: Government Printing Office, 1993), iii.

27. John Fletcher, "Evolution of Ethical Debate about Human Gene Therapy," *Human Gene Therapy* 1 (1990): 56–57.

28. Eric T. Juengst, "Germ-Line Gene Therapy: Back to Basics," *Journal of Medicine and Philosophy* 16 (1991): 587–92.

29. Max Weber, *The Protestant Ethic and the Spirit of Capitalism* (New York: Charles Scribner's Sons, 1958), 181.

30. Brubaker, *Limits of Rationality,* 44.

31. Frank Dobbin, "Cultural Models of Organization: The Social Construction of Rational Organizing Principles," in *The Sociology of Culture,* ed. Diana Crane (Cambridge, Mass.: Blackwell, 1994), 119.

32. This is generally known as the "production of culture" perspective, which has also been incorporated into neo-institutionalist explanations of cultural change. See Richard A. Peterson, "The Production of Culture: A Prolegomenon," in *The Production of Culture,* ed. Richard A. Peterson (Beverly Hills, Calif.: Sage Publications, 1976), 7–22; Robert Wuthnow, *Communities of Discourse: Ideology and Social Structure in the Reformation, the Enlightenment and European Socialism* (Cambridge, Mass.: Harvard University Press, 1989); Paul DiMaggio, "Cultural Entrepreneurship in Nineteenth-century Boston, Part 1," *Media, Culture and Society* 4 (1982): 33–50; Paul DiMaggio and Walter W. Powell, "Introduction," in *The New Institutionalism in Organizational Analysis,* ed. Powell and DiMaggio (Chicago: University of Chicago Press, 1991), 1–40.

33. Wuthnow, *Communities of Discourse.*

34. Although I lack comprehensive data on how difficult it is to publish texts in this area, it appears that there is a home for almost any text. Influential and uninfluential texts alike are found in the most obscure places. For example, some of the most influential texts in the early debate were published in a volume

edited by the rector of the Washington Cathedral and published by a religious publishing house. Scores of texts that appear never to have been looked at again appear in similar venues.

35. The boundaries of the environment where producers compete for limited resources is clearly dependent on the topic at hand. A study of the production of culture disseminated through television would have almost the entire citizenry of the United States as the environment, because almost everyone can contribute resources indirectly through watching television advertisements. The HGE debate has a much more limited environment of persons concerned with the ethics of HGE. It is an elite debate, one that the vast majority of the public either is not interested in or does not have time for.

36. For an overview of institutionalization, see Ronald L. Jepperson, "Institutions, Institutional Effects, and Institutionalism," in *New Institutionalism in Organizational Analysis,* ed. Powell and DiMaggio, 143–63.

37. Stout, *Ethics after Babel.*

38. Albert R. Jonsen, "Foreword," in *A Matter of Principles? Ferment in U.S. Bioethics,* ed. Edwin R. DuBose, Ronald P. Hamel, and Laurence J. O'Connell (Valley Forge, Pa.: Trinity Press International, 1994), xii.

39. The perspective on professions here is largely adapted from Abbott's work, although I have changed some of his terms to make them consistent with mine. See Andrew Abbott, *The System of Professions: An Essay on the Division of Expert Labor* (Chicago: University of Chicago Press, 1988).

40. The first two methods of achieving jurisdiction are made by more established professions.

41. The history of professions is filled with examples of competition for jurisdiction. For example, Abbott examines the different professions responsible for the work of helping people with their personal problems. The clergy, neurology, and other professions eventually lost out in the competition for this work to psychiatry and psychology. The professional boundaries between psychiatry and psychology remain contested, with both groups mobilizing resources to fight against the other, as well as to ward off potential new threats. Another clear example is the case of physicians and their many competitors: chiropractors, midwives, acupuncturists, and herbalists.

42. Abbott, *System of Professions,* 9.

43. Ibid., 98.

44. My "means" and "ends" analogies for jurisdictional claims are the same as Abbott's "act" and "purpose." Abbott's other methods of claiming jurisdiction—scene, agent, and agency—are not relevant to this particular jurisdictional competition. See ibid., 99.

45. Thomas F. Gieryn, "Boundary-Work and the Demarcation of Science from Non-Science: Strains and Interests in Professional Ideologies of Scientists," *American Sociological Review* 48 (1983): 781–95; Thomas F. Gieryn, George M. Bevins, and Stephen C. Zehr, "Professionalization of American Scientists: Public Science in the Creation/Evolution Trials," *American Sociological Review* 50 (1985): 392–409.

46. Abbott, *System of Professions,* 75.

47. Daniel Callahan, quoted in Albert R. Jonsen, *The Birth of Bioethics* (New York: Oxford University Press, 1998), 83–84.

48. This categorization is a mix of the ideas of Rosenberg and Callahan. See Charles E. Rosenberg, "Meanings, Policies, and Medicine: On the Bioethical En-

terprise and History," *Daedalus* 128, no. 4 (1999): 39; Daniel Callahan, "The Social Sciences and the Task of Bioethics," *Daedalus* 128, no. 4 (1999): 279.

49. Albert R. Jonsen and Andrew Jameton, "Medical Ethics, History of: The Americas," in *Encyclopedia of Bioethics*, ed. Warren Thomas Reich (New York: Macmillan, 1995).

50. The definition of *influential* is to follow, below.

51. The distribution of all authors by profession is shown in table A-1.

52. U.S. Congress, Office of Technology Assessment, *Biomedical Ethics in U.S. Public Policy*, iii.

53. Gutmann and Thompson, *Democracy and Disagreement.*

54. Brubaker, *Limits of Rationality*, 43; Espeland, *Struggle for Water.*

55. Espeland, *Struggle for Water.* The social construction of rationality in economic and organizational processes has been one of the primary research emphases of neo-institutional organization theory. See Powell and DiMaggio, eds., *New Institutionalism in Organizational Analysis;* Dobbin, "Cultural Models of Organization."

56. I am convinced of the importance, on a theoretical level, of market capitalism as a transformative force. See Habermas, *Theory of Communicative Action, Vol. 2;* Alan Wolfe, *Whose Keeper? Social Science and Moral Obligation* (Berkeley: University of California Press, 1989). Unfortunately, in the HGE debate the influence of the market is undetectable—either because it does not exist, because it is subsumed into the actions of the state, or because it is simply too subtle. Regardless, all explanations are partial, to some extent, and the influence of the market, if it exists, is necessarily underplayed here.

57. Theodore M. Porter, *Trust in Numbers: The Pursuit of Objectivity in Science and Public Life* (Princeton, N.J.: Princeton University Press, 1995), 195. Those familiar with Porter's work will note that he portrays the move to rulelike behavior as a function of the strength of the government functionaries. Functionaries who are in stronger positions can still rely on discretion. The HGE debate is so fraught with controversy that no functionary is ever trusted. Therefore, this critical variable in Porter's book is a constant in this one, due to the particularity of the HGE debate.

58. Richard Hammond, in ibid., 195.

59. One of the primary reasons presidential commissions exist is to offer political cover for difficult decisions. See Terrence R. Tuchings, *Rhetoric and Reality: Presidential Commissions and the Making of Public Policy* (Boulder, Colo.: Westview Press, 1979); David Flitner, *The Politics of Presidential Commissions* (Dobbs Ferry, N.Y.: Transnational Publishers, 1986).

60. For example, in his review of the use of science in modern liberal democracies, Ezrahi found that, unlike a medieval European monarchy with decrees coming from the subjective perspective of a king, the U.S. political system was partly founded on the idea that "politics is transparent, that political agents, political actions, and political power can be viewed." See Yaron Ezrahi, *The Descent of Icarus: Science and the Transformation of Contemporary Democracy* (Cambridge, Mass.: Harvard University Press, 1990), 69.

61. Porter, *Trust in Numbers.*

Chapter 2

1. The years included in this period are those when the texts by the average members of the debate were published. Due to the inevitable lag between initial

publication and subsequent citation, influential texts—those which are frequently cited—may have been published before the time period began. This is especially true for later eras covered in this study, which have fewer years in them.

2. More specifically, this finding comes from a regression equation where the dependent variable is the number of influential authors in a community that an average author in that era cites. The independent variable of interest is the profession of the average author. Models also controlled for the total number of texts that were cited by the average author because different professions have different citation practices. Negative binomial regression is used because the dependent variable is a count. There are instances where a Poisson regression was more appropriate, and was used instead. The details of these analyses are available from the author.

For this community, the coefficient on the scientist dummy variable is .342. This means that being a scientist increases the expected number of influential authors cited in this community by 41 percent, compared to members of other professions, holding the other variables constant (100*[exp(.342) − 1]). See J. Scott Long, *Regression Models for Categorical and Limited Dependent Variables* (Thousand Oaks, Calif.: Sage Publications, 1997), 228. This finding is significant at the $p = .089$ level, with an N of 69 common authors. A high p-value is used due to the low N. Models for each of the other professions under examination show no significant differential representation in this community.

3. Muller is cited by 26 percent of the authors writing about HGE during this era.

4. For details on Muller and his beliefs, see Elof Axel Carlson, *Genes, Radiation and Society: The Life and Work of H. J. Muller* (Ithaca, N.Y.: Cornell University Press, 1981); Carlson, "Eugenics and Basic Genetics in H. J. Muller's Approach to Human Genetics," *History and Philosophy of the Life Sciences* 9 (1987): 57–78; John Beatty, "Weighing the Risks: Stalemate in the Classical/Balance Controversy," *Journal of the History of Biology* 20, no. 3 (1987): 289–319; Diane B. Paul, "'Our Load of Mutations' Revisited," *Journal of the History of Biology* 20, no. 3 (1987): 321–35; James F. Crow, "Muller, Dobzhansky, and Overdominance," *Journal of the History of Biology* 20, no. 3 (1987): 351–80; Crow, "H. J. Muller's Role in Evolutionary Biology," in *The Founders of Evolutionary Genetics,* ed. Sahotra Sarkar (Dordrecht: Kluwer Academic Publishers, 1992), 83–106.

5. It appears that his biggest mistake was sending Josef Stalin a copy of his 1935 eugenics manifesto *Out of the Night,* which contained his eugenical notions of progress, obviously different from the Marxist-Leninist narrative. Stalin was "displeased," which was the beginning of the end for Muller in the USSR. He escaped the Soviet Union and Stalin's persecution of geneticists by volunteering to serve against the fascists in the Spanish Civil War, eventually finding his way to Great Britain. See Carlson, *Genes, Radiation and Society,* 233–35.

6. The "mainline/reform" distinction in eugenics movements, and most of the information in this section, is taken from Kevles's canonical work on the topic. See Daniel Kevles, *In the Name of Eugenics: Genetics and the Uses of Human Heredity* (Berkeley: University of California Press, 1985).

7. Ibid., ix.

8. Ibid., 34.

9. Ibid., 33.

10. Ibid., 44, 47.

11. Ibid., 47.
12. Ibid., 85.
13. Ibid., 97.
14. Ibid., 110–11.
15. Ibid., 164.
16. Ibid., 118.
17. Ibid., 173.
18. Hermann J. Muller, "Our Load of Mutations," *American Journal of Human Genetics* 2, no. 2 (1950): 111–76.
19. Hermann Muller, "Genetic Progress by Voluntarily Conducted Germinal Choice," in *Man and His Future,* ed. Gordon Wolstenholme (London: J. and A. Churchill Ltd., 1963), 11.
20. Ibid.
21. Bruce Wallace, *Fifty Years of Genetic Load: An Odyssey* (Ithaca, N.Y.: Cornell University Press, 1991), 1.
22. Bruce Wallace, *Genetic Load: Its Biological and Conceptual Aspects* (Englewood Cliffs, N.J.: Prentice-Hall, 1970), 1.
23. Howard L. Kaye, *The Social Meaning of Modern Biology: From Social Darwinism to Sociobiology* (New Brunswick, N.J.: Transaction Publishers, 1997), 42. Looking for meaning in evolution goes back at least as far as Darwin, who was reluctant to conclude that evolution was entirely random and purposeless. See also Michael Ruse, *Monod to Man: The Concept of Progress in Evolutionary Biology* (Cambridge, Mass.: Harvard University Press, 1996).
24. Kaye, *Social Meaning of Modern Biology,* 41.
25. C. Kenneth Waters, "Introduction: Revising Our Picture of Julian Huxley," in *Julian Huxley: Biologist and Statesman of Science,* ed. C. Kenneth Waters and Albert Van Helden (Houston: Rice University Press, 1992), 3.
26. Garland E. Allen, "Julian Huxley and the Eugenical View of Human Evolution," in *Julian Huxley,* ed. Waters and Van Helden, 193–222.
27. Waters, "Introduction," in *Julian Huxley,* ed. Waters and Van Helden, 2.
28. Julian Huxley, "The Humanist Frame," in *The Humanist Frame,* ed. Julian Huxley (London: George Allen and Unwin Ltd., 1961), 19.
29. The first quotation is from Marc Swetlitz, "Julian Huxley and the End of Evolution," *Journal of the History of Biology* 28 (1995): 198. The second is from William B. Provine, "Progress in Evolution and Meaning in Life," in *Julian Huxley,* ed. Waters and Van Helden, 166.
30. Muller thought that "because of resistance by antiquated religious traditions and moralities, the American people . . . had failed to incorporate into their society and personalities 'the wonderful world view opened up by Darwin and other Western biologists . . . [which was] the source of the profoundest idealism and hope.' . . . Evolutionary biology would [rescue] . . . American youth from decadence and self-indulgence by imbuing them with a sense of meaning and purpose in life" (Kaye, *Social Meaning of Modern Biology,* 13).
31. Costas B. Krimbas, "The Evolutionary Worldview of Theodosius Dobzhansky," in *The Evolution of Theodosius Dobzhansky,* ed. Mark B. Adams (Princeton, N.J.: Princeton University Press, 1994), 179–94. Like those of Muller and Huxley, Dobzhansky's life, beliefs, and work have developed a secondary literature. See Adams, ed., *Evolution of Theodosius Dobzhansky;* Louis Levine, *Genetics of Natural Populations* (New York: Columbia University Press, 1995); Ruse, *Monod to Man.*

32. Paul, "'Our Load of Mutations' Revisited," 334.

33. Although Dobzhansky's views are present in his most influential book (*Mankind Evolving: The Evolution of the Human Species* [New Haven, Conn.: Yale University Press, 1962]), they are clearer in his later works, such as *The Biology of Ultimate Concern* (New York: New American Library, 1967). In the latter he makes extensive use of Teilhard de Chardin's *Phenomenon of Man* (New York: Harper and Brothers, 1955). Dobzhansky was also the president of the American Teilhard de Chardin Association in 1969 (Krimbas, "Evolutionary Worldview of Theodosius Dobzhansky," 189). Huxley also was an admirer of Teilhard de Chardin, having written a glowing introduction to the English edition of his 1955 book.

34. Dobzhansky, *Biology of Ultimate Concern,* 5. The last two words in this title are clearly taken from Tillich, whom Dobzhansky acknowledges in the introduction as a source for many of his ideas.

35. Dobzhansky, *Mankind Evolving,* 346–47. Directly following this passage, Dobzhansky closes his book by quoting from Teilhard de Chardin, justifying his new purpose and meaning for humankind. Huxley also believed that humans had transcended the evolutionary process because they had achieved consciousness and had an obligation to now direct their own evolution as well as that of the rest of the species of the world. See Swetlitz, "Julian Huxley and the End of Evolution." The title of Muller's 1959 paper, "The Guidance of Human Evolution" (*Perspectives in Biology and Medicine* 3, no. 1), accurately portrays his (extremely similar) view. He believes that as the first evolved life form to achieve consciousness, human beings must transcend evolution in favor of control: "From now on, evolution is what we make it, provided that we choose the true and the good." If humans do indeed control their evolution, "evolution will become, for the first time, a conscious process. . . . [It] will be the highest form of freedom that man, or life, can have" (42).

36. James F. Crow, "Mechanisms and Trends in Human Evolution," *Daedalus* 90, no. 3 (1961): 430 (emphasis added). Another member of this community, microbiologist René Dubos, who later became an environmental guru of sorts, turned on their heads Dobzhansky and the others who wanted to set genetic ends, but retained a vision of setting the ends for society. Dubos writes less about genetics controlling humans than about humans controlling their destiny through their ability to adapt to the demands of a changing environment. Humans should indeed set ends, because "there is no foreseeable limit to the variety and extent of the social adaptive mechanisms that man can bring to bear on the external world, in order to modify it according to his needs and wishes." See René Dubos, *Man Adapting* (New Haven, Conn.: Yale University Press, 1965), 275. Dubos was debating with Dobzhansky here. Not only is Dubos's title, *Man Adapting,* linguistically linked to Dobzhansky's *Mankind Evolving,* but also Dubos himself notes that this text "bears a close relation" to Dobzhansky's. See Dubos, *Man Adapting,* xvii. Dubos's text, like Dobzhansky's, was first presented in the Silliman Foundation Lectures at Yale, a series designed to showcase arguments about science replacing traditional religion. The series was described as an "annual course of lectures designed to illustrate the presence and providence of God as manifested in the natural and moral world. . . . It was the belief of the testator that any orderly presentation of the facts of nature or history contributed to this end more effectively than dogmatic or polemical theology, which should therefore be excluded from the scope of the lectures."

37. See Paul, "'Our Load of Mutations' Revisited," 328, 332.

38. Julian Huxley, "Eugenics in Evolutionary Perspective," *Perspectives in Biology and Medicine* (1963): 173.

39. Crow, "Mechanisms and Trends in Human Evolution," 429.

40. Huxley, "Eugenics in Evolutionary Perspective," 163. Muller had a similar list: "When one considers how much the world owes to single individuals of the order of capability of an Einstein, Pasteur, Descartes, Leonardo, or Lincoln, it becomes evident how vastly society would be enriched if they were to be manifolded. Moreover, those who repeatedly proved their worth would surely be called upon to reappear age after age until the population in general had caught up with them" ("Guidance of Human Evolution," 35). Lenin was included in his 1935 book but not his 1959 text. See Dobzhansky, *Mankind Evolving,* 328.

41. Muller, "Guidance of Human Evolution," 15.

42. Ibid., 26.

43. Huxley, "Eugenics in Evolutionary Perspective," 183. Huxley's voluntarism was not perfect, and the mainline creed was not entirely dead, because while the people with the "good" genes would have a choice, the people with "bad" genes might not be able to see reason. For example, Huxley thought that the "so-called social problem group"—those persons who live in poverty and squalor, and who are "genetically subnormal" in qualities such as "initiative, pertinacity, general exploratory urge and interest, energy, emotional intensity and will-power"—might have to face "compulsory or semi-compulsory" birth control. See Huxley, "Eugenics in Evolutionary Perspective," 177.

44. Elof Axel Carlson, "Eugenics Revisited: The Case for Germinal Choice," *Stadler Genetics Symposia* 5 (1973): 22. The account just given papers over a huge debate about whether eugenics *could* achieve its ends. The most persistent critic of Muller's schemes was Dobzhansky, whose criticisms were embedded in a dispute about the nature of genes in populations, later referred to as the "classical/balance" controversy. The "classical" position, held by Muller and Crow, was that there was a "normal" allele at most loci on the genome. Deviations from the "normal" state of affairs was considered bad, a part of the "load." The "balance" position, held by Dobzhansky and microbiologist Peter Medawar, was that it was not meaningful to speak of "normal" alleles; rather, there was variation at each locus, and that variation was good. In an overstatement of the case, Dobzhansky would often accuse Muller of wanting a world of Einsteins, of believing in the one true and perfect genotype, the "ideal man, or the ideal woman," and of believing that "the entire population of the world, the whole of mankind, [should] carry this ideal genotype" (Dobzhansky, *Mankind Evolving,* 329). Dobzhansky himself, in contrast, believed that it was the diversity in the population that was the strength of the human race, acting "as a leaven of creative effort in the past and will to act in the future" (ibid., 330). As Medawar would point out, if the classical view was false, and variation was important to evolution (the balance view), then Muller's eugenic schemes would be disastrous (P. B. Medawar, *The Future of Man* [New York: Basic Books, 1960], chap. 3). This does not mean, however, that Dobzhansky was opposed to eugenics in general or to the end that humans should control their genetic future: he concurred with Muller's opinion that "the problem of the management of human evolution should not be postponed until the conjectural time when directed mutation in man will have been discovered" (Dobzhansky, *Mankind Evolving,* 333). According to Beatty, Dobzhansky was opposed to Muller's eugenics programs not

because they were eugenics programs, but because they were based on the wrong theory of genetic inheritance. See Beatty, "Weighing the Risks," 315–17. On this debate see also Crow, "Muller, Dobzhansky, and Overdominance"; and Ruse, *Monod to Man,* 400–401.

45. Kevles, *In the Name of Eugenics,* 237.

46. This is poignantly illustrated by the description of a 1959 meeting discussing the new successes in molecular biology, where Muller and a fellow fruit-fly geneticist were sitting by themselves, "rather forlorn, largely ignored and recognized, if at all, as dim figures from a period of classical genetics which had long since seen its best days." After commenting on the bad manners of the new young crowd, Muller said, "I had always known that someday there would be a chemical basis for the structure and function of the gene, but I never believed that I would live to see it in our lifetime." Quoted in Carlson, *Genes, Radiation and Society,* 392.

47. Robert Bud, *The Uses of Life* (New York: Cambridge University Press, 1993), 168. In addition to this meeting, there was a very similar meeting in the same year in England. The proceedings of both these meetings are highly influential texts in this community. See Gordon Wolstenholme, ed., *Man and His Future* (London: J. and A. Churchill, Ltd., 1963); Tracy M. Sonneborn, ed., *The Control of Human Heredity and Evolution* (New York: Macmillan, 1965); Joshua Lederberg, "Biological Future of Man," in *Man and His Future,* ed. Wolstenholme, 263–73.

48. Sonneborn, ed., *Control of Human Heredity and Evolution.*

49. In his address, Muller does not approve of the move to the new biology, because these techniques are in the future and his germinal-choice scheme is available now: "For any group of people who have a rational attitude toward matters of reproduction, and who also have a genuine sense of their own responsibility to the next and subsequent generations, the means exist right now of achieving a much greater, speedier, and more significant genetic improvement of the population, by the use of selection, than could be effected by the most sophisticated methods of treatment of the genetic material that might be available in the twenty-first century" (Muller, "Means and Aims in Human Betterment," in *The Control of Human Heredity and Evolution,* ed. T. M. Sonneborn [New York: Macmillan, 1965], 100).

50. S. E. Luria, "Directed Genetic Change: Perspectives from Molecular Genetics," in *Control of Human Heredity and Evolution,* ed. Sonneborn, 4.

51. G. Pontecorvo, "Prospects for Genetic Analysis in Man," in *Control of Human Heredity and Evolution,* ed. Sonneborn, 80–81.

52. Sonneborn, ed., *Control of Human Heredity and Evolution,* viii.

53. Ibid., viii.

54. Luria, "Directed Genetic Change," 3, 17.

55. These authors are science journalists Gordon Rattray Taylor (*The Biological Time Bomb* [New York: World Publishing, 1968]) and Albert Rosenfeld (*The Second Genesis: The Coming Control of Life* [Englewood Cliffs, N.J.: Prentice Hall, 1969]). Rosenfeld's book begins: "Coming: the control of life. All of life, including human life. With man himself at the controls. Also coming: a new Genesis—The Second Genesis. The creator, this time around—man." Like Rosenfeld's book, Taylor's text reflects the substantively rational tendency to consider problems as representative of deeper or larger problems. He connects the many technological developments of the previous ten to fifteen years into one

deeper problem. The blurb on the back cover asks a series of questions: "Test-tube babies? Postponing death? Mind control? The semi-artificial man? Brain without a body? Genetic warfare? Creation of life? Is sex necessary?" The final question, echoing the Sonneborn conference, is, "Where are we going?" The answer, clearly, lies in the title of the first chapter: we are going wherever "the biologists [are] taking us."

56. An oocyte is a cell derived from the ovary that transforms itself into an ovum.

57. R. G. Edwards, B. D. Bavister, and P. C. Steptoe, "Early Stages of Fertilization In Vitro of Human Oocytes Matured In Vitro," *Nature* 221 (1969): 633.

58. Aldous Huxley, *Brave New World* (London: Chatto and Windus, 1932), 3–4.

59. This section describes the debate in the middle community in figure 2. It is not dominated by any one profession.

60. R. G. Edwards and Ruth E. Fowler, "Human Embryos in the Laboratory," *Scientific American* 223, no. 6 (1970): 54.

61. Glass had been president of the American Association for the Advancement of Science, the American Society of Human Genetics, Phi Beta Kappa, and the American Association of University Professors. After spending eighteen years at Johns Hopkins University, after 1965 he was academic vice president and professor of biology at the State University of New York, Stony Brook.

62. Bentley Glass, *Science and Ethical Values* (Chapel Hill: University of North Carolina Press, 1965), vii; and Glass, *The Timely and the Timeless* (New York: Basic Books, 1970), 37.

63. Glass, *Science and Ethical Values*, 59–60.

64. Ibid., 61. By 1971 Glass was still talking about the necessity of man "direct[ing] the course of his own evolution," but he now looked to Edwards's research with "expectant attention." The reason was that although "Edwards cautiously limits the application of his developing techniques to the provision of a healthy embryo for a woman whose oviducts are blocked . . . [i]t should be obvious that the technique can be quickly and widely extended" (Bentley Glass, "Science: Endless Horizons or Golden Age?" *Science* 171 [1971]: 28).

65. Glass, "Science: Endless Horizons or Golden Age?" 28–29.

66. Krishna R. Dronamraju, "Introduction," in *Haldane's Daedalus Revisited*, ed. Dronamraju (Oxford: Oxford University Press, 1995), 12.

67. Amitai Etzioni, *Genetic Fix* (New York: Macmillan, 1973), 59.

68. Kelly Moore, "Organizing Integrity: American Science and the Creation of Public Interest Organizations, 1955–1975," *American Journal of Sociology* 101, no. 6 (1996): 1601.

69. Sheldon Krimsky, *Genetic Alchemy: The Social History of the Recombinant DNA Controversy* (Cambridge, Mass.: MIT Press, 1982), chap. 1; Moore, "Organizing Integrity"; Susan Wright, *Molecular Politics: Developing American and British Regulatory Policy for Genetic Engineering, 1972–1982* (Chicago: University of Chicago Press, 1994), 36–49.

70. After Jonathan Beckwith and his team isolated the first gene in 1969, he and colleague James Shapiro publicly discussed the frightening possibilities raised by the discovery, with Shapiro deciding to leave science due to his concern that the research might be used for immoral purposes. See Sheldon Krimsky, *Biotechnics and Society: The Rise of Industrial Genetics* (New York: Praeger, 1991), 159. Others, such as Ethan Singer of MIT, saw the new science through a

New Left lens. Singer apparently scandalized the Thirtieth Symposium of the Society for Developmental Biology in 1971 by reading a paper—thanking Noam Chomsky for his critical comments—in which he concluded that "the way science is used is determined by the interests and orientation of the dominant social sectors, and there is no reason to expect this to be any less true for gene manipulation than other research" (Ethan Singer, "Gene Manipulation: Progress and Prospects," in *Macromolecules Regulating Growth and Development*, ed. Elizabeth D. Hay, Thomas J. King, and John Papaconstantinou [New York: Academic Press, 1974], 229).

71. Wright, *Molecular Politics*, 37.

72. All of Glass's texts considered here have large sections on the ethics of science. Although appearing to desire consultation with society, in the final analysis he appears to be more or less a technocrat. For example, in his "scientific commandments," scientists have an "obligation" to communicate to the general public "the great new revelations of science, the important advances, the noble syntheses of scientific knowledge." The "third commandment" of science "requires fearlessness in the defense of intellectual freedom, for science cannot prosper where there is constraint upon daring thinking, where society dictates what experiments may be conducted, or where the statement of one's conclusions may lead to loss of livelihood, imprisonment or even death. . . . No doors must be barred to its inquiries, except by reason of its own limitations" (Glass, *Science and Ethical Values*, 95, 90).

73. Ward Madden, "Foreword," in *The Timely and the Timeless: The Interrelationships of Science, Education and Society* (New York: Basic Books, 1970), v–vii.

74. Philip Handler, "Science's Continuing Role," *BioScience* 20, no. 20 (1970): 1102.

75. Ibid., 1103.

76. Although only two of the four influential authors in this community were theologians, among the common authors being a theologian increases the expected number of influential authors cited by 151 percent, compared to other professions, holding other variables constant. This finding is significant at the $p = .047$ level. Also in this community, being a scientist decreases the expected number of influential authors cited by 70 percent, compared to other professions, holding other variables constant. This finding is significant at the $p = .007$ level. The N for both analyses is 69.

77. The theologians involved with the HGE debate have been mainline Protestants and Catholics of a more liberal bent. The extremely limited evangelical voice in these debates did not begin until the mid-1980s. There were only a few Jewish theologians among the almost 100 common authors who were theologians. Although by today's standards Ramsey would probably be considered a conservative theologian, at the time he was within the mainstream of liberal Protestantism.

78. Robert Wuthnow, *The Restructuring of American Religion* (Princeton, N.J.: Princeton University Press, 1988), 145; Harold Quinley, *The Prophetic Clergy: Social Activism among Protestant Ministers* (New York: John Wiley and Sons, 1974); Jeffrey K. Hadden, *The Gathering Storm in the Churches* (Garden City, N.Y.: Doubleday, 1969).

79. Before turning to medical issues, Ramsey was well known for basic theology and just-war theory. There are many summaries of Ramsey's work. See

D. Stephen Long, *Tragedy, Tradition, Transformism: The Ethics of Paul Ramsey* (Boulder, Colo.: Westview Press, 1993); David Attwood, *Paul Ramsey's Political Ethics* (Lanham, Md.: Rowman and Littlefield, 1992); Charles E. Curran, *Politics, Medicine, and Christian Ethics: A Dialogue with Paul Ramsey* (Philadelphia: Fortress Press, 1973); David H. Smith, "On Paul Ramsey: A Covenant-centered Ethic for Medicine," in *Theological Voices in Medical Ethics,* ed. Allen Verhey and Stephen E. Lammers (Grand Rapids, Mich.: William B. Eerdmans, 1993), 7–29; James B. Tubbs, *Christian Theology and Medical Ethics* (Dordrecht: Kluwer Academic Publishers, 1996).

80. Paul Ramsey, *Fabricated Man: The Ethics of Genetic Control* (New Haven, Conn.: Yale University Press, 1970), 138–45. The phrase "playing God" is often misunderstood in bioethical debates. It is often assumed to be a simple prohibitionary statement: those who are "playing God" are doing wrong. This assumes a view of theology that most theologians do not share. Ramsey, for one, wanted human beings to "play God," but only as God plays God, not as a scientistic understanding of the world would have people act. See Allen Verhey, "'Playing God' and Invoking a Perspective," *Journal of Medicine and Philosophy* 20 (1995): 357. For example, in his influential book *The Patient as Person* ([New Haven, Conn.: Yale University Press, 1970], 256) Ramsey claims that because human lives cannot be compared, if only some can survive due to the shortage of a medical resource, "men should then 'play God' in the correct way: he makes his sun rise upon the good and the evil and sends rain upon the just and the unjust alike." Thus, human beings should treat all people, good or bad, the same—as God does.

81. Leon R. Kass, "New Beginnings in Life," in *The New Genetics and the Future of Man,* ed. Michael Hamilton (Grand Rapids, Mich.: William B. Eerdmans, 1972), 52.

82. Leon R. Kass, "The New Biology: What Price Relieving Man's Estate?" *Science* 174 (1971): 785.

83. Beecher's 1966 article critiquing the exclusive use of this end included descriptions of research such as this: "Live cancer cells were injected into 22 human subjects as part of a study of immunity to cancer. According to a recent review, the subjects (hospitalized patients) were 'merely told they would be receiving "some cells"' . . . the word cancer was entirely omitted" ("Ethics and Clinical Research," *New England Journal of Medicine* 274, no. 24 [1966]: 1354–60). A similar and more famous case was an experiment that lasted from the mid-1930s to the early 1970s, in which poor black men in Tuskegee, Alabama, who had syphilis were left untreated by U.S. Public Health Service workers—even after the discovery in 1945 that penicillin would cure the disease—to study the effects of advanced syphilis. See David J. Rothman, *Strangers by the Bedside: A History of How Law and Bioethics Transformed Medical Decision Making* (New York: Basic Books, 1991), 183.

84. Ramsey, *Patient as Person;* Leon R. Kass, "Babies by Means of In Vitro Fertilization: Unethical Experiments on the Unborn?" *New England Journal of Medicine* 285, no. 21 (1971): 1174–79. Both Kass and Ramsey also develop an informed-consent argument against the IVF techniques of Edwards, opposing them not only because they may lead to the violation of the proper role for humans in creation, but also because IVF research is experimenting on not yet existing people without their consent, and in the absence of anyone who could con-

sent for them. See Kass, "Babies by Means of In Vitro Fertilization"; Ramsey, *Fabricated Man,* 134.

85. Fletcher's first work in this area (*Morals and Medicine* [1954]) is what provoked Ramsey to enter these debates for the first time, writing a critical review article on Fletcher's book in 1956. See LeRoy Walters, "Religion and the Renaissance of Medical Ethics in the United States," in *Theology and Bioethics,* ed. E. E. Shelp (Boston: D. Reidel Publishing, 1985), 5. One of Fletcher's most influential books in this era can be read as an extended reaction to the views articulated by Ramsey (Joseph Fletcher, *Situation Ethics: The New Morality* [Philadelphia: Westminster Press, 1966]).

86. Kass, "New Beginnings in Life," 60.

87. Joseph Fletcher, "Ethical Aspects of Genetic Controls," *New England Journal of Medicine* 285, no. 14 (1971): 780–81.

88. The texts of Beecher indicate no interaction with theological ideas. Ramsey is a theologian throughout the period, Fletcher begins the era as a theologian and ends up a bioethicist, and Kass is a scientist. Although Kass is not a theologian and denies being a "student of religion," he makes extensive use of the theology of the time, particularly when compared to other nontheologians in the debate. His interlocutors are not only Ramsey and Fletcher, but C. S. Lewis and Karl Rahner (whom Kass labels a "theologian-turned-technocrat"). Moreover, Kass is "well aware of the debt [he owes] . . . to the ideas and moral teachings of the great religious traditions which have informed our civilization and on which we are nourished, wittingly or not. Only those men stripped of noble sentiment and good sense, men for whom the truth comes only in differential equations or on semi-log paper, will deny that great truth and wisdom can be and often are conveyed in what is for them, at best, myth and story. It is irrational to ignore what reasonableness these stories contain" (Kass, "New Beginnings in Life," 60).

Readers aware of the canonical histories of the bioethics profession will want to label these four the first bioethicists. However, during this era they still belonged to other professions, since bioethics did not yet exist as a profession according to my definition (that is, it did not yet have its own form of argumentation). Moreover, with the exception of Fletcher, none of these authors identified himself with the bioethics profession even after its creation in the late 1970s. For histories of the bioethics profession see Rothman, *Strangers by the Bedside;* Daniel Callahan, "Why America Accepted Bioethics," *Hastings Center Report* (November–December 1993): S8–S9; Callahan, "Bioethics," in *Encyclopedia of Bioethics,* ed. Warren Thomas Reich (New York: Macmillan, 1995), 247–56; Albert R. Jonsen, *The Birth of Bioethics* (New York: Oxford University Press, 1998); Albert R. Jonsen and Andrew Jameton, "Medical Ethics, History of: The Americas," in *Encyclopedia of Bioethics,* ed. Reich; Warren Thomas Reich, "The Word 'Bioethics': Its Birth and the Legacies of Those Who Shaped It," *Kennedy Institute of Ethics Journal* 4, no. 4 (1994): 319–35; Reich, "The Word 'Bioethics': The Struggle over Its Earliest Meanings," *Kennedy Institute of Ethics Journal* 5, no. 1 (1995): 19–34; Walters, "Religion and the Renaissance of Medical Ethics in the United States"; U.S. Congress, Office of Technology Assessment, *Biomedical Ethics in U.S. Public Policy—Background Paper* (Washington, D.C.: Government Printing Office, 1993).

89. Kass, "New Beginnings in Life," 14.

90. Ibid., 16.

91. Ibid. Elsewhere, reacting to one of Muller's phrases, Kass wants to "inquire into the meaning of phrases such as the 'betterment of mankind.'" To assume "betterment," we must know "what is a good man, what is a good life for man, [and] what is a good community" (Kass, "New Biology," 779).

92. Ramsey, *Fabricated Man,* 81.

93. Long, *Tragedy, Tradition, Transformism,* 125–26.

94. Letter, Karen Lebacqz to Alexander Capron, 28 December 1981. Archives of the President's Commission for the Study of Ethical Problems in Medicine and Biomedical and Behavioral Research, National Reference Center for Bioethics Literature, Georgetown University.

95. Smith, "On Paul Ramsey," 8.

96. Fletcher's 1966 *Situation Ethics* is essentially an argument that decisions in medicine and science should follow a consequentialist maximization of agape. There he states that the two principles to be maximized come from the apostle Paul: "The written code kills, but the Spirit gives life" (2 Cor. 3:6), and "For the whole law is fulfilled in one word, 'You shall love your neighbor as yourself'" (Gal. 5:14) (Fletcher, *Situation Ethics,* 30). The first principle refers to Fletcher's rejection of rules and the second to agape. Fletcher would become the first bioethicist in these data. When *Situation Ethics* appeared in 1966, he was still at Episcopal Theological School, but by the time of publication of his 1971 text he was in the Department of Ethics in the School of Medicine at the University of Virginia. Even a cursory examination of the citations in his texts across time shows his removal from theological debate. In a short memoir published after his death, he discusses how in the period between these two texts he had "de-Christianized" himself. Facing "the final realization that social justice was not going to get any significant help from Christian social ethics," he took "a hard look at Christian doctrine itself, on its own merits" and concluded that "the whole thing was weird and untenable." An additional insight is that "along with Christianity went all ideology of any kind," and that he had "reached what is expressed in the title of a book by Daniel Bell, *The End of Ideology.*" Joseph Fletcher, "Memoir of an Ex-Radical," in *Joseph Fletcher: Memoir of an Ex-Radical,* ed. Kenneth Vaux (Louisville, Ky.: Westminster/John Knox, 1993), 85. This embrace at this time of technocracy in the political sense of the word—the use of value-free, objective criteria for decision making, free of politics and ideology—as if his replacement system were somehow neutral, is a telling portent of the technocratic end and value-free direction that the HGE debate would follow.

97. Attwood, *Paul Ramsey's Political Ethics,* 30.

98. Ibid., 29.

99. Ibid., 69. *Deontology* is the philosophical term for what I am calling the substantively rational link between means and ends. In a deontological argument, an act is right or wrong due to the intrinsic qualities of the act itself, not due to its consequences. The contrasting term in philosophy is *consequentialism,* the belief that no act is right or wrong in and of itself, but rather that acts are wrong only if they do not maximize the given ends. *Consequentialism* is then the philosophical term for what I am calling a formally rational link between means and ends.

100. Long, *Tragedy, Tradition, Transformism,* 106.

101. Fletcher, "Ethical Aspects of Genetic Controls," 779. It could be said that Fletcher's entire project in this era was not so much the ethics of HGE and

medical decision making, but rather use of the issue to oppose Ramsey, Kass, and other substantively rational thinkers. All of his texts use the distinction between substantive and instrumental rationality as their guiding source, typically beginning with a claim such as this: "[when] we tackle right-wrong or good-evil or desirable-undesirable questions there are fundamentally two alternative lines of approach. The first one supposes that whether any act or course of action is right or wrong depends on its consequences. The second approach supposes that our actions are right or wrong according to whether they comply with general moral principles or prefabricated rules of conduct. . . . The first approach is consequentialist, the second is a priori. This is the rock-bottom issue, and it is also . . . the definitive question in the ethical analysis of genetic control" (Fletcher, "Ethical Aspects of Genetic Controls," 777).

102. Kass, "New Beginnings in Life," 35.

103. Ramsey, *Fabricated Man,* 22. This chapter in Ramsey's book is reprinted from the proceedings of that conference.

104. Ibid., 27. This is a reference to a traditional Christian eschatology, according to which the Second Coming and the establishment of the Kingdom of God on Earth would "kill" the human species as we know it, re-creating humankind as beings who presumably would not have genetic disease.

105. Ibid., 30 (emphasis in original).

106. For example, he thought that Muller's germinal-choice scheme violated the God-given covenant of marriage. Ibid., 32.

107. Beecher, Kass, and Ramsey all spent a good amount of time writing on the ethics of research with human subjects. They argued that the means should be evaluated not by whether they maximized the ends, but by whether they were consistent with the ends—in this case the end that was eventually called autonomy, or "respect for persons." From this end would flow the requirement of the informed consent of research subjects first argued for in the post-1960s era by Beecher.

108. Here Ramsey inserts a footnote stating that for an example of someone who thinks that ethics should provide the rationalization for what scientists want to do, the reader should see a particular page in the Wolstenholme volume. On that page, Julian Huxley is commenting on the proposal by Lederberg for various forms of biological control. Huxley, addressing the problem that the ethics of the population is out of touch with the scientists' means, states that "at the moment the population certainly wouldn't tolerate compulsory eugenic or sterilization measures, but if you start some experiments, including some voluntary ones, and see that they work and if you make a massive attempt at educating people and making them understand what is at issue, you might be able, within a generation, to have an effect on the general population. After all, our moral values evolve like everything else and they evolve largely on the basis of the knowledge we have and share" (Wolstenholme, *Man and His Future,* 290).

109. Ramsey, *Fabricated Man,* 122–23. Kass held a similar view: "To deny that rationality might dictate that there are some things we can do that we must never do . . . can only be regarded as the height of folly. . . . In the absence of that 'ultimate wisdom,' we can be wise enough to know that we are not wise enough. When we lack sufficient wisdom to do, wisdom consists in not doing" ("New Beginnings in Life," 61–62).

110. On these scientists as popularizers, see Ruse, *Monod to Man,* 408. Note, for example, that Medawar's most influential book was actually first read

over the radio in Great Britain. Many of the rest of the texts considered in this chapter were written for public symposia at universities. For example, the most influential books of Dobzhansky and Dubos were first given at a public symposium at Yale; Glass's texts were presented to the National Society of College Teachers of Education and as a lecture series at the University of North Carolina.

111. See the quotation from Huxley in note 108, above.

112. Bentley Glass, "The Ethical Basis of Science," *Science* 150 (1965): 1255.

113. Sonneborn, ed., *Control of Human Heredity and Evolution,* viii.

114. Fletcher, "Ethical Aspects of Genetic Controls," 782; Kass, "Babies by Means of In Vitro Fertilization," 1178.

115. Michael Hamilton, *The New Genetics and the Future of Man* (Grand Rapids, Mich.: William B. Eerdmans, 1972).

Chapter 3

1. Ramsey was cited 25 times out of a total of 71 texts. That is, fully 35 percent of the texts referred to his work. Muller was cited 14 times in this period by the 71 texts. See table A-3.

2. The members of this theological community average 18.3 citations per author, while the other communities average 14.7, 13.8, 12, and 16 citations per author. Being a theologian increases the expected number of influential authors cited in this community by 91 percent, compared to members of other professions, holding the other variables constant. This finding is significant at the p = .014 level. Being a scientist decreases the expected number of influential authors cited in this community by 68 percent, compared to members of other professions, holding the other variables constant. This finding is significant at the p = .016 level. These analyses have an N of 71 common authors. Models for each of the other professions under examination show no significant differential representation in this community by other professions.

3. Examination of the texts from the common members of the community reinforces the reading of the texts from the influential members. Claims about the content of the texts of the common members of a community are made using an analysis of keywords that were coded for each text by a team of reference librarians for the National Institute of Medicine. This type of data, when used by sociologists of science, is called qualitative scientometrics or co-word analysis. See Michel Callon, John Law, and Arie Rip, *Mapping the Dynamics of Science and Technology: Sociology of Science in the Real World* (London: Macmillan, 1986); J. Law et al., "Policy and the Mapping of Scientific Change: The Co-Word Analysis of Research into Environmental Acidification," *Scientometrics* 14, nos. 3–4 (1988): 251–64; John Whittaker, "Creativity and Conformity in Science: Titles, Keywords and Co-Word Analysis," *Social Studies of Science* 19 (1989): 473–96; Jean-Pierre Courtial and John Law, "A Co-Word Analysis of Artificial Intelligence," *Social Studies of Science* 19 (1989): 301–11. Further discussion of how the keywords are used in this study can be found in the appendix.

Compared to the authors in other communities, the common members disproportionately discussed "informed consent," the means for forwarding the end of "respect for persons." This claim is made using a very similar method to that used to demonstrate that a community is disproportionately composed of members of a particular profession. The dependent variable is the number of influential authors in a community whom a common author cites. The independent variable of interest is whether the text that is doing the citing has the keyword coded onto it; the negative binomial regression models control for the total number of texts that were cited and the total number of keywords coded onto the text, because both of these measures are structured by profession. Models are Poisson, when appropriate. Keywords are coded 0, 1, or 2 (0 if the topic is not found in the text, 1 if it is present but of lesser importance to the text, 2 if the topic is central to the text). Thus, for this analysis, a one-point difference in this keyword coding scale for "informed consent" means that the expected number of influential authors in this community that the text will cite increases 47 percent (100*[exp(.383) − 1]). The finding is significant at the $p = .056$ level, with N of 66.

The common members also disproportionately were *not* using "risk/benefit analysis," the formally rational method of determining whether the means maximize the end. A one-point difference in this keyword coding scale for "risk/benefit analysis" means that the expected number of influential authors in this community that the text will cite decreases 37 percent. The finding is significant at the $p = .047$ level, with N of 66. The details of these analyses are available from the author.

4. See figure 3. The community has as its most influential authors R. Edwards, M. Lappé, W. F. Anderson, T. Friedmann, B. Davis, L. Kass, and J. Watson.

5. Bernard D. Davis, "Prospects for Genetic Intervention in Man," *Science* 170 (1970): 1279. He cites Medawar's *Future of Man,* the *Daedalus* issue with Muller's and Crow's texts, Wolstenholme's edited volume, Lederberg's articles, Huxley's *Essays of a Humanist,* and Sonneborn's edited volume.

6. Diseases such as Tay-Sachs are the result of a mistake in one gene that leads to a particular enzyme needed for cellular function not being produced. The cells therefore do not work properly, resulting in the manifestations of these dysfunctions we call disease—in the case of Tay-Sachs, the destruction of the brain. Other traits that are supposedly the result of genes such as "intelligence" are probably not the result of one gene. Lack of "intelligence" and the other traits that concerned the reform eugenicists would presumably be due to the lack of some other proteins in the cells of the brain, which combine in some manner to enhance "intelligence." How this would work is not clear, even today.

7. This "objective" analysis is his move toward considering only monogenetic traits as relevant to the HGE debate. The second half of the paper is the now commonplace defense of scientific freedom against the challengers; to "defend vigorously the value of objective and verifiable knowledge, especially when it comes into conflict with political, theological or sociological dogmas." See Davis, "Prospects for Genetic Intervention in Man," 1279, 1283.

8. Bernard Davis, "Novel Pressures on the Advance of Science," *Annals of the New York Academy of Sciences* 265 (1976): 195.

9. Ibid., 198.

10. The sense of threat to the home jurisdiction pervades this community. For example, Robert Edwards, the IVF pioneer discussed in chapter 2, writes about the threats to "major social values." After considering the first value—"the sanctity of life"—he focuses the rest of the paper on the other major social value, "independence of scientific inquiry," which scientists must go on the counter-attack to defend. The implication here is that scientists should not heed the will of the people, but should become active in politics "in the hope that the

attitudes of society . . . will mature at a rate not too far behind the transition of scientific discovery into technological achievement." He states that "what is to be feared is that if the biologists do not invent a method of taking counsel of mankind, society will thrust its advice on biologists and other scientists and probably in a manner or form seriously hampering to science." See Robert G. Edwards and David J. Sharpe, "Social Values and Research in Human Embryology," *Nature* 231 (14 May 1971): 89–90.

11. Theodore Friedmann and Richard Roblin, "Gene Therapy for Human Genetic Disease?" *Science* 175 (1972): 949–55.

12. H. Vasken Aposhian, "The Use of DNA for Gene Therapy—The Need, Experimental Approach, and Implications," *Perspectives in Biology and Medicine* 14 (1970): 98–108. Sinsheimer also uses the term *gene therapy* in a 1970 article. See Robert L. Sinsheimer, "Genetic Engineering: The Modification of Man," *Impact of Science on Society* 20, no. 4 (1970): 279–91.

13. Aposhian, "Use of DNA for Gene Therapy," 106–7.

14. Anderson's attempt at HGE occurred in 1990. See LeRoy Walters and Julie Gage Palmer, *The Ethics of Human Gene Therapy* (New York: Oxford University Press, 1997), 18. The first attempt at HGE occurred in the early 1970s with techniques that would be considered naive by today's standards. See Paul Ramsey, "Genetic Therapy: A Theologian's Response," in *The New Genetics and the Future of Man,* ed. Michael Hamilton (Grand Rapids, Mich.: William B. Eerdmans, 1972), 162. The second attempt was in 1980 by Martin Cline at UCLA who, after being refused permission to conduct the experiment by the UCLA Institutional Review Board, went to Israel and Italy to conduct his experiments, which apparently had no negative or positive effect. Cline was sanctioned by NIH, which was funding the experiments, for both ignoring UCLA officials and not demonstrating the safety and efficacy of his technique with animals first. See President's Commission, *Splicing Life: A Report on the Social and Ethical Issues of Genetic Engineering with Human Beings* (Washington, D.C.: Government Printing Office, 1983), 44.

15. W. French Anderson, "Genetic Therapy," in *New Genetics and the Future of Man,* ed. Hamilton, 109.

16. Ramsey, "Genetic Therapy," 158–59.

17. Ibid., 159.

18. Ibid.

19. For example, one of his fundamental ends that Ramsey wants to debate is whether science should perfect humanity. As might be expected from what we have seen of his arguments so far, Ramsey says no. Consider this lively passage: "Dostoevsky's hope—against environmental perfectionists who would rationalize all suffering away—was that an undergroundling would come to stick out his tongue at living in a Crystal Palace or being reduced to the handle on a hurdy-gurdy or a key on the keyboard of a piano even if they played a merry social tune. Against genetic perfectionists who would rationalize human suffering away, the health of the human race may require an undergroundling who will stick out his tongue at genetic screening that aims to tell us all the mistakes that we are" (ibid., 173).

20. Ibid., 158.

21. Friedmann and Roblin, "Gene Therapy for Human Genetic Disease?" 953.

22. See note 3, above, for an explanation of this method. The more one is em-

bedded in this community, the more likely he is to discuss "gene therapy," "human experimentation," and "informed consent." A one-point difference in the coding scale for these keywords means that the expected number of influential authors in this community that the text will cite increases 50 percent, 44 percent, and 72 percent, respectively. The findings are significant at the $p = .015$, .062 and .036 levels, respectively. *N* is 66. The keyword "risk/benefit analysis," indicating the formally rational maximization of beneficence, fell just shy of significance.

23. Leon Kass, whose work was discussed in chapter 2, is now found in the community under discussion. In his 1971 text critiquing Edwards's work, he suggests that although scientists might first try self-regulation, "one must question the wisdom of leaving the decision to go ahead for the private judgement of a team of physicians and scientists. . . . or even for the collective judgement of the medical and scientific community. Is this not a matter that deserves broader public deliberation and, in the end, might be one for public decision?" See Leon R. Kass, "Babies by Means of In Vitro Fertilization: Unethical Experiments on the Unborn?" *New England Journal of Medicine* 285, no. 21 (1971): 1178.

24. Albert R. Jonsen, *The Birth of Bioethics* (New York: Oxford University Press, 1998), 90.

25. Ibid., 91.

26. Ibid., 93.

27. Ibid., 94.

28. "To protect rights and ensure policy responsiveness, contemporary liberal democracy endows elected authorities with powers and devices sufficiently strong to control administrative and regulatory agencies. In the twentieth century, legislative and executive authority over an agency's budget—the proverbial 'power of the purse'—has emerged as one of the most forceful and frequently exercised tools of political control" (Daniel P. Carpenter, "Adaptive Signal Processing, Hierarchy, and Budgetary Control in Federal Regulation," *American Political Science Review* 90, no. 2 [1996]: 283).

29. Edwards and Sharpe, "Social Values and Research in Human Embryology," 89, 90 (emphasis added).

30. Moreover, Anderson is opposed to a public regulatory body because "such a concentrated localization of authority could not be used wisely in the field of genetics at this time." See Anderson, "Genetic Therapy," 120. However, as Ramsey points out, if the problem is the inability of a small group to make such decisions, this would also seem to exclude the small coterie of scientists at the National Institutes of Health and the National Science Foundation who currently make such decisions, and who would continue to have the power to make such decisions under Anderson's laissez-faire system. See Ramsey, "Genetic Therapy," 165.

31. Anderson, "Genetic Therapy," 122. Edwards as well thinks that an advisory commission would "probably take a fairly conservative line as lay attitudes struggle to catch up with what the scientists can do" (Edwards and Sharpe, "Social Values and Research in Human Embryology," 90).

32. Ramsey, "Genetic Therapy," 166.

33. David Dickson, *The New Politics of Science* (New York: Pantheon Books, 1984), 220.

34. See chapter 2, note 83.

35. For other scholars' accounts of the birth of bioethics, see David J. Roth-

man, *Strangers by the Bedside: A History of How Law and Bioethics Transformed Medical Decision Making* (New York: Basic Books, 1991); Daniel Callahan, "Why America Accepted Bioethics," *Hastings Center Report* (November–December 1993): S8–S9; Albert R. Jonsen, "The Birth of Bioethics," *Hastings Center Report* (November–December 1993): S1–S4; Jonsen, "Foreword," in *A Matter of Principles? Ferment in U.S. Bioethics,* ed. Edwin R. DuBose, Ronald P. Hamel, and Laurence J. O'Connell (Valley Forge, Pa.: Trinity Press International, 1994), ix–xvii; Jonsen, "American Moralism and the Origin of Bioethics in the United States," *Journal of Medicine and Philosophy* 16 (1991): 113–30; Albert R. Jonsen and Andrew Jameton, "Medical Ethics, History of: The Americas," in *Encyclopedia of Bioethics,* ed. Warren Thomas Reich (New York: Macmillan, 1995); Jonsen, *Birth of Bioethics;* U.S. Congress, Office of Technology Assessment, *Biomedical Ethics in U.S. Public Policy—Background Paper* (Washington, D.C.: Government Printing Office, 1993); Warren Thomas Reich, "The Word 'Bioethics': Its Birth and the Legacies of Those Who Shaped It," *Kennedy Institute of Ethics Journal* 4, no. 4 (1994): 319–35; Reich, "The Word 'Bioethics': The Struggle over Its Earliest Meanings," *Kennedy Institute of Ethics Journal* 5, no. 1 (1995): 19–34; LeRoy Walters, "Religion and the Renaissance of Medical Ethics in the United States," in *Theology and Bioethics,* ed. E. E. Shelp (Boston: D. Reidel Publishing, 1985), 3–16; Daniel Callahan, "Bioethics," in *Encyclopedia of Bioethics,* ed. Reich, 247–56.

36. Rothman, *Strangers by the Bedside,* 188.

37. Conversation with Charles McCarthy, director of the Ethics Advisory Board, 1978–1980, at the "Belmont Revisited" conference, Charlottesville, Va., 16 April 1999. See also Tom L. Beauchamp, "How the Belmont Report Was Written," paper presented at the "Belmont Revisited" conference, 21.

38. Cited in Jonsen, "Foreword," xiv.

39. Ibid., xvi (emphasis added).

40. Jonsen, *Birth of Bioethics,* 103. "Respect for persons" evolved into "autonomy." See Karen Lebacqz, "Twenty Years Older But Are We Wiser?" paper presented at the "Belmont Revisited" conference.

41. National Commission for the Protection of Human Subjects of Biomedical and Behavioral Research, *The Belmont Report: Ethical Principles and Guidelines for the Protection of Human Subjects of Research* (Washington, D.C.: Government Printing Office, 1978), 10.

42. Jonsen, *Birth of Bioethics,* 103.

43. The principles are often thought to be a compromise between two different ways of linking means and ends. Autonomy and justice are taken to be deontological, substantively rational ends, which any means must be consistent with. Beneficence and nonmaleficence are taken to be consequentialist, formally rational ends, where means that maximize these ends are selected. Bioethicists are working on the problem that the different ends in their form of argumentation conflict, and there is no set manner of working out these conflicts. In essence, they are trying to make their system more calculable. For a summary of the positions in the debate over how to make the principles commensurable with each other, see Robert M. Veatch, "Resolving Conflicts among Principles: Ranking, Balancing, and Specifying," *Kennedy Institute of Ethics Journal* 5, no. 3 (1995): 199–218.

44. This is the interpretation of James Laney, "The New Morality and the Re-

ligious Communities," *Annals of the American Academy of Political and Social Science* 387 (1970): 18.

45. Ibid., 19.

46. Neo-scholasticism in Catholic ethics produced a stream of moral manuals that, according to one account, have "a great air of security and certainty," reflecting "a confident understanding of the identity of Christian morality." To oversimplify, the whole point is to make sure people know the rules to get to heaven. See Vincent MacNamara, *Faith and Ethics: Recent Roman Catholicism* (Washington, D.C.: Georgetown University Press, 1985), 9, 10.

47. Josef Fuchs, writing in 1970, quoted in ibid., 40.

48. Gustav Ermecke, writing in 1972, quoted in ibid., 55.

49. Edward LeRoy Long, *A Survey of Recent Christian Ethics* (New York: Oxford University Press, 1982), 3.

50. Courtney S. Campbell, "On James F. Childress: Answering That God in Every Person," in *Theological Voices in Medical Ethics,* ed. Allen Verhey and Stephen E. Lammers (Grand Rapids, Mich.: William B. Eerdmans, 1993), 127 (emphasis in original).

51. Ramsey denied the legitimacy of this universalizing move in a letter to theologian Stanley Hauerwas in 1978: "I certainly do not believe that 'it is still possible to do Christian ethics for those who do not share Christian convictions' (was it ever?); or that 'there is no difference between the best in our culture and Christianity'" (quoted in D. Stephen Long, *Tragedy, Tradition, Transformism: The Ethics of Paul Ramsey* (Boulder, Colo.: Westview Press, 1993), 207.

52. Jonsen, "Foreword," xii.

53. Stephen Toulmin, "How Medicine Saved the Life of Ethics," *Perspectives in Biology and Medicine* 25, no. 4 (1982): 736–50.

54. Robert J. Henle, S.J., in Warren Thomas Reich, "Revisiting the Launching of the Kennedy Institute: Re-visioning the Origins of Bioethics," *Kennedy Institute of Ethics Journal* 6, no. 4 (1996): 324.

55. Ibid., 324; Reich, "The Word 'Bioethics': The Struggle over Its Earliest Meanings," 30–31.

56. Callahan, "Why America Accepted Bioethics," S8. See also Stephen E. Lammers, "The Marginalization of Religious Voices in Bioethics," in *Religion and Medical Ethics: Looking Back, Looking Forward,* ed. Allen Verhey (Grand Rapids, Mich.: William B Eerdmans, 1996), 19–43.

57. H. Tristram Engelhardt, *The Foundations of Bioethics* (New York: Oxford University Press, 1986), 5.

58. U.S. Department of Health and Human Services, "Protection of Human Subjects," *Code of Federal Regulations,* Part 46, Subtitle A (1981).

59. John C. Fletcher, "Evolution of Ethical Debate about Human Gene Therapy," *Human Gene Therapy* 1 (1990): 58; Fletcher, "Moral Problems and Ethical Issues in Prospective Human Gene Therapy," *Virginia Law Review* 69 (1983): 524.

60. Jonsen, *Birth of Bioethics,* 103–4.

61. Reich, "The Word 'Bioethics': The Struggle over Its Earliest Meanings," 22.

62. Tom L. Beauchamp and James F. Childress, *Principles of Biomedical Ethics* (New York: Oxford University Press, 1979).

63. K. Danner Clouser and Bernard Gert, "A Critique of Principlism," *Journal of Medicine and Philosophy* 15 (1990): 219.

64. Edwin R. DuBose, Ronald P. Hamel, and Laurence J. O'Connell, "Introduction," in *A Matter of Principles? Ferment in U.S. Bioethics,* ed. DuBose, Hamel, and O'Connell (Valley Forge, Pa.: Trinity Press International, 1994), 1.

65. Clouser and Gert, "Critique of Principlism," 219. Ironically, Clouser and Gert are advocating an even more formally rational system that has only one universal commensurable end.

66. Robert M. Veatch, *A Theory of Medical Ethics* (New York: Basic Books, 1981); Engelhardt, *Foundations of Bioethics.*

67. For a summary of these two approaches as varieties of principlism, see Jonsen, *Birth of Bioethics,* 329–31.

68. For example, Raymond Devettere talks about the "misuse of principles," noting that "unfortunately, the following caricature is all too widespread in clinical ethics: 'The ethics best suited for medicine and health care is applied normative ethics. Applied normative ethics is the formulation and defense of a system of norms—the principles and rules—to determine which actions are right and which actions are wrong. The application of these action-guides to various ethical problems allows people to determine the morally right action, and this determination establishes a moral obligation.' How did it happen that so many believe this is just about all there is to health care ethics? No serious moral philosopher or theologian advocating principles and rules reduces ethics to this—neither philosophers such as the Stoics, Cicero, Philo, Aquinas, Kant, and Mill, or such bioethicists as Beauchamp, Childress, Veatch, Engelhardt, Brody, Pellegrino and Brock. All claim that normative principles and rules are only one aspect of ethics." Devettere thinks that the average bioethicist thinks that principles are all that matter because theorists do a poor job of explaining the other parts of ethics, because principles are simple to apply, because some people like the certainty of rulelike principles, and because, as I identify, the government has adopted principlism. See Raymond Devettere, "The Principled Approach: Principles, Rules and Actions," in *Meta Medical Ethics: The Philosophical Foundations of Bioethics,* ed. Michael A. Grodin (Dordrecht: Kluwer Academic Publishers, 1995), 35–39.

69. Rothman, *Strangers by the Bedside,* 89, 107. "Patients' rights" are derived from autonomy or respect for persons.

70. Daniel Callahan, "At the Center," *Hastings Center Report* (June 1982): 4.

71. Reich, "The Word 'Bioethics': The Struggle over Its Earliest Meanings," 22.

72. Ibid., 23. Clearly, Reich also believes that the charisma of this first director was partially responsible for this success, and not simply that the principles were particularly well suited for the state and other bureaucratic forms.

73. Daniel M. Fox, "View the Second," *Hastings Center Report* 23, no. 6 (1993): S13.

74. The form of argumentation developed at Georgetown "introduced a notion of bioethics that would deal with *concrete medical dilemmas* restricted to three issue-areas: (1) the rights and duties of patients and health professionals; (2) the rights and duties of research subjects and researchers; and (3) the formulation of public policy guidelines for clinical care and biomedical research" (Reich, "The Word 'Bioethics': The Struggle over Its Earliest Meanings," 20).

75. This is the second community from the bottom in figure 3. Being a bioethicist increases the expected number of influential authors cited in this community by 100 percent, compared to members of other professions, holding the other variables constant. This finding is significant at the $p = .032$ level, with N

of 71 common authors. Models for each of the other professions under examination show no significant differential representation in this community by other professions.

That Muller and Dobzhansky are leaders in this community indicates that the concern about genetic load, the gene pool, and control of the genetic destiny of the population was still being debated here. It is also telling that they are not co-cited enough with the authors in the community discussing "gene therapy" to be in that community, which in turn shows how quickly those authors were moving away from the arguments used by Muller, Dobzhansky, and their colleagues.

The more a common author in this era is embedded in this community, the more likely he or she is to discuss "genetic counseling" and "genetic disorders." A one-point difference in the coding scale for these keywords means that the expected number of influential authors in this community that the text will cite increases 65 percent and 69 percent respectively. The findings are significant at the $p = .008$ and .001 levels, respectively. N is 66.

76. The screening and counseling movement really gained adherents with the eclipse of probabilistic predictions about potential offspring and the rise of the near certain predictions made available through amniocentesis. By the late 1960s amniocentesis was no longer an experimental procedure and was in widespread use. By the mid-1970s parents could know with a high degree of certainty whether or not their child would have Tay-Sachs or twenty-three other metabolic disorders, as well as Down syndrome and a hundred other chromosomal disorders. With the liberalization of abortion laws in the late 1960s to include fetal disorders, and the placing of the decision in the hands of the pregnant woman following the Supreme Court's *Roe v. Wade* decision in 1973, amniocentesis became a real possibility in the United States for the selection of the genetic qualities of a person's offspring. In 1960 there were between thirty and forty clinics and counseling centers in the United States; by 1974 there were about four hundred. See Daniel Kevles, *In the Name of Eugenics: Genetics and the Uses of Human Heredity* (Berkeley: University of California Press, 1985), 255–57.

77. At the time he wrote the texts under examination here, Fletcher was an Episcopal priest, the director of an experimental seminary in Washington, D.C., promoting experience-based training for clergy, and by my categorization a theologian. His writing has more the feel of sociology of religion than theology, in that his empirical research is concerned with medical interaction, with an open ear toward the participants' religious concerns. His work is highly universalistic, making almost no references to any religious sources. Like many of the early theologians in these debates whose writings are not tightly connected to their particular tradition, by the next time period under consideration Fletcher had become a bioethicist. This is not to say that everyone who switched from theology to bioethics was no longer religious or no longer saw himself as serving the theological tradition. Like Joseph Fletcher, however, John Fletcher had a secularization experience after becoming a bioethicist. As he told a conference in 1998, "in the late eighties, after thirty-five years of trying to hold together the beliefs of Christian theology and modern biology, I gave up the struggle and resigned from the Episcopal ministry and became a friendly critic of religion" (Gregory Stock and John Campbell, *Engineering the Human Germline* [New York: Oxford University Press, 2000], 74).

78. John Fletcher, “The Brink: The Parent-Child Bond in the Genetic Revolution,” *Theological Studies* 33, no. 3 (1972): 484.

79. Ibid., 480.

80. Clearly, autonomy was rising as an end not only within bioethics but in society more generally, as exemplified by the *Griswold v. Connecticut* (1965) and *Roe v. Wade* (1973) decisions of the Supreme Court, which held autonomy to be the principal end to forward in decisions about birth control and abortion. It is then easy to see why the National Commission would think that autonomy was a basic, universal end of society, and that adopting this end would give bioethicists an advantage. Of the influential authors in this time period, it was the members of the theological debating community who were the most skeptical of autonomy as a universal end for all decision making.

81. Susan Wright, *Molecular Politics: Developing American and British Regulatory Policy for Genetic Engineering, 1972–1982* (Chicago: University of Chicago Press, 1994), 71.

82. Ibid., 75.

83. Ibid., 130.

84. Both letters are cited in Wright’s canonical book on this debate. Ibid., 135.

85. Ibid., 136.

86. Ibid., 137.

87. Paul Berg, “Potential Biohazards of Recombinant DNA Molecules,” *Science* 185 (1974): 303.

88. Wright, *Molecular Politics,* 140.

89. Organizers tried to limit press coverage by allowing only eight reporters, requiring that reporters not file reports until after the end of the meeting, and restricting release of the official tapes of the meeting until the year 2000. After threats of a lawsuit by the *Washington Post,* restrictions were eased. Several of the reporters present at the meeting later concluded that “Asilomar was intended to avoid public involvement rather than to encourage it” (Wright, *Molecular Politics,* 146, 151).

90. Sheldon Krimsky, *Biotechnics and Society: The Rise of Industrial Genetics* (New York: Praeger, 1991), 100.

91. Wright, *Molecular Politics,* 153.

92. Robert Sinsheimer, “Troubled Dawn for Genetic Engineering,” *New Scientist* 16 (1975): 150. Between his first statements in the 1960s extolling the glorious possibilities of HGE and the mid-1970s, Sinsheimer appears to have changed his mind. He became one of the few scientific voices to question the ends to which HGE was being put, and apparently paid for this position by being shunned by his colleagues in later years. See Robert Sinsheimer, *The Strands of a Life: The Science of DNA and the Art of Education* (Berkeley: University of California Press, 1994).

93. Quoted in Wright, *Molecular Politics,* 148–49.

94. Sinsheimer, “Troubled Dawn for Genetic Engineering,” 150.

95. Dickson, *New Politics of Science,* 246.

96. Wright, *Molecular Politics,* 165.

97. Within the HGE debate, a separate debating community arose, focused on the relationship between HGE and the Asilomar controversy, with Berg’s 1974 letter as the most influential text. Along with Berg in this community are Lederberg and Sinsheimer. Sinsheimer’s texts are critical of the decisions made at

Asilomar. Lederberg's most influential texts precede this controversy, but the sum of his more prevalent and less influential texts—which are supportive of the Asilomar consensus—result in his being placed in this debating community rather than another. A separate community consisting only of science journalist Nicholas Wade is considered simultaneously because all of Wade's texts are popular descriptions of the events in the biohazard controversy. While the community that has only Wade as its influential author is not structured by profession, the community with Lederberg, Sinsheimer, and Berg is disproportionately composed of scientists. Being a scientist increases the expected number of influential authors cited in this community by 96 percent, compared to members of other professions, holding the other variables constant. This finding is significant at the $p = .091$ level, with N of 71 common authors. Models for each of the other professions under examination show no significant differential representation in this community by other professions.

98. The interested reader is referred to Wright's text or the others for description of this process. See Wright, *Molecular Politics;* Sheldon Krimsky, *Genetic Alchemy: The Social History of the Recombinant DNA Controversy* (Cambridge, Mass.: MIT Press, 1982); Clifford Grobstein, *A Double Image of the Double Helix: The Recombinant DNA Debate* (San Francisco: W. H. Freeman, 1979); Dickson, *New Politics of Science,* 243–60.

Chapter 4

1. The leaders were Rev. Dr. Claire Randall (National Council of Churches), Rabbi Bernard Mandelbaum (Synagogue Council of America), and Bishop Thomas Kelly (U.S. Catholic Conference).

2. The full text of their letter can be found in President's Commission, *Splicing Life: A Report on the Social and Ethical Issues of Genetic Engineering with Human Beings* (Washington, D.C.: Government Printing Office, 1983), app. B.

3. It has been found that one of the primary reasons presidential commissions exist is to offer political cover for difficult decisions. See Terrence R. Tuchings, *Rhetoric and Reality: Presidential Commissions and the Making of Public Policy* (Boulder, Colo.: Westview Press, 1979); David Flitner, *The Politics of Presidential Commissions* (Dobbs Ferry, N.Y.: Transnational Publishers, 1986).

4. The National Commission ended operation in 1978. The statutory authority for the President's Commission was enacted in 1978; it first met in January 1980. The legislation creating the President's Commission mandated that there be eleven volunteer commissioners, appointed by the president, who were "distinguished" in a variety of professions. Three members were to be from biomedical and behavioral research, three from medicine, and five from the following fields: ethics, theology, law, social science, humanities, health administration, government, public affairs, and nonbiomedical or behavioral natural science. See President's Commission, *Summing Up* (Washington, D.C.: Government Printing Office, 1983), 4. The commission was mandated by Congress to prepare reports on topics such as the definition of death, informed consent for medical care, and the availability of health care services. HGE was not a statutorily required topic, but one taken on by the commission after it had begun to deliberate other issues. In all, the commission produced nine substantive reports and one summary report that were printed by the Government Printing Office and widely distributed. See Bradford H. Gray, "Bioethics Commissions: What Can We Learn from Past Successes and Failures?" in *Society's Choices: Social and*

Ethical Decision Making in Biomedicine, ed. Ruth Ellen Bulger, Elizabeth Meyer Bobby, and Harvey V. Fineberg (Washington, D.C.: National Academy Press, 1995), 262.

The commissioners met a total of twenty-eight times over a period of a little more than three years, typically monthly, to review the work of staff and paid consultants in the preparation of reports and to debate the topics under study. See President's Commission, *Summing Up,* 9. The commission had approximately nine professional staff members during its tenure, led by the executive director, who was chosen by the chair. Unlike the predecessor commission, the President's Commission had a "strong staff" that "probably had stronger professional credentials than did the commission itself" for examining the topics under consideration. This has been attributed to the hiring of lawyer Alexander Capron as the executive director; Capron then hired staff members who had made careers of studying the issues on the commission's agenda. See Gray, "Bioethics Commissions," 269.

The commissioners had limited terms, and many who were appointed by President Carter were replaced as their terms expired by new commissioners appointed by President Reagan. There is a small academic literature that analyzes the decisions of the commission and its processes, although the HGE report is generally not discussed in these texts. There is consensus that the Reagan appointees transformed the positions of the commission on a few issues, most notably in the report on access to health care, which was made more conservative. See Ronald Bayer, "Ethics, Politics, and Access to Health Care: A Critical Analysis of the President's Commission for the Study of Ethical Problems in Medicine and Biomedical and Behavioral Research," *Cardozo Law Review* 6 (1984): 303–20. There was also consensus among the persons whom I interviewed that HGE was not a priority for the Reagan commissioners and that they did not try to change it because it did not have clear right-left ideological overtones. Examination of the meeting transcripts reveals that newly appointed Reagan commissioners said almost nothing about the HGE report, which was almost finished by the time they joined.

5. Capron testimony, Hearings before the Subcommittee on Investigations and Oversight of the Committee on Science and Technology, 97th Cong., 2d sess., 16–18 November 1982, 150.

6. See Alexander Morgan Capron, "The Impact of the Report, Splicing Life," *Human Gene Therapy* 1 (1990): 70 (emphasis added).

7. It received 52 citations from the 180 texts examined in the 12 years from the time it was published in 1983 until 1995.

8. Capron, "Impact of the Report," 70.

9. Although the commission wanted the group to be independent of NIH, this did not occur due to the quick response of people at NIH. See chapter 5, below; also President's Commission, *Splicing Life,* 4.

10. John C. Fletcher, "Evolution of Ethical Debate about Human Gene Therapy," *Human Gene Therapy* 1 (1990): 61. This committee is now called the Human Gene Therapy Subcommittee.

11. LeRoy Walters, "Human Gene Therapy: Ethics and Public Policy," *Human Gene Therapy* 2 (1991): 116.

12. Elizabeth A. Milewski, "Development of a Points to Consider Document for Human Somatic Cell Gene Therapy," *Recombinant DNA Technical Bulletin* 8, no. 4 (1985): 176–77. Moreover, then Rep. Albert Gore held three days of

congressional hearings on the topic of HGE one week after the final approval of *Splicing Life*. The report was read into the congressional record. The testimony of the executive director was a prominent part of the proceedings, as he reiterated the main points of the commission's conclusions. Hearings before the Subcommittee on Investigations and Oversight of the Committee on Science and Technology, 97th Cong., 2d sess., 16–18 November 1982.

13. Alexander Morgan Capron, "Looking Back at the President's Commission," *Hastings Center Report* (October 1983): 10.

14. Ibid., 9–10.

15. Flitner, *Politics of Presidential Commissions,* 17.

16. Jonathan D. Moreno, *Deciding Together: Bioethics and Moral Consensus* (New York: Oxford University Press, 1995), 57–58.

17. Gray, "Bioethics Commissions"; Capron, "Looking Back at the President's Commission"; Morris B. Abram and Susan M. Wolf, "Public Involvement in Medical Ethics," *New England Journal of Medicine* 310 (1984): 627–32. Consensus was also sought by the President's Commission because, like most presidential commissions, it actually had no power. See Flitner, *Politics of Presidential Commissions;* Tuchings, *Rhetoric and Reality.* In the words of its chair, Morris Abram, the commission had only the "power of persuasion," and without coercive powers, it could not be "persuasive without internal agreement." See Abram and Wolf, "Public Involvement in Medical Ethics," 629.

18. Moreno, *Deciding Together,* 45.

19. Capron, "Looking Back at the President's Commission," 8.

20. "It was, remember, the beginning of the commission, people were geared up [and were] really intellectually interested, excited about doing things, tons of energy to work, ready to take on the world. And so I think it was seen as very interesting" (Schapiro interview, 11). (See appendix, p. 214, for the list of interviews conducted.)

21. Schapiro interview, 7.

22. Meeting no. 3 transcript, 369, Archives of the President's Commission for the Study of Ethical Problems in Medicine and Biomedical and Behavioral Research, National Reference Center for Bioethics Literature, Georgetown University (hereafter cited as President's Commission Archives).

23. Motulsky interview, 10.

24. Meeting no. 3 transcript, 372. Whether "calming the fears" of the public was in the public's own interest at that time will never be known because the commission took on the case contingent on the conclusion being reassuring.

25. Schapiro interview, 8.

26. Alan Weisbard, assistant director for law (Weisbard interview, 8). Motulsky himself often talked in this way. For example, in the fourth meeting of the commission he countered a claim about genetic effects by starting his argument with "speaking for the biomedical community" (meeting no. 4 transcript, 296).

27. Weisbard interview, 8.

28. Schapiro interview, 11.

29. Schapiro interview, 11. Weisbard shared Schapiro's belief, stating that "it was unlikely that the ultimate report would be approved without Arno being comfortable with it" (Weisbard interview, 8). How could Motulsky, who was only one of the three statutorily assigned biomedical or behavioral researchers, and only one of eleven commissioners, have such influence? First, given the desire of this commission to operate under consensus, any strong-willed commis-

sioner would have to be deferred to. In particular, a dissent on a genetics report from the one geneticist on the commission would have damaged the credibility and influence that the commission strove for. Second, he was the only commissioner who had written about the ethics of genetic issues. And indeed "the commissioners really looked to him as the expert. Very few commissioners, possibly even none, felt like they were expert in this issue. Either the science of it, or the more social issues of it. And they had enormous respect for Arno [Motulsky] as a world-class geneticist. . . . This was a case where the commission just turned their heads to Arno whenever a decision was made—and he had that kind of influence" (Schapiro interview, 11). An example of this deference occurred when the commissioners opened the discussion of HGE at their eleventh meeting: the chair asked Motulsky if he "would care to start the discussion because of [his] expertise" and said that "it would be helpful to get the benefit of [his] wisdom" (meeting no. 11 transcript, 8).

30. Since the distinction is of little importance for this project, I would like to avoid debates on whether scientific "facts" are "facts" in some objective sense. I will simply say that there are some claims about nature that are so highly institutionalized that even if they are not "facts" in an objective sense, they function as if they are, at least for the purposes of these debates.

31. Meeting no. 3 transcript, 377.

32. Steven Lukes, *Power* (London: Macmillan, 1974).

33. Meeting no. 3 transcript, 370–72. At first it was not accepted by the commissioners that they could not speculate. Immediately after the statement by Motulsky, another commissioner responds—producing laughter among his colleagues—"I'd like to think that thinking ahead is helpful. Sometimes it is" (meeting no. 3 transcript, 373).

34. Executive director Capron hired a consultant, Tabitha Powledge, who had worked on similar issues in the past, to write a first draft of a report on short notice. This first draft was ten pages long, and was written in less than a month, so as to be ready for the next meeting.

35. "Splicing Life," draft no. 1, 7 (emphasis added), President's Commission Archives.

36. Meeting no. 4 transcript, 366–67 (emphasis added).

37. Ibid., 370.

38. Ibid., 367–68.

39. As staff ethicist Allen Buchanan put it in a 1996 interview, "it was thought that these were concerns that were not just being raised by this particular group, but that this group's communication represented a much larger constituency of ordinary citizens and other people who were concerned by these issues and that it was appropriate to address them for that reason" (Buchanan interview, 3).

40. Meeting no. 4 transcript, 371–72.

41. Ibid., 372–73 (emphasis added).

42. Public Law 95–622, 9 November 1978. An agency was not bound to follow these recommendations, but had to respond in the *Federal Register* if it decided not to adopt the recommended policies.

43. It is clear that among those persons I interviewed, Jonsen took most seriously the idea that the commission should primarily suggest policy. When asked to identify the task of the two government advisory commissions of which he had been a member, he responded: "The task is quite literally the task given to

them by the Congress. One is reminded again and again, when you serve on a government body, that you do what you're told to do. You can think of a whole lot of other purposes and you can imagine a whole lot of other purposes, like a broad effect on the education of young people or changing the minds of doctors or whatever, getting legislation. You can imagine those things; but you literally do what you're told to do in the mandate" (Jonsen interview, 6).

44. President's Commission, *Summing Up*, 3 (emphasis added).

45. It is, however, quite common for presidential commissions to be assigned or to adopt educational functions. See Flitner, *Politics of Presidential Commissions.*

46. Meeting no. 4 transcript, 372.

47. Ibid., 373–74.

48. Ibid., 384.

49. Ibid., 385.

50. The draft had been lightly rewritten. Consultants were described by Capron as representing "mixtures of disciplines and attitudes on the subject" (letter from Capron to the commissioners, 10 June 1981, President's Commission Archives). Possible consultants were described as "antis" or "middle-roaders" (letter from Tabitha Powledge to Capron, 6 November 1980, President's Commission Archives), or as "neutral observers," "critics," or "harsh" (letter from Schapiro to Capron, 3 December 1980, President's Commission Archives). One person was described as having "mellowed a bit . . . And [now] sees the need for compromise" (ibid.). Another was described as "exactly right" for his role because he was so "*very* pro" that "*scientists* [have taken] him to task for not being critical enough" (Powledge, memorandum, n.d. [emphasis in original], President's Commission Archives).

51. For this and the next round of consultant comments, I cannot determine from the archives whether I have the entire set of persons who responded, a random sample, or only the responses that were held to be the most useful. It was not uncommon for the commission to request comments from many people who did not respond. The letters I have are from Peter Hutt (former lawyer for the Food and Drug Administration), Maxine Singer (biochemistry, NIH), David Baltimore (cancer research, MIT), Ruth Hubbard (biology, Harvard), Michael Shapiro (law, University of Southern California), and Sheldon Krimsky (urban and environmental policy, Tufts).

52. Although in the previous meeting the vague issues were to be described but not resolved, with Krimsky's framework they would not be dealt with at all. The next draft was written as an "amalgam of suggestions from [staff member] Renie [Schapiro] & Krimsky plus the exchange between Alex [Capron] and Al Jonsen" at meeting no. 4 (memo from Tabitha Powledge to Capron, n.d., President's Commission Archives).

53. Letter from Sheldon Krimsky to Morris Abram, chair of the commission, 31 December 1980, President's Commission Archives.

54. The data here are a bit sketchy—the meetings were not transcribed, interviewees who participated in the meetings could not recall details, and only the brief notes of executive director Capron exist in the archives. However, when combined with a few follow-up letters from participants summarizing the positions they had taken at the meetings, the meetings can be adequately reconstructed.

55. In a letter to one of the participants, Capron hopes that the next draft

will "be able to convey the 'vaguer' concerns about values" (letter from Capron to Leon Kass, 17 April 1981, President's Commission Archives).

56. Why Kass's voice was taken to represent this view is unclear, but several possibilities exist. First, he was widely respected in debates on these issues at the time. Second, he was a physician, biologist, and philosopher, and therefore understood, and was respected by, people from many professions. Third, he actually wrote back with a long follow-up letter on his comments in the meeting and provided a marked-up draft of the report. Fourth, as his supporters and opponents would probably agree, he is a gifted writer.

57. Letter from Kass to Capron, 7 April 1981 (emphasis in original).

58. Ibid. Although he does not use the terms, Kass provides here an excellent description of formal rationality, as well as the problem with it from a substantively rational point of view.

59. Ibid. (emphasis in original).

60. Memo from Powledge to Capron, 26 May 1981, President's Commission Archives.

61. "Splicing Life," draft no. 4, 6 October 1981, p. 4, President's Commission Archives.

62. Meeting no. 11 transcript, 29. In an interview in 1997 Jonsen said, referring to an early draft, that he "felt that the draft didn't meet the letter's questions very accurately; because it really left out the broader issues about human nature and destiny. We had to address those in order to be responsive, and also because those were questions that people were really asking" (Jonsen interview, 3).

63. Meeting no. 11 transcript, 36–37.

64. Ibid., 27.

65. Ibid., 29.

66. Ibid., 64–65 (emphasis added).

67. Letter from Schapiro to general secretaries of the U.S. Catholic Conference, National Council of Churches, and Synagogue Council of America, 28 August 1981, President's Commission Archives.

68. The commission received papers from Fr. John Connery, S. J., of the National Conference of Catholic Bishops, and J. Robert Nelson, writing for the National Council of Churches. A short article written by Rabbi Seymour Siegel, who later became a commissioner, was submitted on behalf of the Synagogue Council of America so late in the process that it did not play a role in this round of consultations.

69. Connery's paper was less scrutinized by the commission and its consultants—perhaps because it pointed out problems the commission had already considered—but it did analyze the issues using a fairly formally rational type of argumentation. With the exception of a few references to papal documents, a reader would have had to be versed in Catholic theology to recognize the theological reasoning behind many of the points. In reviewing the paper for the commission, Catholic theologian Charles Curran described its reasoning as "based almost entirely on human reason and natural law." Even the official Church teachings to which it appeals "claim to be based on human reason and human nature" (letter from Charles Curran to Capron, 5 January 1982). Note that this is basically the "autonomous ethic" common in Roman Catholic theology—that natural law theology should be similar to universalist secular reasoning.

70. J. Robert Nelson, "Theological Concerns about Genetic Engineering," n.d., 15 (emphasis in original), President's Commission Archives.

71. "The initial intent would doubtless be therapeutic for the human organism in correcting a potentially debilitating genetic disease; and the secondary intent would be the protection of future progeny. However, in practice it currently seems unlikely that sufficient knowledge is available even to theorize accurately about the long-term generational effects. Empirical data could only be received, obviously, from observations over many years. For the present, then, modifying germ cells should not be practiced at all" (ibid.).

72. Ibid., 9.

73. These included the following: Robert Brungs, James Childress, Charles Curran, Margaret Farley, James Gustafson, Karen Lebacqz, Donald McCarthy, Kevin O'Rourke, Paul Ramsey, Richard Roche, Harmon Smith, and Gabriel Vahanian. Letters from Brungs, Childress, Farley, Gustafson, and McCarthy are not in the archives and could not be located. All the consultants were Protestant or Catholic. A list of Jewish theological consultants was prepared, but was apparently not used, probably because the Jewish paper came too late in the process for comments.

74. Letter from Capron to Paul Ramsey, 19 December 1981, President's Commission Archives.

75. She specifically refers to this as the perspective of theologian Stanley Hauerwas, who is the most famous proponent of this view of Christian ethics. Hauerwas's ethics may be described as substantively rational, as opposed to formally rational. Instead of Christians asking themselves, "Does this act maximize a Christian principle?" they are instead to act in ways that are consistent with the entire cluster of Christian ends (what Hauerwas would call the Christian narrative). It should be noted that Lebacqz was on the National Commission that wrote the *Belmont Report*. Is it then contradictory that she seems to have advocated a thin list of universal, commensurable ends in one context and a thicker notion of ends in another? It seems likely that if she had reflected on her theological beliefs during the Belmont process, created theological ends, and translated them to secular language, she would have ended up with the Belmont principles. However, most theologians, when encountering a different issue, would start this process over again, and possibly end up with a different set of ends. It is also clear that she is not entirely happy with how the Belmont ends have been applied to human experimentation itself, since she argues for a different set of ends. See Karen Lebacqz, "Twenty Years Older But Are We Wiser?" paper presented at the "Belmont Revisited" conference, Charlottesville, Va., 18 April 1999.

76. Both the quotations in these two paragraphs are drawn from the same source: Letter from Karen Lebacqz to Capron, 28 December 1981, President's Commission Archives.

77. Letter from Paul Ramsey to Capron, 7 January 1982 (emphasis in original), President's Commission Archives.

78. Letter from Vahanian to Capron, 21 December 1981, President's Commission Archives.

79. Memorandum from Jonsen to Schapiro, n.d. (emphasis added), President's Commission Archives.

80. Ibid. (emphasis added).

81. President's Commission, *Splicing Life,* 54.

82. Buchanan interview, 3.

83. Ibid., 4. He expressed a similar motive for breaking down slogans during

the commission deliberations: "It is clear in the report that in deflating the rhetoric, one of the things we are responding to is that kind of extreme position that this whole endeavor should be stopped" (meeting no. 26 transcript, 72).

84. Buchanan interview, 4.

85. Ibid., 6. Buchanan is no doubt sincere in his assessment that the theological input made no sense to the staff. To reiterate, from the perspective of someone trained as an analytic philosopher or a bioethicist, using the formally rational type of argumentation, the theologians' claims would indeed make no sense, since they violate the very assumptions of reasoned debate. Furthermore, much of the input was written as if it were to be read only by other theologians. For example, Lebacqz stated in her letter, without further explanation, that "without disagreeing with the understanding of sin [in Nelson's document] as a disposition towards egoism, I believe that this definition needs to be expanded . . . [using a concept] that provides a corrective to the kind of individualist emphasis given to the notion of sin in these pages. The Fall is not merely an individual or simply human occurrence but is the Fall of the entire created order" (letter from Karen Lebacqz to Capron, 28 December 1981). What is entailed by "The Fall" and what this would imply for the topic at hand is undoubtedly unclear for someone not trained in Christian theology.

86. Buchanan interview, 10. These were not second-rate theologians, but the top theologians in the field who had written about HGE. The consultants for the commission were generally selected by Capron, who "recognized quality and went after it" in the selection process (Weisbard interview, 12).

87. Buchanan interview, 11.

88. President's Commission, *Splicing Life,* 19–20 (emphasis added).

89. The commission had been engrossed in the other projects and was under great pressure to complete the congressionally mandated reports by year-end, when the commission's authorization would expire. The terms of many of the original commissioners appointed by President Carter had expired, and they had been replaced by Reagan appointees, although the commissioners who had been the most involved—Motulsky and Jonsen—were still on the commission. What was essentially the final draft was distributed at meeting no. 22, and the document was ultimately approved after a shorter discussion a few months later at meeting no. 26. Very little of substantive interest occurred at meeting no. 26 where the draft was given final approval. The most interesting moments in that meeting were Motulsky and others trying to remove language from the report, making the argument that some concerns, although debunked in the report, might through their very mention cause the public to become fearful.

90. Meeting no. 22 transcript, 8.

91. Ibid., 9.

92. Ibid., 13–14.

93. Commissioner George Dunlop (ibid., 44). Shortly afterward, Wade commented, "Surely it's not the purpose of the commission to try and reassure the public, or to guarantee the scientists' ability to work, as Dr. Dunlop so eloquently put it" (p. 49). In meeting no. 26 Dunlop said that he hoped "that there is nothing in this draft that would antagonize the scientific community or that would take positions that are not in keeping with the acceptable positions of the scientific community" (transcript, p. 20).

94. Meeting no. 22 transcript, 156.

95. Ibid., 49–54.

96. The final report consisted of four chapters: an introduction, a technical chapter, a "social and ethical issues" chapter, and a chapter discussing possible continued oversight. The introduction was generally an overview of the contents. Chapter 2 outlined basic scientific information necessary to understand human genetics, such as the nature of DNA and the difference between somatic and germline engineering. It is here that one finds the calming of the fears of the public through providing accurate scientific information. I will focus on the "social and ethical issues" chapter (chapter 3) because this is where most of the analysis of ethical arguments occurs.

97. President's Commission, *Splicing Life,* 51–52.

98. Ibid., 52 (emphasis added). "As such" means without regard to consequences, or substantively rational.

99. Ibid.

100. This is my phrase, not the commission's. The insight here draws upon the work of Allen Verhey, "'Playing God' and Invoking a Perspective," *Journal of Medicine and Philosophy* 20 (1995): 347–64.

101. President's Commission, *Splicing Life,* 54.

102. Ibid., 55.

103. Ibid., 56.

104. Ibid.

105. Ibid., 57.

106. Ibid. (emphasis added).

107. Ibid., 59.

108. Ibid., 58.

109. Ibid., 22.

110. Ibid., 66.

111. Ibid., 65.

112. Ibid., 68.

113. President's Commission, *Summing Up,* 66.

114. President's Commission, *Splicing Life,* 77.

115. Verhey, "'Playing God' and Invoking a Perspective," 348.

116. Ibid., 353.

117. For example, Capron interprets "playing God" as being "a way of saying that this is knowledge which man ought not to seek" (meeting no. 22 transcript, 54). That is, HGE falls into one of the "gaps" that is God.

118. President's Commission, *Splicing Life,* 53.

119. Even in the hands of its most famous proponent, Paul Ramsey, the phrase "playing God" is "not immediately identified with a particular moral rule or principle." Nor does the phrase as used by Ramsey refer to a "God of the Gaps"; rather, it invokes a perspective opposed to "an 'attitude,' an 'outlook' [and] certain 'operating unspoken premises' at work in Western scientific culture." See Verhey, "'Playing God' and Invoking a Perspective," 355–56.

120. The latter claim is in the letter from Lebacqz to the commission (Karen Lebacqz to Capron, 28 December 1981).

121. Karen Lebacqz, "The Ghosts Are on the Wall: A Parable for Manipulating Life," in *The Manipulation of Life,* ed. Robert Esbjornson (San Francisco: Harper and Row, 1984), 33–34.

122. The NCC is an ecumenical council, which at the time was made up of thirty-two Protestant and Eastern Orthodox denominations. The Protestant denominations include the traditionally African American denominations and lib-

eral and moderate Protestant denominations. Evangelical denominations are generally not part of the NCC.

123. National Council of Churches, Panel on Bioethical Concerns, *Genetic Engineering: Social and Ethical Consequences* (New York: Pilgrim Press, 1984).

124. C. Keith Boone, "Splicing Life, with Scalpel and Scythe," *Hastings Center Report* (April 1983): 8.

125. Not all the participants would, in their religious communities, have called themselves theologians, a term that has a more specific meaning within religious groups than my more general definition. They were clergy who worked for the NCC or who represented their denomination to the NCC. However, for the purposes of a sociological project concerned with general trends, and following my definition of a theologian in the appendix, all these persons can be considered theologians.

126. The ends are "the worth of human life," "the interdependence of life systems," "the growth process," "knowledge and the pursuit of truth," "responsible scientific inquiry," "participation in decisions that may affect personal or community well-being," "diversity," and "distributive justice" (National Council of Churches, Panel on Bioethical Concerns, *Genetic Engineering,* 34–35).

127. Meeting no. 22 transcript, 127.

128. Ibid., 125–30.

129. John Robertson, "Genetic Alteration of Embryos: The Ethical Issues," in *Genetics and the Law III,* ed. Aubrey Milunsky and George Annas (New York: Plenum Press, 1985), 117 (emphasis added).

Chapter 5

1. Alexander Morgan Capron, "The Impact of the Report, Splicing Life," *Human Gene Therapy* 1 (1990): 70.

2. LeRoy Walters, "Human Gene Therapy: Ethics and Public Policy," *Human Gene Therapy* 2 (1991): 117.

3. LeRoy Walters, "The Ethics of Human Gene Therapy," *Nature* 320 (1986): 226.

4. For the data on the changing distribution of professions among both the common and influential authors in the debate, see figure 1. For the list of the texts of the influential authors in this era, see table A-4.

5. They are placed in separate communities because the scientists use a large amount of their texts to describe the scientific techniques of how HGE would actually occur. There are actually two distinct but closely related communities of scientists in this era: one with Kantoff, Miller, Anderson, and Friedmann as the influential authors, and another with Palmiter and Weatherall as the influential authors (see fig. 4). They will be discussed together. All these influential authors are scientists, and being a scientist increases the expected number of influential authors cited in the first community by 130 percent, compared to members of other professions, holding the other variables constant. This finding is significant at the $p < .001$ level. This analysis has an N of 79 common authors. Models for each of the other professions under examination show no significant differential representation in this community by other professions.

We can get a sense of what the common authors are debating by examining the keywords in use. The more one is embedded in the first community, the more likely one is to discuss "gene therapy," "germ cells," and "human experimentation." A one-point difference in the coding scale for these keywords means that the expected number of influential authors in this community that the text will cite increases 216 percent, 29 percent, and 73 percent, respectively. The findings are significant at the $p < .001$, $p = .084$, and $p < .001$ levels, respectively. The second community of scientists is disproportionately discussing genetic disorders (57 percent; $p = .073$). Contrary to the theologians, who tend to discuss many means simultaneously, the first of these communities is disproportionately not discussing abortion, cloning, genetic screening, prenatal diagnosis, and reproductive technologies. A one-point difference in the coding scale for these keywords means that the expected number of influential authors in this community that the text will cite decreases 58 percent, 64 percent, 38 percent, 36 percent, and 51 percent, respectively. The findings are significant at the $p =$ .020, .072, .044, .072, and .007 levels, respectively. N is 84.

6. A one-point difference in the coding scale of the "risk/benefit analysis" keyword in the first and second community means that the expected number of influential authors in this community that the text will cite decreases 42 percent and 72 percent, respectively. The findings are significant at the $p = .015$ and .042 levels, respectively. N is 84.

7. The texts by A. Dusty Miller, Philip Kantoff, and Richard Palmiter are all purely scientific reports of how the most recent experiments in the development of HGE were progressing. During this era scientific work was being done on developing viral vectors to deliver strands of DNA into cells. For example, Miller's 1984 text describes using a retrovirus with a human gene sequence in it to transfer this human gene to a mouse (A. Dusty Miller, Douglas J. Jolly, and Inder M. Verma, "Expression of a Retrovirus Encoding Human HPRT in Mice," *Science* 225 [1984]: 630–32). Palmiter's experiment was to insert a growth gene from a rat into a mouse, resulting in giant mice—an act of germline engineering in that the trait of this new transgenic animal was expressed in its offspring (Richard D. Palmiter et al., "Dramatic Growth of Mice That Develop from Eggs Microinjected with Metallothionein-Growth Hormone Fusion Genes," *Nature* 300 [1982]: 611–15).

8. W. French Anderson, "Prospects for Human Gene Therapy," *Science* 226 (1984): 402.

9. Theodore Friedmann, "Progress toward Human Gene Therapy," *Science* 244 (1989): 1275.

10. The first quotation is from D. J. Weatherall, *The New Genetics and Clinical Practice* (New York: Oxford University Press, 1985), 181. The second is from Friedmann, "Progress toward Human Gene Therapy," 1280.

11. W. French Anderson and John C. Fletcher, "Gene Therapy in Human Beings: When Is It Ethical to Begin?" *New England Journal of Medicine* 303 (1980): 1295.

12. Ibid., 1293.

13. Ibid.

14. Anderson, "Prospects for Human Gene Therapy," 408.

15. LeRoy Walters and Julie Gage Palmer, *The Ethics of Human Gene Therapy* (New York: Oxford University Press, 1997), 40.

16. The current regulations of institutional review boards are found in U.S. Department of Health and Human Services, "Protection of Human Subjects," *Code of Federal Regulations,* Part 46, Subtitle A (1981). The "Points to Con-

sider" document is Human Gene Therapy Subcommittee, Recombinant DNA Advisory Committee, "Points to Consider in the Design and Submission of Human Somatic Cell Gene Therapy Protocols," *Recombinant DNA Technical Bulletin* 8, no. 4 (1985): 182.

17. Walters, "Ethics of Human Gene Therapy," 226.

18. Human Gene Therapy Subcommittee, Recombinant DNA Advisory Committee, "Points to Consider in the Design and Submission of Human Somatic Cell Gene Therapy Protocols," 186.

19. The "Points to Consider" document was revised in 1989.

20. These events were momentous enough that the common members of the HGE debate had a distinct debate about them. The community in figure 4 containing scientist Bernard Davis, science journalist Barbara Culliton, and Rifkin represents this discussion. Culliton's articles are news stories in *Science* magazine about the first somatic gene therapy experiments and the advisory commissions that govern them. Rifkin, who will be more extensively discussed in chapter 6, was the author of several texts critical of HGE and genetic engineering more generally. Examination of common members of this debate reveals that they are discussing Rifkin's texts to explain why he would sue to stop the experiments that Culliton is reporting on. Examination of the work of Davis, the ardent defender of science whom we first met in chapter 3, shows that authors are discussing him in relation to the Asilomar controversy (the subject of one of his most influential texts). As shown in figure 4, he is most closely related to Rifkin in this community, a finding that makes sense given that Rifkin's most influential text during this era (*Who Should Play God?*) is a retelling of the Asilomar controversy in somewhat spectacular terms. (Although Rifkin's book was written with Ted Howard, the references to it were compressed into Rifkin's other works.) Common authors may summarize the Asilomar debate by referring to Rifkin and Davis, who were antagonists. For example, Davis was clearly motivated to write his 1987 text by "demagogic attacks" against his field. "Indeed, the whole science of bacterial genetics might have been aborted if [the] pioneers had had a Rifkin in the wings, demanding absolute guarantees of safety" (Bernard D. Davis, "Bacterial Domestication: Underlying Assumptions," *Science* 235 [1987]: 1329, 1335). Rifkin is also clearly in his mind when he concludes that "the agenda has been set for too long by apocalyptic activists" and that "the scientific community must take initiative in helping the public and decision-makers to distinguish reasonable probabilities from remote fantasies" (ibid., 1335). Rifkin is thus the linchpin to this particular community, which, besides its focus on Rifkin, is otherwise of little substantive interest.

21. Human Gene Therapy Subcommittee, Recombinant DNA Advisory Committee, "Points to Consider in the Design and Submission of Human Somatic Cell Gene Therapy Protocols," 182 (emphasis added).

22. President's Commission, *Splicing Life: A Report on the Social and Ethical Issues of Genetic Engineering with Human Beings* (Washington, D.C.: Government Printing Office, 1983), 45.

23. Friedmann, "Progress toward Human Gene Therapy," 1280. By "efficient disease control" is meant not having to somatically engineer every generation that is born. Reference to "the need to prevent damage early in development" means that some genetic diseases do their damage in the fetal or infant stage, before one could somatically engineer the patient. Reference to "inaccessible cells" means that some somatic cells are inaccessible once the body has grown. How-

ever, if a one-celled fertilized egg is engineered before it multiplies into an entire body, the genetic problems are fixed throughout the entire body.

24. W. French Anderson, "Human Gene Therapy: Scientific and Ethical Considerations," *Journal of Medicine and Philosophy* 10 (1985): 275–76 (emphasis added).

25. Ibid., 285–87.

26. Ibid.

27. Ibid., 288.

28. President's Commission, *Splicing Life,* 47.

29. Being a bioethicist increases the expected number of influential authors cited in this community by 108 percent, compared to members of other professions, holding the other variables constant. This finding is significant at the $p = .003$ level. This analysis has an N of 79 common authors. Models for each of the other professions under examination show no significant differential representation in this community by other professions.

A one-point difference in the coding scale of the "germ cell" and "gene therapy" keywords means that the expected number of influential authors in this community that the text will cite increases 43 percent in each case. The findings are significant at the $p = .004$ and .015 levels, respectively. N is 84.

30. A one-point difference in the coding scale of the "informed consent" keyword means that the expected number of influential authors in this community that the text will cite increases 53 percent. The findings are significant at the $p = .053$ level. N is 84.

31. At this point, for the first time, keywords representing various government mechanisms are significant. This community is disproportionately discussing government advisory commissions, presumably the RAC. A one-point difference in the coding scale of the "government advisory commission" keyword means that the expected number of influential authors in this community that the text will cite increases 52 percent. The findings are significant at the $p = .026$ level. N is 84.

32. At this time John Fletcher was the assistant for bioethics at NIH, which would ultimately approve any act of HGE. LeRoy Walters, a bioethicist at Georgetown University, was during this period the chair of the subcommittee of the RAC that evaluated HGE protocols. The government entities were the President's Commission and the Office of Technology Assessment, whose reports were to be used by policy makers (see appendix).

33. Clifford Grobstein and Michael Flower, "Gene Therapy: Proceed with Caution," *Hastings Center Report* 14, no. 2 (1984): 13; Walters, "Ethics of Human Gene Therapy," 225; John C. Fletcher, "Ethical Issues in and beyond Prospective Clinical Trials of Human Gene Therapy," *Journal of Medicine and Philosophy* 10, no. 3 (1985): 293.

34. Grobstein and Flower, "Gene Therapy," 16.

35. Taking Walters as an example, the only ethical questions in HGE that he sees are those derived from the *Belmont Report.* Primary concerns are the risk/benefit analysis of safety and efficacy (beneficence and nonmaleficence). "Other ethical issues" concern the selection of candidates (justice), "informed consent," and "privacy and confidentiality" (autonomy). See Walters, "Ethics of Human Gene Therapy."

36. A one-point difference in the coding scale of the "abortion" and "euthanasia" keywords means that the expected number of influential authors in this

community that the text will cite decreases 37 percent and 334 percent respectively. The findings are significant at the $p = .047$ and .004 levels, respectively. N is 84.

37. Fletcher, "Ethical Issues in and beyond Prospective Clinical Trials of Human Gene Therapy," 307.

38. Clearly, their intellectual worlds intersect. Besides their shared employer and workplace, they cowrote the 1980 article that is Anderson's second most influential text in this era. Their collaboration highlights the decreasing difference between scientists and bioethicists.

39. Reduction is the argument that the tasks of the profession under challenge are reducible to the core jurisdiction of the challenging profession. See Andrew Abbott, *The System of Professions: An Essay on the Division of Expert Labor* (Chicago: University of Chicago Press, 1988), 98.

40. John C. Fletcher, "Moral Problems and Ethical Issues in Prospective Human Gene Therapy," *Virginia Law Review* 69 (1983): 534.

41. Ibid., 537 (emphasis added).

42. Fletcher summarizes Kass and Ramsey at this point as being opposed to "the use of positive eugenics and genetic engineering to attempt improvements in complex traits like intelligence, memory, and life span" (Fletcher, "Moral Problems and Ethical Issues in Prospective Human Gene Therapy," 537).

43. Leon R. Kass, "New Beginnings in Life," in *The New Genetics and the Future of Man,* ed. Michael Hamilton (Grand Rapids, Mich.: William B. Eerdmans, 1972), 39.

44. Fletcher, "Moral Problems and Ethical Issues in Prospective Human Gene Therapy," 538.

45. Ibid., 544.

46. Fletcher, "Ethical Issues in and beyond Prospective Clinical Trials of Human Gene Therapy," 303–4, 307.

47. A one-point difference in the coding scale of the "gene therapy," "human experimentation," and "risk/benefit analysis" keywords means that the expected number of influential authors in this community that the text will cite decreases 38 percent, 46 percent, and 38 percent, respectively. The findings are significant at the $p = .005$, .002, and .012 levels, respectively. N is 84.

48. A one-point difference in the coding scale of the "abortion" and "artificial insemination" keywords means that the expected number of influential authors in this community that the text will cite increases 59 percent and 73 percent, respectively. The findings are significant at the $p = .070$ and .063 levels, respectively. N is 84.

49. Being a scientist decreases the expected number of influential authors cited in this community by 58 percent, compared to members of other professions, holding the other variables constant. This finding is significant at the $p = .017$ level. This analysis has an N of 79 common authors. Models for each of the other professions under examination show no significant differential representation in this community by other professions.

50. A one-point difference in the coding scale of the "eugenics" keyword means that the expected number of influential authors in this community that the text will cite increases 39 percent. This finding is significant at the $p = .091$ level. N is 84.

51. Marc Lappé, "Moral Obligations and the Fallacies of 'Genetic Control,'" *Theological Studies* 33, no. 3 (1972): 413.

52. Ibid., 415.

53. Daniel Kevles, *In the Name of Eugenics: Genetics and the Uses of Human Heredity* (Berkeley: University of California Press, 1985). More technically, Kevles and the other influential members of this community are co-cited by the common members. See the appendix for more details on this method of analysis.

54. The Human Genome Project is a government-sponsored effort that officially began in the fall of 1990 with the aim of mapping and sequencing all of the estimated three billion chemical base pairs of the human genome, and thus all fifty thousand to one hundred thousand genes in the genome. This is the first step toward determining where on the genome every genetic disease or trait lies. See Robert Cook-Deegan, *Gene Wars: Science, Politics, and the Human Genome* (New York: W. W. Norton, 1990); James D. Watson, "The Human Genome Project: Past, Present and Future," *Science* 248 (1990): 44–49.

55. George J. Annas, "Who's Afraid of the Human Genome?" *Hastings Center Report* 19, no. 4 (July–August 1989): 20 (emphasis in original).

56. Ibid., 21.

57. The authors in this community (see fig. 4) average 13.4 citations each, while the authors in the most influential scientist community and the bioethicist community average 22.8 and 20.8, respectively. That is, this community receives 59 percent and 64 percent of the amount of citations compared to the more influential communities.

58. Annas, "Who's Afraid of the Human Genome?" 19.

Chapter 6

1. Stuart H. Orkin and Arno G. Motulsky, *Report and Recommendations of the Panel to Assess the NIH Investment in Research on Gene Therapy* (1995). This report is available on-line at http://www4.od.nih.gov/oba/panelrep.htm.

2. Comparing the 1985–91 and 1992–95 time periods, 48.2 percent vs. 39.6 percent of texts discussed public policy, 40.2 percent vs. 31.7 percent discussed government regulation, 23.3 percent vs. 16.5 percent discussed advisory commissions. An F test of the difference in mean values for these variables in the two periods reveals that the difference in discussion of public policy is significant at .005, government regulations at .024, and advisory commissions not quite significant at .142. The lack of significance of the final comparison is the result of a substantial number of authors in the last year of the data calling on the RAC to begin deliberations to set regulations and policy for germline HGE. See table A-5 for a list of the texts of the influential authors in this period.

3. It did, however, begin to discuss germline HGE in 1995, just as the data used for this study end.

4. LeRoy Walters, "Human Gene Therapy: Ethics and Public Policy," *Human Gene Therapy* 2 (1991): 118–19 (emphasis added).

5. Robert M. Veatch, "Resolving Conflicts among Principles: Ranking, Balancing, and Specifying," *Kennedy Institute of Ethics Journal* 5, no. 3 (1995): 199–218.

6. Ibid.

7. On these cases, see David J. Garrow, *Liberty and Sexuality: The Right to Privacy and the Making of* Roe v. Wade (New York: Macmillan/Lisa Drew, 1994).

8. Although it would take a separate book to defend this thesis, I believe that

the abortion debate is more substantively rational than the HGE debate, primarily because the arguments are being produced for public consumption and not government advisory commissions. To take just one example, consider the absolutist "rights" language that permeates the debate. Abortion is either a priori wrong, regardless of consequences, or restriction of women's autonomy is wrong. Consequentialism is a minor part of the debate. Moreover, while the end pursued on the pro-choice side is largely autonomy, there are also arguments about population control, women's health, and defense of the establishment clause of the Constitution. On the pro-life side there is a much wider spectrum of ends. The abortion debate is about these ends: "life" vs. "choice."

9. Authors in the community containing LeRoy Walters, Burke Zimmerman, Marc Lappé, Eric Juengst, W. French Anderson, John Fletcher, and the President's Commission average 19.1 citations each, with other communities averaging 11.6, 13.3, 10, and 13.5 (see fig. 5). Being a bioethicist increases the expected number of influential authors cited in this community by 66 percent, compared to members of other professions, holding the other variables constant. This finding is significant at the $p = .023$ level. This analysis has an N of 64 common authors. Models for each of the other professions under examination show no significant differential representation in this community by other professions.

10. Scientist W. F. Anderson is the most influential author in this period, and his work is perceived by common authors to be the most similar to that of John Fletcher and the President's Commission in this community (see fig. 5).

11. The influential authors are personally still involved with state policy making on this topic. Walters was chair of the Human Gene Therapy Subcommittee of the RAC during this era. Eric Juengst was program director of the Ethical, Legal and Social Implications Program of the Human Genome Project at NIH. The President's Commission was a government agency. In his earlier texts in this era John Fletcher was director of bioethics at NIH, but when he wrote the later texts he had become director of the Center for Biomedical Ethics at the University of Virginia. While Anderson continued as the leading HGE researcher at NIH, Zimmerman and Lappé had no clear connections to state policy making on this issue.

12. We can get a sense of what the common authors are debating by examining the keywords in use. The more one is embedded in this community, the more likely one is to discuss "gene therapy," "germ cells," and "genetic disorders." A one-point difference in the coding scale for these keywords means that the expected number of influential authors in this community that the text will cite increases 130 percent, 78 percent, and 57 percent, respectively. The findings are significant at the $p < .001$, $p < .001$, and $p < .001$ levels, respectively. A one-point difference in the coding scale for the keyword "cloning" means that the expected number of influential authors in this community that the text will cite decreases 67 percent. This finding is significant at the $p = .026$ level. N is 68.

13. The more one is embedded in this community, the more likely one is to discuss "autonomy," "informed consent," and "risk/benefit analysis." A one-point difference in the coding scale for these keywords means that the expected number of influential authors in this community that the text will cite increases 49 percent, 59 percent, and 49 percent, respectively. The findings are significant at the $p = .058$, .022, and .001 levels, respectively. N is 68.

14. Eric T. Juengst, "Germ-Line Gene Therapy: Back to Basics," *Journal of*

Medicine and Philosophy 16 (1991): 587. Also critiqued are theologians Bernard Haring and Ken Vaux as well as W. Schirmacher. Although the first two authors were referred to in the early debate, neither was influential enough to make the cutoff for consideration in chapter 2. The third author was not found in the 16,000 citations gathered.

15. Juengst, "Germ-Line Gene Therapy," 587 (emphasis added).

16. Ibid., 589 (emphasis added).

17. Ibid.

18. Walters, "Human Gene Therapy," 118 (emphasis added).

19. We should reflect on what a "neutral and objective" ethical argument is. If we accept that there is no natural or objectively true ethical system that an author could be biased against, then what does this mean? It is simply a reiteration of the legitimacy claim of the bioethics profession: that its ethics are "neutral and objective" because they are shared by all Americans.

20. John C. Fletcher and W. French Anderson, "Germ-Line Gene Therapy: A New Stage of the Debate," *Law, Medicine and Health Care* 20 (1992): 26.

21. Ibid., 28.

22. Evidence for their close relationship can be seen in figure 5: the two communities are more like each other than they are like the other debating communities in this era. The first community has Annas, Robertson, Capron, Suzuki, and Watson as influential authors. The more one is embedded in this community, the more likely one is to discuss "abortion," "genetic counseling," "genetic screening," and "prenatal diagnosis." A one-point difference in the coding scale for these keywords means that the expected number of influential authors in this community that the text will cite increases 150 percent, 121 percent, 139 percent, and 104 percent, respectively. The findings are significant at the $p = .025$, $p = .056$, $p < .001$, and $p = .029$ levels, respectively. The second community has Holtzmann, Wertz, Kevles, and the Office of Technology Assessment as the influential authors. The more one is embedded in this community, the more likely one is to discuss "genetic counseling" and "genetic screening." A one-point difference in the coding scale for these keywords means that the expected number of influential authors in this community that the text will cite increases 73 percent and 55 percent, respectively. The findings are significant at the $p = .032$ and $.018$ levels, respectively.

23. Being a lawyer increases the expected number of influential authors cited in this community by 330 percent, compared to members of other professions, holding the other variables constant. This finding is significant at the $p = .058$ level. This analysis has an N of 64 common authors.

24. John Robertson, "Genetic Alteration of Embryos: The Ethical Issues," in *Genetics and the Law III,* ed. Aubrey Milunsky and George Annas (New York: Plenum Press, 1985), 118. To reiterate a methodological point made earlier, the time period for each chapter is determined by the data from the citing authors. These citing authors by logical necessity cite texts from the past. Therefore, the texts from the most influential authors in any period may have been written significantly before the era under examination.

25. Ibid., 119–21.

26. Ibid., 124–25.

27. The influential authors in this debating community are philosophers Peter Singer and Jonathan Glover. Being a philosopher increases the expected number of influential authors cited in this community by 271 percent, compared to

members of other professions, holding the other variables constant. This finding is significant at the $p = .007$ level. This analysis has an N of 64 common authors. The concern with more speculative questions is found among the common authors who, more than the members of other communities, are interested in cloning. A one-point difference in the coding scale for the keyword "cloning" means that the expected number of influential authors in this community that the text will cite increases 458 percent. This finding is significant at the $p = .001$ level. N is 68.

28. In utilitarianism, unlike the bioethics profession's form of argumentation, there are not four ends but one—human happiness. This end is justified because it is universal. This is expressed in several places: "From ancient times, philosophers and moralists have expressed the idea that ethical conduct is acceptable from a point of view that is somehow universal"; "ethics requires us to go beyond 'I' and 'you' to the universal law, the universalizable judgement, the standpoint of the impartial spectator or ideal observer, or whatever we choose to call it" (Peter Singer, *Practical Ethics* [New York: Cambridge University Press, 1979], 10, 11). The link between the means and this one end is not a priori, as Ramsey advocated, but rather is a selection of which means will maximize the ends. Singer is perhaps best known for his book *Animal Liberation,* which argues from a utilitarian position that animals should be largely treated like humans. See Peter Singer, *Animal Liberation* (New York: Random House, 1975). Some of Singer's influence in this community is due to this text. People refer to Singer's book in order to argue about the rights of animals, which would be the subject of engineering or experimentation before HGE experiments would begin. More prevalent are references to Singer's arguments about utilitarianism. See Singer, *Practical Ethics.*

29. "There is a widespread view that any project for the genetic improvement of the human race ought to be ruled out: that there are fundamental objections of principle. The aim of this discussion is to sort out some of the main objections. . . . The debate on human genetic engineering should become like the debate on nuclear power: one in which large possible benefits have to be weighed against big problems and the risks of great disasters" (Jonathan Glover, *What Sort of People Should There Be?* [New York: Penguin, 1984], 25).

30. Ibid., 47–48.

31. Burke K. Zimmerman, "Human Germ-Line Therapy: The Case for Its Development and Use," *Journal of Medicine and Philosophy* 16 (1991): 597.

32. Gregory Fowler, Eric T. Juengst, and Burke K. Zimmerman, "Germ-Line Therapy and the Clinical Ethos of Medical Genetics," *Theoretical Medicine* 10 (1989): 163 (emphasis added). Although Zimmerman refers to the earlier work, this text itself falls just shy of inclusion in this community.

33. Ibid., 159.

34. Dorothy C. Wertz and John C. Fletcher, "Fatal Knowledge? Prenatal Diagnosis and Sex Selection," *Hastings Center Report* 19 (1989): 21–22. Wertz is a member of the second community discussed above, which is debating genetic screening and counseling.

35. Ibid., 21.

36. Ibid., 24.

37. Ibid., 21.

38. Ibid., 26.

39. Ibid., 26.

40. On how this impulse already exists, see Marque-Luisa Miringoff, *The Social Costs of Genetic Welfare* (New Brunswick, N.J.: Rutgers University Press, 1991); Troy Duster, *Backdoor to Eugenics* (New York: Routledge, 1990).

41. Certainly, as many authors have pointed out, autonomy has already brought about negative eugenics through prenatal screening and abortion. It is simply that abortion for fetal abnormalities such as Tay-Sachs and Down syndrome is accepted as legitimate by the public.

42. Zimmerman, "Human Germ-Line Therapy," 606–7.

43. Glover, *What Sort of People Should There Be?* 32.

44. Being a theologian increases the expected number of influential authors cited in this community by 101 percent, compared to members of other professions, holding the other variables constant. This finding is significant at the $p = .026$ level. This analysis has an N of 64 common authors.

45. The more one is embedded in this community, the more likely one is to discuss "cloning," "euthanasia," and "reproductive technologies." A one-point difference in the coding scale for these keywords means that the expected number of influential authors in this community that the text will cite increases 111 percent, 410 percent, and 54 percent, respectively. The findings are significant at the $p = .046$, .007, and .028 levels, respectively. N is 68.

46. D. Stephen Long, *Tragedy, Tradition, Transformism: The Ethics of Paul Ramsey* (Boulder, Colo.: Westview Press, 1993), 103. Long is not part of the data, but is a secondary source on Ramsey's influence in the 1990s more generally.

47. Susan Wright, *Molecular Politics: Developing American and British Regulatory Policy for Genetic Engineering, 1972–1982* (Chicago: University of Chicago Press, 1994), 224.

48. Sheldon Krimsky, *Biotechnics and Society: The Rise of Industrial Genetics* (New York: Praeger, 1991), 109.

49. Ibid., 122–23.

50. Barbara J. Culliton, "Gene Test Begins," *Science* 244 (1989): 913.

51. The "guerrilla theater" depiction comes from Krimsky, *Biotechnics and Society,* 109.

52. Ted Howard and Jeremy Rifkin, *Who Should Play God? The Artificial Creation of Life and What It Means for the Future of the Human Race* (New York: Dell, 1977), 10.

53. Krimsky, *Biotechnics and Society,* 120.

54. For example, the Contemporary Authors Online review of Rifkin's career (not a complete source) mentions reviews of *Algeny* in *Esquire,* the *Los Angeles Times, Maclean's, Discover,* and *Time,* and an interview with Rifkin about the book in the *Detroit Free Press.*

55. Krimsky, *Biotechnics and Society,* 109.

56. John C. Fletcher, "Evolution of Ethical Debate about Human Gene Therapy," *Human Gene Therapy* 1 (1990): 62.

57. Stephen Budiansky, "Churches against Germ Changes," *Nature* 303 (1983): 563. See also Jeremy Rifkin, *The Theological Letter concerning the Moral Arguments against Genetic Engineering of the Human Germline Cells* (Washington, D.C.: Foundation on Economic Trends, 1983).

58. Budiansky, "Churches against Germ Changes."

59. Ibid., 563. The Nobel laureate was George Wald.

60. Colin Norman, "Clerics Urge Ban on Altering Germline Cells," *Science* 220 (1983): 1360.

61. Ibid.

62. The canonical text on this topic is by Hannah Pitkin, who calls the "delegate" view the "mandate," and the "trustee" view, the "independence." See Hanna Fenichel Pitkin, *The Concept of Representation* (Berkeley: University of California Press, 1967).

63. For the trustee motivation in bioethics and government advisory commissions, recall the comment of the staff member who wrote *Splicing Life:* that the point of the document was to improve the quality of the HGE debate among both policy makers and the public. See chapter 4.

64. The claim that bioethics is a "philosophy of the people" comes from Albert R. Jonsen and Andrew Jameton, "Medical Ethics, History of: The Americas," in *Encyclopedia of Bioethics,* ed. Warren Thomas Reich (New York: Macmillan, 1995).

65. Fletcher, "Evolution of Ethical Debate about Human Gene Therapy," 63. I believe that Ramsey, were he alive today, would not have welcomed the comparison because as an academic he believed that arguments are to be won through the power of reasoning, not through an appeal to the emotions of the reader or an attempt to force the issue through a lawsuit. However, I would describe Rifkin's arguments, compared to those of the other authors in this period, as similar to Ramsey's with the nuances removed.

66. In addition to the example of his organizing the broad group of religious leaders to sign his 1983 declaration, some religious leaders wrote blurbs for his texts and others are thanked in the acknowledgments. For example, in his 1983 text he acknowledges the helpful advice of Wes Michelson, who in later years as Rev. Wes Granberg-Michelson worked for the National Council of Churches on the HGE issue. That same text carries a blurb by United Methodist bishop A. James Armstrong, who was then the president of the National Council of Churches.

67. Ted Howard is not a mainline Protestant, like most of the other theologians in this debate, but is either an evangelical or a fundamentalist. I use the term *fundamentalist* in its technical sense. Put very simply, fundamentalists are a stricter subset of evangelicals. See James Hunter, *American Evangelicalism: Conservative Religion and the Quandary of Modernity* (New Brunswick, N.J.: Rutgers University Press, 1983). Howard's affiliations seem to indicate that he is a fundamentalist. Howard is an ordained minister in the Pentecostal Free-Will Baptist Church. He received his D. D. degree from the Fundamental Christian College in 1957, and pastored a Pentecostal church in Virginia starting in the 1960s. The only membership listed for him in Contemporary Authors Online is the Association of Fundamental Ministers and Churches. See the web site: http://galenet.gale.com.

68. Howard and Rifkin, *Who Should Play God?* 229.

69. Ibid.

70. Ronald Cole-Turner, *The New Genesis: Theology and the Genetic Revolution* (Louisville, Ky.: Westminster/John Knox Press, 1993).

71. Jeremy Rifkin, *Algeny* (New York: Penguin Books, 1983), 248.

72. Alexander M. Capron, "The Latest in Genes," *Commonweal,* 16 December 1983, 697.

73. Rifkin, *Algeny,* 50.

74. Ibid., 250.

75. John C. Fletcher, "Ethical Issues in and beyond Prospective Clinical Trials of Human Gene Therapy," *Journal of Medicine and Philosophy* 10, no. 3 (1985): 306.

76. Cole-Turner, *New Genesis,* 61.

77. Ibid., 45.

78. Ibid., 108.

Chapter 7

1. "Pluralistic" and "overlapping consensus" are from Bruce Jennings, "Possibilities of Consensus: Toward Democratic Moral Discourse," *Journal of Medicine and Philosophy* 16, no. 4 (1991): 447–63.

2. John C. Fletcher, "Evolution of Ethical Debate about Human Gene Therapy," *Human Gene Therapy* 1 (1990): 55–68; Eric T. Juengst, "Germ-Line Gene Therapy: Back to Basics," *Journal of Medicine and Philosophy* 16 (1991): 587–92.

3. Rogers Brubaker, *The Limits of Rationality: An Essay on the Social and Moral Thought of Max Weber* (Boston: George Allen and Unwin, 1984), 44.

4. Juengst, "Germ-Line Gene Therapy."

5. Albert R. Jonsen, *The Birth of Bioethics* (New York: Oxford University Press, 1998), 310.

6. Ruth Ellen Bulger, Elizabeth Meyer Bobby, and Harvey V. Fineberg, *Society's Choices: Social and Ethical Decision Making in Biomedicine* (Washington, D.C.: National Academy Press, 1995), 95.

7. John C. Fletcher and W. French Anderson, "Germ-Line Gene Therapy: A New Stage of the Debate," *Law, Medicine and Health Care* 20 (1992): 34.

8. Susan Wright, *Molecular Politics: Developing American and British Regulatory Policy for Genetic Engineering, 1972–1982* (Chicago: University of Chicago Press, 1994); Sheldon Krimsky, *Genetic Alchemy: The Social History of the Recombinant DNA Controversy* (Cambridge, Mass.: MIT Press, 1982); Nicholas Wade, *The Ultimate Experiment: Man-Made Evolution* (New York: Walker, 1977).

9. The Religious Coalition for Abortion Rights was founded in 1973 by a coalition of social action agencies from mainline Protestant denominations and Jewish movements. Founded to defeat constitutional amendments designed to overturn the *Roe v. Wade* decision, it exists to this day. For a history of this organization, see John H. Evans, "Multi-organizational Fields and Social Movement Organization Frame Content: The Religious Pro-Choice Movement," *Sociological Inquiry* 67, no. 4 (1997): 451–69.

10. At least, the HGE debate that I am examining. I acknowledge that the HGE debate has occurred in other media, such as oral discussion. I have focused on the textual debate for pragmatic reasons.

11. Malcolm Spector and John I. Kitsuse, "Social Problems: A Reformulation," *Social Problems* 21 (1973): 145–59; Joseph Gusfield, *The Culture of Public Problems* (Chicago: University of Chicago Press, 1981).

12. Aldon Morris and Carol McClurg Mueller, *Frontiers of Social Movement Theory* (New Haven, Conn.: Yale University Press, 1992).

13. Robert N. Bellah et al., *Habits of the Heart: Individualism and Commitment in American Life* (New York: Harper and Row, 1985); Robert Wuthnow,

Acts of Compassion (Princeton, N.J.: Princeton University Press, 1991); Anthony Giddens, *Modernity and Self-Identity* (Stanford, Calif.: Stanford University Press, 1991); Christopher Lasch, *The Culture of Narcissism: American Life in an Age of Diminishing Expectations* (New York: Warner Books, 1979).

14. See also Michael Pusey, *Economic Rationalism in Canberra* (New York: Cambridge University Press, 1991).

15. Wendy Espeland, *The Struggle for Water: Politics, Rationality and Identity in the American Southwest* (Chicago: University of Chicago Press, 1998), 251.

16. In 1994 the RAC received a protocol for the treatment of rheumatoid arthritis. Claudia Mickelson, "Meeting Overview and Historical Background: Human Gene Transfer Research in the U.S.," 5, paper presented at NIH First Gene Therapy Policy Conference, "Human Gene Transfer: Beyond Life-Threatening Disease," Bethesda, Md., 11 September 1997.

17. Ibid., 12 (emphasis added).

18. Ibid., 27.

19. Ibid., 10.

20. Quoted in Rick Weiss, "Science on the Ethical Frontier: Engineering the Unborn: The Code of Cross-Generation Cures," *Washington Post,* 22 March 1998, A1.

21. "Clinical efficacy has not been definitively demonstrated at this time in any gene therapy protocol, despite anecdotal claims of successful therapy and the initiation of more than 100 . . . protocols" (Stuart H. Orkin and Arno G. Motulsky, *Report and Recommendations of the Panel to Assess the NIH Investment in Research on Gene Therapy* [1995]). This report is available on-line at http://www/nih.gov/od/orda/panelrep.htm.

22. "One of the things that's enticing from a biological point of view about germline therapy is, that it's actually much simpler than somatic gene therapy. Modalities that we would never think would be possible with somatic gene therapy would be quite easy to do with germline gene therapy" (geneticist Mario Capecchi, speaking at a conference held at UCLA, 20 March 1998).

23. Gina Kolata, "Scientists Brace for Changes in Path of Human Evolution," *New York Times,* 20 March 1998, A1.

24. Ibid.

25. Meredith Wadman, "Germline Gene Therapy 'Must Be Spared Excessive Regulation,'" *Nature* 392 (26 March 1998): 317.

26. "When the time comes that [germline engineering] is truly safe, that we understand how human cells work, how the brain works, and so on, which I think will take centuries—when that time comes, I have no objection to enhancements. But the fundamental point I'm trying to make is that we don't know enough about what the consequences would be, from a medical point of view, to attempt anything at this point but treatment of serious disease" (Anderson, in Gregory Stock and John Campbell, "Summary Report: Engineering the Human Germline Symposium," University of California, Los Angeles, 20 March 1998. Available on-line at http://www.ess.ucla.edu:80/huge/report.html).

27. Sharon Begley, "Designer Babies," *Newsweek,* 9 November 1998. Available on-line at http://www.ess.ucla.edu/huge/Nature.html.

28. Ibid.

29. Stock and Campbell, "Summary Report." These policy recommendations were actually from the program at UCLA that sponsored the conference. They

did not seem to be the result of any decision-making process among the conferees. At the subsequent RAC meeting, they were portrayed as the opinions of some of the more influential scientists at the meeting.

30. Wadman, "Germline Gene Therapy 'Must Be Spared Excessive Regulation.'"

31. Ibid.

32. Summary minutes, RAC meeting, 18–19 June 1998. Available on-line at http://www.nih.gov/od/orda.

33. Ibid.

34. Erik Parens, "Tools from and for Democratic Deliberations," *Hastings Center Report* 27, no. 5 (1997): 20–22.

35. Susan Cohen, "A House Divided," *Washington Post Magazine,* 12 October 1997, 12.

36. Courtney S. Campbell, "Prophecy and Policy," *Hastings Center Report* 27, no. 5 (1997): 17.

37. President Clinton, in James F. Childress, "The Challenges of Public Ethics: Reflections on NBAC's Report," *Hastings Center Report* 27, no. 5 (1997): 11.

38. See National Bioethics Advisory Commission, *Cloning Human Beings: Report and Recommendations of the National Bioethics Advisory Commission* (Rockville, Md.: NBAC, 1997), 7.

39. Childress, "Challenges of Public Ethics," 11.

40. Leon Kass, testimony to National Bioethics Advisory Commission, 14 March 1997. Reprinted as Leon R. Kass, "The Wisdom of Repugnance," *New Republic,* 2 June 1997, 17–26.

41. Alexander M. Capron, "Inside the Beltway Again: A Sheep of a Different Feather," *Kennedy Institute of Ethics Journal* 7, no. 2 (1997): 178.

42. Childress, "Challenges of Public Ethics," 10.

43. Ibid.

44. Lee Silver, quoted in Meredith Wadman, "Cloned Mice Fail to Rekindle Ethics Debate," *Nature* 394 (30 July 1998): 408–9.

45. In 1998 the American Society for Bioethics and Humanities had 1,500 members. See Jonsen, *Birth of Bioethics,* xiii.

46. Charles L. Bosk and Joel Frader, "Institutional Ethics Committees: Sociological Oxymoron, Empirical Black Box," in *Bioethics and Society: Constructing the Ethical Enterprise,* ed. Raymond DeVries and Janardan Subedi (Upper Saddle River, N.J.: Prentice Hall, 1998), 94–116.

47. The conference was the "Belmont Revisited" conference, held at the University of Virginia, Charlottesville, 16–18 April 1999. An example of a critical text is Edwin R. DuBose, Ronald P. Hamel, and Laurence J. O'Connell, *A Matter of Principles? Ferment in U.S. Bioethics* (Valley Forge, Pa.: Trinity Press International, 1994).

48. Albert R. Jonsen and Stephen Toulmin, *The Abuse of Casuistry: A History of Moral Reasoning* (Berkeley: University of California Press, 1988); Edmund D. Pellegrino, "Toward a Virtue-based Normative Ethics for the Health Professions," *Kennedy Institute of Ethics Journal* 5, no. 3 (1995): 253–77.

49. This literature is becoming quite extensive. See Renee C. Fox and Judith P. Swazey, *The Courage to Fail: A Social View of Organ Transplants and Dialysis* (Chicago: University of Chicago Press, 1974); Fox and Swazey, "Medical Morality Is Not Bioethics—Medical Ethics in China and the United States," *Per-*

spectives in Biology and Medicine 27, no. 3 (1984): 336–60; Charles Bosk, *Forgive and Remember: Managing Medical Failure* (Chicago: University of Chicago Press, 1979); Bosk, *All God's Mistakes: Genetic Counseling in a Pediatric Hospital* (Chicago: University of Chicago Press, 1992); Robert Zussman, *Intensive Care: Medical Ethics and the Medical Profession* (Chicago: University of Chicago Press, 1992); Renee R. Anspach, *Deciding Who Lives: Fateful Choices in the Intensive-Care Nursery* (Berkeley: University of California Press, 1993); Daniel F. Chambliss, *Beyond Caring: Hospitals, Nurses and the Social Organization of Ethics* (Chicago: University of Chicago Press, 1996); Raymond DeVries and Janardan Subedi, *Bioethics and Society: Constructing the Ethical Enterprise* (Upper Saddle River, N.J.: Prentice Hall, 1998).

50. John H. Evans, "A Sociological Account of the Growth of Principlism," *Hastings Center Report* 30, no. 5 (2000): 31–38.

51. Andrew Abbott, *The System of Professions: An Essay on the Division of Expert Labor* (Chicago: University of Chicago Press, 1988), 102.

52. Daniel Callahan, "Cloning: The Work Not Done," *Hastings Center Report* 27, no. 5 (1997): 18–20.

53. James M. Gustafson, "Theology Confronts Technology and the Life Sciences," *Commonweal,* 16 June 1978, 392.

54. D. Stephen Long, *Tragedy, Tradition, Transformism: The Ethics of Paul Ramsey* (Boulder, Colo.: Westview Press, 1993), 207–8.

55. Michael Walzer, *Thick and Thin* (Notre Dame, Ind.: Notre Dame Press, 1994), 16–17.

56. Jonathan D. Moreno, *Deciding Together: Bioethics and Moral Consensus* (New York: Oxford University Press, 1995), 56.

57. Jennings, "Possibilities of Consensus," 458.

58. Ibid.

59. Amy Gutmann and Dennis Thompson, "Deliberating about Bioethics," *Hastings Center Report* 27, no. 3 (1997): 41.

60. "Techno-Utopianism," *New York Times,* 28 August 2000, A11 (advertisement purchased by the Turning Point Project).

61. James D. Davidson, Ralph E. Pyle, and David V. Reyes, "Persistence and Change in the Protestant Establishment," *Social Forces* 74, no. 1 (1995): 157–75.

62. Amy Gutmann and Dennis Thompson, *Democracy and Disagreement* (Cambridge, Mass.: Harvard University Press, 1996), 143.

63. Ibid.; Jennings, "Possibilities of Consensus."

64. The first two quotations are from the address by Claudia Mickelson, chair of the RAC, to the NIH First Gene Therapy Policy Conference, 11 September 1997. The third is from the report of the Ad Hoc Review Committee on the RAC, 8 September 1995. Available on-line at http://www4.od.nih.gov/oba/adhoc-re.htm.

65. American Association for the Advancement of Science, "Human Inheritable Genetic Modifications: Assessing Scientific, Ethical, Religious, and Policy Issues" (Washington, D.C.: AAAS, 2000).

66. George Annas, "Why We Should Ban Human Cloning," *New England Journal of Medicine* 339, no. 2 (1998): 124.

67. Tom L. Beauchamp and James F. Childress, *Principles of Biomedical Ethics,* 4th ed. (New York: Oxford University Press, 1994), 275–76.

68. For the interpretation of new social movements as defenders of substan-

tive rationality, see Jean Cohen, "Strategy or Identity: New Theoretical Paradigms and Contemporary Social Movements," *Social Research* 52 (1985): 663–716; Claus Offe, "New Social Movements: Challenging the Boundaries of Institutional Politics," *Social Research* 52 (1985): 663–716.

Appendix: Methods and Tables

1. Institute of Society, Ethics and the Life Sciences, *Bibliography of Society Ethics and the Life Sciences* (Hastings on Hudson, N.Y.: Institute of Society, Ethics and the Life Sciences, 1973), 1.

2. In 1973, under the heading of "birth and the biological revolution," I selected the categories of "general material"; "future questions: engineering, in vitro fertilization, cloning"; and "population genetics: the future of the gene pool," resulting in 52 items (ibid.). In 1974, under the heading of "genetics, fertilization and birth," I selected "general material"; "future questions: engineering, in vitro fertilization, cloning"; and "population genetics: the future of the gene pool," resulting in 70 items (Institute of Society, Ethics and the Life Sciences, *Bibliography of Society Ethics and the Life Sciences 1974* [Hastings on Hudson, N.Y.: Institute of Society, Ethics and the Life Sciences, 1974]). In 1975, 1976–77, 1979–80, and 1984, under the heading of "genetics, fertilization and birth," I selected the categories of "general readings"; "gene therapy and genetic engineering"; and "the future of the gene pool," resulting in 97, 109, 144, and 128 cases, respectively.

3. James R. Sorenson, *Social and Psychological Aspects of Applied Human Genetics: A Bibliography* (Washington, D.C.: Government Printing Office, 1973). Items in the sections on "genetics and philosophy" and "genetic engineering" were used.

4. U.S. Library of Congress, *Genetic Engineering: Evolution of a Technological Issue* (Washington, D.C.: Government Printing Office, 1973). Items under the headings of "general references" and "clonal propagation" were used.

5. This distribution of years (truncated to the final two digits) has a mean of 82.075, a median of 83, a standard deviation of 9.3, and a skew of −.226. The skew toward later years is attributable to the increase in publishing in general, and in bioethics in particular, from the 1950s to the present.

6. LeRoy Walters, *Bibliography of Bioethics* (Washington, D.C.: Georgetown University, 1975), x.

7. For example, items were included from the *Journal of the Indiana State Medical Association, Social Theory and Practice,* and a South African medical journal that I was never able to obtain even through interlibrary loan. Moreover, the Library of Congress bibliographic source had probably the most extensive collection of items in the world to search from at the time.

8. LeRoy Walters and Tamar Joy Kahn, *Bibliography of Bioethics* (Washington, D.C.: Georgetown University, 1995).

9. Walters, *Bibliography of Bioethics.*

10. Blais Cronin, "The Need for a Theory of Citing," *Journal of Documentation* 37, no. 1 (1981): 16–24.

11. Harriet Zuckerman, "Citation Analysis and the Complex Problem of Intellectual Influence," *Scientometrics* 12, nos. 5–6 (1987): 333; M. H. MacRoberts and Barbara R. MacRoberts, "Testing the Ortega Hypothesis: Facts and Artifacts," *Scientometrics* 12, nos. 5–6 (1987): 293–95; G. Nigel Gilbert, "Referencing as Persuasion," *Social Studies of Science* 7 (1977): 113–22; Ter-

rence A. Brooks, "Private Acts and Public Objects: An Investigation of Citer Motivations," *Journal of the American Society for Information Science* 36, no. 4 (1985): 223–29; Brooks, "Evidence of Complex Citer Motivations," *Journal of the American Society for Information Science* 37, no. 1 (1986): 34–36; Stephane Baldi, "Normative versus Social Constructivist Processes in the Allocation of Citations: A Network-Analytic Model," *American Sociological Review* 63, no. 6 (1998): 829–46.

12. Jonathan R. Cole and Stephen Cole, "The Ortega Hypothesis," *Science* 178 (1972): 368–75.

13. Gilbert, "Referencing as Persuasion"; D. Lindsey, "Using Citations Counts as a Measure of Quality in Science," *Scientometrics* 15 (1989): 189.

14. Brooks, "Private Acts and Public Objects," 227.

15. Nicholas C. Mullins et al., "The Group Structure of Cocitation Clusters: A Comparative Study," *American Sociological Review* 42 (1977): 552–62; Zuckerman, "Citation Analysis and the Complex Problem of Intellectual Influence."

16. Diana Crane and Henry Small, "American Sociology since the Seventies: The Emerging Identity Crisis in the Discipline," in *Sociology and Its Publics,* ed. Terence Halliday and Morris Janowitz (Chicago: University of Chicago Press, 1992), 202.

17. Henry G. Small and Diana Crane, "Specialties and Disciplines in Science and Social Science: An Examination of Their Structure Using Citation Indexes," *Scientometrics* 1, nos. 5–6 (1979): 445–61.

18. Items that were never published or were self-published (primarily dissertations) were also excluded. As a practical matter, citations that were not in English were not counted toward the threshold of four, nor were they compiled with the other citations (see below). Dutch, German, French, and Italian were the most common languages other than English. Due to the inherently limited audience of these citations in the U.S. debate, I am certain that none of these would reach the threshold for use in the citation study. The pre-1973 universe items also contained some texts that were dropped because, upon examination, they were genetics textbooks with no discussion of moral, social, or ethical issues. These were apparently placed in the bibliographies by their compilers to give the reader basic scientific information. As a practical matter, if they had been included, they would have had no influence on the results of the study because their hundreds of citations to the technical literature would not have been found in any of the other items in the sample—resulting in every one of these citations being below the citation thresholds used for analysis.

19. When a randomly drawn item referred not to a single chapter in an edited volume but to the edited volume itself, one of the chapters was randomly selected to represent that random draw. The other option—to include all chapters—was not chosen because it would have been impossible to describe the characteristics of the author.

20. Of the items examined but not included in the sample, 18 were textbooks; 9, proposed laws; 7, unpublished documents from nongovernmental associations; 6, works in a foreign language; 5, unpublished master's theses or Ph.D. dissertations; 4, television transcripts, movies, or filmstrips; 3, items that could not be located. Three hundred sixty-eight items had too few citations.

21. While citations were being compiled, some items were not collected and were removed from the data for each citing text: citations to newspaper or magazine articles without an author's name, unpublished citations, "conversations

with . . . ," "in the files of . . . ," manuscript collections, and citations in a foreign language. A cited item was coded only once per case (i.e., multiple citations of one item were not collected). Different editions of the same book were coded as identical. Books, chapters, and articles with the same titles, but in two different places, were coded as identical under the assumption that no author would write two different items with exactly the same title. Many of the most highly cited articles were later reprinted in collected volumes, which were then cited rather than the original source.

22. The primary logistical challenge here is that, unlike the citation studies using the citation indexes of the ISI, each of the citations in my case is not in a uniform format. In my data, citing texts used endnotes, footnotes, and references. Citations were in every conceivable format (or what often seemed to be made-up formats). Further, articles in scientific outlets generally use initials rather than authors' first names. An example of two citations that required the same, unique citation number is the following: "Weber, Max. 1958. *The Protestant Ethic and the Spirit of Capitalism.* New York: Charles Scribner." and "see the interesting thesis of M. Weber in his *Protestant Ethic and the Spirit of Capitalism.*" This required that my program search by last name and first initial, and bring all of that author's unique citations to my screen (in probabilistic order), and that I then tell the program whether the citation I was looking at was the same as the one I was coding. Since the program brought up all information from a citation (title, publisher, year, and so forth) there was little ambiguity. Given that I added this level of (time-consuming) verification, I believe that the data collection was very accurate.

23. The program that assisted in this had a similar logic to the citation-level data program. The program gathered unique citations of authors with the same last name and first initial from the list produced in the earlier analysis. It then queried me as to whether the unique citations were all from the same author. An inevitable source of error here is that two persons may have the same first initial and last name (most obvious to observers of this literature is the example of John Fletcher and Joseph Fletcher). This problem—which is built into all studies of the ISI data as well—is minimized in my case because I had a function on the citation-level program that replaced the information on a unique citation with subsequent information if I determined that it was a better description of the citation, which usually meant having an entire first name and/or a more complete title. One source of bias I share with studies using the ISI data is that only the first author in multiply authored items was compiled, due to the pervasive use of "et al.," especially among physical scientists. This is a minor problem for my data compared to the ISI data because compared to social scientists—and especially physical scientists—people writing about HGE by and large work alone.

24. Although both individual texts and texts aggregated by author have been used to study the structure of literatures, focusing on authors has a few advantages. First, by downplaying the particularity of the individual items written by the same author, one can highlight the general ideas of that author and downplay the particularities. See Katherine W. McCain, "Mapping Authors in Intellectual Space: A Technical Overview," *Journal of the American Society for Information Science* 41, no. 6 (1990): 433–43; Alan E. Bayer, John C. Smart, and Gerald W. McLaughlin, "Mapping Intellectual Structure of a Scientific Subfield through Author Cocitations," *Journal of the American Society for Information*

Science 41, no. 6 (1990): 444–52. Moreover, many authors in the data never produced a "seminal" work—or perhaps produced too much representative work—so that citing authors cite interchangeably many different pieces of the cited author's work. This is exemplified by Hermann Muller, who has four individual writings during the 1959–74 period in the top thirty-five most cited works: the tenth, eleventh, twenty-second, and thirty-first most cited works. However, *as an author* during the same period he was cited by the most citing texts from 1959 to 1974 ($N = 24$) of any author. This is due to his production of many highly popular items, all of which basically make the same claims and seem to be used by citing texts to represent the same ideas. The contrary case for focusing on cited items is made by the few bench scientists who also write ethical pieces, most notably W. French Anderson. However, the citations to Anderson's lab work are few compared to citations of his ethical work.

25. Thus in 1959–74 the cutoff for being an influential author is 10 or more citations from the citing texts in that period, resulting in a total of 19 authors from 69 remaining citing texts, with the maximum citations for one author being 24. In 1975–84 the cutoff is 11 or more citations (21 authors, 71 citing texts, maximum 25 citations). In 1985–91 the cutoff is 10 or more citations (20 authors, 79 citing texts, maximum 45 citations). In 1992–95 the cutoff is 10 or more citations (22 authors, 64 citing texts, maximum 33 citations). Using these thresholds does lower the N because in some cases all the citations in a citing text will be below the threshold, and thus the case will be excluded.

26. Mark S. Aldenderfer and Roger K. Blashfield, *Cluster Analysis* (Newbury Park, Calif.: Sage Publications, 1984).

27. Ibid., 29.

28. Ibid.; Maurice Lorr, *Cluster Analysis for Social Scientists* (San Francisco: Jossey-Bass, 1983); Marija J. Norusis, *SPSS Base System User's Guide* (Chicago: SPSS, Inc., 1990).

29. This is actually a "dissimilarity" measure. Although this results in identical structure in cluster analyses, multidimensional scaling can accept only dissimilarity measures. When my research was designed, I intended also to use multidimensional scaling, but later decided this was superfluous. The formula for the relationship between two cases is then

$$\text{Binary Lance / Williams} = \frac{b+c}{2a+b+c},$$

where a is the number of items that both cases cite; b is the number of cases cited by the first case but not the second; and c is the number of items cited by the second case but not the first. As stated above, the total number of items not cited by either case is not included in this formula. Note that other options have undesirable features. For example, the Russell and Rao measure (simply the number of co-citations over the total number of citations in the two cases) results in highly cited articles clustering together solely on the basis of the number of citations they receive.

30. In the social sciences there are four commonly used clustering algorithms for hierarchical cluster analysis: Ward's method, single linkage, complete linkage, and average linkage. The more conceptually difficult Ward's method is not useful with binary data because in practice the clusters are organized by the number of citations. In my research, this would result in the most highly cited items clustering together regardless of whether they are often cited together. In

the single linkage method cases are joined together if at least one member of an existing cluster is at the same level of similarity as the case under consideration for inclusion. This method is not used in citation studies because of its tendency to create long, elongated clusters. That is, in practice one cluster is formed and cases are added—one by one—to this growing cluster until the nonuseful result is obtained of two clusters, one with just one case and the other with all the rest. The average linkage method adds an entity to a cluster if it is within a certain level of similarity to the *average* member of that cluster. In the complete linkage method, a candidate for inclusion in a cluster must be within a certain level of similarity to *all* members of that cluster. See Aldenderfer and Blashfield, *Cluster Analysis,* 40.

31. Ibid., 57.

32. Bibliographic resources, such as the *National Faculty Directory, American Men and Women of Science,* and *Who's Who in America,* and the online Biography and Genealogy Master Index, provided the necessary data for the majority of the remaining items. See *National Faculty Directory* (Detroit: Gale Research, 1990); *American Men and Women of Science* (New York: R. R. Bowker, 1990); *Who's Who in America* (Chicago: Marquis, 1990). If all these sources failed, I looked for the author in the rest of the 52,000 items in the Bioethicsline data. If the author's name was found, I would examine the item closest in year to the citing item in question and use that data. In sum, of the 345 citing authors in the citation study, profession data on 13 could not be found. The top frame of figure 1 (which corresponds to the data in the top half of table A-1) is based on gathering profession data from the entire post-1973 population, whether I gathered their citations or not, using the same methods. Of these 719 cases, the professions of 23 authors could not be found.

33. Items with corporate authors or no stated authors were coded by publisher. Thus, such authorless items published in the *Journal of the American Medical Association* were assigned the "medical doctor" designation; items published by the Hastings Center received a "bioethics profession" designation; and items published by the National Council of Churches were given a "theology" designation.

34. Since what one is attempting to measure here is the profession most strongly identified with, authors who use their religious titles (Reverend, Rabbi, S. J., and so forth) were also placed in the "theology" category.

35. Matthew B. Miles and A. Michael Huberman, *Qualitative Data Analysis,* 2d ed. (Thousand Oaks, Calif.: Sage Publications, 1994), 57.

36. Michel Callon, John Law, and Arie Rip, *Mapping the Dynamics of Science and Technology: Sociology of Science in the Real World* (London: Macmillan, 1986); J. Law et al., "Policy and the Mapping of Scientific Change: The Co-Word Analysis of Research into Environmental Acidification," *Scientometrics* 14, nos. 3–4 (1988): 251–64; John Whittaker, "Creativity and Conformity in Science: Titles, Keywords and Co-Word Analysis," *Social Studies of Science* 19 (1989): 473–96; Jean-Pierre Courtial and John Law, "A Co-Word Analysis of Artificial Intelligence," *Social Studies of Science* 19 (1989): 301–11.

37. Bioethics Information Retrieval Project, *Bioethics Thesaurus* (Washington, D.C.: Kennedy Institute of Ethics, Georgetown University, 1995).

38. Ibid. Keywords with ambiguous definitions in the thesaurus were clarified through an interview with the senior bibliographer of the Bioethicsline database. The frequency of use in the universe for each keyword was then investi-

gated, with many potentially interesting keywords having to be eliminated from consideration because they did not meet my threshold of having a mean of .10 in the sample. This threshold was selected based on the intuitive notion that, since most of my analyses were predicting variance, there needed to be some variance to predict.

39. Keywords in this category are: "gene therapy"; "gene pool"; "germ cells" (including narrower terms "ovum" and "sperm"); "human experimentation" (including narrower terms "behavioral research," "embryo research," "fetal research," "nontherapeutic research," and "therapeutic research"); "genetic disorders" (including narrower terms "cystic fibrosis," "muscular dystrophy," "hemophilia," "Huntington's disease," "late-onset disorders," "phenylketonuria," "sickle-cell disease," "Tay-Sachs," "thalassemia"); "genetic screening"; "cloning"; "abortion" (including "abortion on demand," "illegal abortion," "selective abortion," and "therapeutic abortion"); "prenatal diagnosis" (including narrower terms "amniocentesis," "chorionic villi sampling," "preimplantation diagnosis," and "sex determination"); "eugenics"; "reproductive technologies"; "artificial insemination"; and "genetic counseling."

40. *Autonomy* is defined as "the freedom of an individual or group to make decisions and to choose a pattern of life." *Freedom* is defined as "the absence of external constraints on the individual's right and ability to act and make decisions."

41. A regression equation reveals that the number of keywords per article is increasing at the rate of .36 per year.

42. J. Scott Long, *Regression Models for Categorical and Limited Dependent Variables* (Thousand Oaks, Calif.: Sage Publications, 1997), 228.

Works Cited

Abbott, Andrew. 1988. *The System of Professions: An Essay on the Division of Expert Labor.* Chicago: University of Chicago Press.

Abram, Morris B., and Susan M. Wolf. 1984. "Public Involvement in Medical Ethics." *New England Journal of Medicine* 310:627–32.

Adams, Mark B., ed. 1994. *The Evolution of Theodosius Dobzhansky.* Princeton, N.J.: Princeton University Press.

Aldenderfer, Mark S., and Roger K. Blashfield. 1984. *Cluster Analysis.* Newbury Park, Calif.: Sage Publications.

Allen, Garland E. 1992. "Julian Huxley and the Eugenical View of Human Evolution." Pp. 193–222 in *Julian Huxley: Biologist and Statesman of Science,* edited by C. Kenneth Waters and Albert Van Helden. Houston: Rice University Press.

American Association for the Advancement of Science. 2000. "Human Inheritable Genetic Modifications: Assessing Scientific, Ethical, Religious, and Policy Issues." Washington, D.C.: AAAS.

American Men and Women of Science. 1990. New York: R. R. Bowker.

Anderson, W. French. 1972. "Genetic Therapy." Pp. 109–24 in *The New Genetics and the Future of Man,* edited by Michael Hamilton. Grand Rapids, Mich.: William B. Eerdmans.

———. 1984. "Prospects for Human Gene Therapy." *Science* 226:401–9.

———. 1985. "Human Gene Therapy: Scientific and Ethical Considerations." *Journal of Medicine and Philosophy* 10:275–91.

Anderson, W. French, and John C. Fletcher. 1980. "Gene Therapy in Human Beings: When Is It Ethical to Begin?" *New England Journal of Medicine* 303:1293–1300.

Annas, George. 1989. "Who's Afraid of the Human Genome?" *Hastings Center Report* 19, no. 4:19–21.

———. 1998. "Why We Should Ban Human Cloning." *New England Journal of Medicine* 339, no. 2:122–25.

Anspach, Renee R. 1993. *Deciding Who Lives: Fateful Choices in the Intensive-Care Nursery.* Berkeley: University of California Press.

Aposhian, H. Vasken. 1970. "The Use of DNA for Gene Therapy—The Need, Experimental Approach, and Implications." *Perspectives in Biology and Medicine* 14:98–108.

Attwood, David. 1992. *Paul Ramsey's Political Ethics.* Lanham, Md.: Rowman and Littlefield.

Baldi, Stephane. 1998. "Normative versus Social Constructivist Processes in the Allocation of Citations: A Network-Analytic Model." *American Sociological Review* 63, no. 6:829–46.

Bayer, Alan E., John C. Smart, and Gerald W. McLaughlin. 1990. "Mapping Intellectual Structure of a Scientific Subfield through Author Cocitations." *Journal of the American Society for Information Science* 41, no. 6:444–52.

Bayer, Ronald. 1984. "Ethics, Politics, and Access to Health Care: A Critical Analysis of the President's Commission for the Study of Ethical Problems in Medicine and Biomedical and Behavioral Research." *Cardozo Law Review* 6:303–20.

Beatty, John. 1987. "Weighing the Risks: Stalemate in the Classical/Balance Controversy." *Journal of the History of Biology* 20, no. 3:289–319.

Beauchamp, Tom L. 1999. "How the Belmont Report Was Written." Paper presented at the "Belmont Revisited" conference, University of Virginia, Charlottesville, 16 April.

Beauchamp, Tom L., and James F. Childress. 1979. *Principles of Biomedical Ethics*. New York: Oxford University Press.

———. 1994. *Principles of Biomedical Ethics*. 4th ed. New York: Oxford University Press.

Beecher, Henry K. 1966. "Ethics and Clinical Research." *New England Journal of Medicine* 274, no. 24:1354–60.

Begley, Sharon. 1998. "Designer Babies." *Newsweek,* 9 November, 61.

Bellah, Robert N., et al. 1985. *Habits of the Heart: Individualism and Commitment in American Life*. New York: Harper and Row.

Berg, Paul. 1974. "Potential Biohazards of Recombinant DNA Molecules." *Science* 185:303.

Bioethics Information Retrieval Project. 1995. *Bioethics Thesaurus*. Washington, D.C.: Kennedy Institute of Ethics, Georgetown University.

Boone, C. Keith. 1983. "Splicing Life, with Scalpel and Scythe." *Hastings Center Report* (April): 8–10.

Bosk, Charles. 1979. *Forgive and Remember: Managing Medical Failure*. Chicago: University of Chicago Press.

———. 1992. *All God's Mistakes: Genetic Counseling in a Pediatric Hospital.* Chicago: University of Chicago Press.

Bosk, Charles, and Joel Frader. 1998. "Institutional Ethics Committees: Sociological Oxymoron, Empirical Black Box." Pp. 94–116 in *Bioethics and Society: Constructing the Ethical Enterprise,* edited by Raymond DeVries and Janardan Subedi. Upper Saddle River, N.J.: Prentice Hall.

Brooks, Terrence A. 1985. "Private Acts and Public Objects: An Investigation of Citer Motivations." *Journal of the American Society for Information Science* 36, no. 4:223–29.

———. 1986. "Evidence of Complex Citer Motivations." *Journal of the American Society for Information Science* 37, no. 1:34–36.

Brubaker, Rogers. 1984. *The Limits of Rationality: An Essay on the Social and Moral Thought of Max Weber.* Boston: George Allen and Unwin.

Bud, Robert. 1993. *The Uses of Life*. New York: Cambridge University Press.

Budiansky, Stephen. 1983. "Churches against Germ Changes." *Nature* 303:563.

Bulger, Ruth Ellen, Elizabeth Meyer Bobby, and Harvey V. Fineberg. 1995. *Society's Choices: Social and Ethical Decision Making in Biomedicine*. Washington, D.C.: National Academy Press.

Callahan, Daniel. 1982. "At the Center." *Hastings Center Report* (June): 4.

———. 1993. "Why America Accepted Bioethics." *Hastings Center Report* (November–December): S8–S9.

———. 1995. "Bioethics." Pp. 247–56 in *Encyclopedia of Bioethics,* edited by Warren Thomas Reich. New York: Macmillan.

———. 1997. "Cloning: The Work Not Done." *Hastings Center Report* 27, no. 5:18–20.

———. 1999. "The Social Sciences and the Task of Bioethics." *Daedalus* 128, no. 4:275–94.

Callon, Michel, John Law, and Arie Rip. 1986. *Mapping the Dynamics of Science and Technology: Sociology of Science in the Real World.* London: Macmillan.

Campbell, Courtney S. 1993. "On James F. Childress: Answering That God in Every Person." Pp. 127–56 in *Theological Voices in Medical Ethics,* edited by Allen Verhey and Stephen E. Lammers. Grand Rapids, Mich.: William B. Eerdmans.

———. 1997. "Prophecy and Policy." *Hastings Center Report* 27, no. 5:15–17.

Capron, Alexander M. 1983. "The Latest in Genes." *Commonweal,* 16 December, 696–97.

———. 1983. "Looking Back at the President's Commission." *Hastings Center Report* (October): 7–10.

———. 1990. "The Impact of the Report, Splicing Life." *Human Gene Therapy* 1:69–71.

———. 1997. "Inside the Beltway Again: A Sheep of a Different Feather." *Kennedy Institute of Ethics Journal* 7, no. 2:171–79.

Carlson, Elof Axel. 1973. "Eugenics Revisited: The Case for Germinal Choice." *Stadler Genetics Symposia* 5:13–34.

———. 1981. *Genes, Radiation and Society: The Life and Work of H. J. Muller.* Ithaca, N.Y.: Cornell University Press.

———. 1987. "Eugenics and Basic Genetics in H. J. Muller's Approach to Human Genetics." *History and Philosophy of the Life Sciences* 9:57–78.

Carpenter, Daniel P. 1996. "Adaptive Signal Processing, Hierarchy, and Budgetary Control in Federal Regulation." *American Political Science Review* 90, no. 2:283–302.

Chambliss, Daniel F. 1996. *Beyond Caring: Hospitals, Nurses and the Social Organization of Ethics.* Chicago: University of Chicago Press.

Childress, James F. 1997. "The Challenges of Public Ethics: Reflections on NBAC's Report." *Hastings Center Report* 27, no. 5:9–11.

Clouser, K. Danner, and Bernard Gert. 1990. "A Critique of Principlism." *Journal of Medicine and Philosophy* 15:219–36.

Cohen, Jean. 1985. "Strategy or Identity: New Theoretical Paradigms and Contemporary Social Movements." *Social Research* 52:663–716.

Cohen, Susan. 1997. "A House Divided." *Washington Post Magazine,* 12 October, 12.

Cole, Jonathan R., and Stephen Cole. 1972. "The Ortega Hypothesis." *Science* 178:368–75.

Cole-Turner, Ronald. 1993. *The New Genesis: Theology and the Genetic Revolution.* Louisville, Ky.: Westminster/John Knox Press.

Cook-Deegan, Robert. 1990. *Gene Wars: Science, Politics, and the Human Genome.* New York: W. W. Norton.

Courtial, Jean-Pierre, and John Law. 1989. "A Co-Word Analysis of Artificial Intelligence." *Social Studies of Science* 19:301–11.

Crane, Diana, and Henry Small. 1992. "American Sociology since the Seventies:

The Emerging Identity Crisis in the Discipline." Pp. 197–234 in *Sociology and Its Publics,* edited by Terence Halliday and Morris Janowitz. Chicago: University of Chicago Press.
Cronin, Blais. 1981. "The Need for a Theory of Citing." *Journal of Documentation* 37, no. 1:16–24.
Crow, James F. 1961. "Mechanisms and Trends in Human Evolution." *Daedalus* 90, no. 3:416–31.
———. 1987. "Muller, Dobzhansky, and Overdominance." *Journal of the History of Biology* 20, no. 3:351–80.
———. 1992. "H. J. Muller's Role in Evolutionary Biology." Pp. 83–106 in *The Founders of Evolutionary Genetics,* edited by Sahotra Sarkar. Dordrecht: Kluwer Academic Publishers.
Culliton, Barbara J. 1989. "Gene Test Begins." *Science* 244:913.
Curran, Charles E. 1973. *Politics, Medicine, and Christian Ethics: A Dialogue with Paul Ramsey.* Philadelphia: Fortress Press.
Davidson, James D., Ralph E. Pyle, and David V. Reyes. 1995. "Persistence and Change in the Protestant Establishment." *Social Forces* 74, no. 1:157–75.
Davis, Bernard. 1970. "Prospects for Genetic Intervention in Man." *Science* 170:1279–83.
———. 1976. "Novel Pressures on the Advance of Science." *Annals of the New York Academy of Sciences* 265:193–202.
———. 1987. "Bacterial Domestication: Underlying Assumptions." *Science* 235:1329–35.
Devettere, Raymond. 1995. "The Principled Approach: Principles, Rules and Actions." Pp. 27–48 in *Meta Medical Ethics: The Philosophical Foundations of Bioethics,* edited by Michael A. Grodin. Dordrecht: Kluwer Academic Publishers.
DeVries, Raymond, and Janardan Subedi. 1998. *Bioethics and Society: Constructing the Ethical Enterprise.* Upper Saddle River, N.J.: Prentice Hall.
Dickson, David. 1984. *The New Politics of Science.* New York: Pantheon Books.
DiMaggio, Paul. 1982. "Cultural Entrepreneurship in Nineteenth-century Boston, Part 1." *Media, Culture and Society* 4:33–50.
DiMaggio, Paul, and Walter W. Powell. 1991. "Introduction." Pp. 1–40 in *The New Institutionalism in Organizational Analysis,* edited by Walter W. Powell and Paul J. DiMaggio. Chicago: University of Chicago Press.
Dobbin, Frank. 1994. "Cultural Models of Organization: The Social Construction of Rational Organizing Principles." Pp. 117–42 in *The Sociology of Culture,* edited by Diana Crane. Cambridge, Mass.: Blackwell.
Dobzhansky, Theodosius. 1962. *Mankind Evolving: The Evolution of the Human Species.* New Haven, Conn.: Yale University Press.
———. 1967. *The Biology of Ultimate Concern.* New York: New American Library.
Dronamraju, Krishna R. 1995. "Introduction." Pp. 1–22 in *Haldane's Daedalus Revisited,* edited by Krishna R. Dronamraju. Oxford: Oxford University Press.
Dubos, René. 1965. *Man Adapting.* New Haven, Conn.: Yale University Press.
DuBose, Edwin R., Ronald P. Hamel, and Laurence J. O'Connell. 1994. "Introduction." Pp. 1–17 in *A Matter of Principles? Ferment in U.S. Bioethics,*

edited by Edwin R. DuBose, Ronald P. Hamel, and Laurence J. O'Connell. Valley Forge, Pa.: Trinity Press International.
———. 1994. *A Matter of Principles? Ferment in U.S. Bioethics*. Valley Forge, Pa.: Trinity Press International.
Duster, Troy. 1990. *Backdoor to Eugenics*. New York: Routledge.
Edwards, R. G., B. D. Bavister, and P. C. Steptoe. 1969. "Early Stages of Fertilization In Vitro of Human Oocytes Matured In Vitro." *Nature* 221:632–35.
Edwards, R. G., and Ruth E. Fowler. 1970. "Human Embryos in the Laboratory." *Scientific American* 223, no. 6:44–54.
Edwards, R. G., and David J. Sharpe. 1971. "Social Values and Research in Human Embryology." *Nature* 231:87–91.
Engelhardt, H. Tristram. 1986. *The Foundations of Bioethics*. New York: Oxford University Press.
Espeland, Wendy. 1998. *The Struggle for Water: Politics, Rationality and Identity in the American Southwest*. Chicago: University of Chicago Press.
Espeland, Wendy, and Mitchell L. Stevens. 1998. "Commensuration as a Social Process." *Annual Review of Sociology* 24:313–31.
Etzioni, Amitai. 1973. *Genetic Fix*. New York: Macmillan.
Evans, John H. 1997. "Multi-organizational Fields and Social Movement Organization Frame Content: The Religious Pro-Choice Movement." *Sociological Inquiry* 67, no. 4:451–69.
———. 2000. "A Sociological Account of the Growth of Principlism." *Hastings Center Report* 30, no. 5:31–38.
Ezrahi, Yaron. 1990. *The Descent of Icarus: Science and the Transformation of Contemporary Democracy*. Cambridge, Mass.: Harvard University Press.
Fletcher, John. 1972. "The Brink: The Parent-Child Bond in the Genetic Revolution." *Theological Studies* 33, no. 3:457–85.
———. 1983. "Moral Problems and Ethical Issues in Prospective Human Gene Therapy." *Virginia Law Review* 69:515–46.
———. 1985. "Ethical Issues in and beyond Prospective Clinical Trials of Human Gene Therapy." *Journal of Medicine and Philosophy* 10, no. 3:293–309.
———. 1990. "Evolution of Ethical Debate about Human Gene Therapy." *Human Gene Therapy* 1:55–68.
Fletcher, John C., and W. French Anderson. 1992. "Germ-Line Gene Therapy: A New Stage of the Debate." *Law, Medicine and Health Care* 20:26–39.
Fletcher, Joseph. 1966. *Situation Ethics: The New Morality*. Philadelphia: Westminster Press.
———. 1971. "Ethical Aspects of Genetic Controls." *New England Journal of Medicine* 285, no. 14:776–83.
———. 1993. "Memoir of an Ex-Radical." Pp. 55–92 in *Joseph Fletcher: Memoir of an Ex-Radical*, edited by Kenneth Vaux. Louisville, Ky.: Westminster/John Knox.
Flitner, David. 1986. *The Politics of Presidential Commissions*. Dobbs Ferry, N.Y.: Transnational Publishers.
Fowler, Gregory, Eric T. Juengst, and Burke K. Zimmerman. 1989. "Germ-Line Therapy and the Clinical Ethos of Medical Genetics." *Theoretical Medicine* 10:151–65.

Fox, Daniel M. 1993. "View the Second." *Hastings Center Report* 23, no. 6: S12–S13.

Fox, Renee C. 1989. *The Sociology of Medicine: A Participant Observer's View.* Englewood Cliffs, N.J.: Prentice Hall.

Fox, Renee C., and Judith P. Swazey. 1974. *The Courage to Fail: A Social View of Organ Transplants and Dialysis.* Chicago: University of Chicago Press.

———. 1984. "Medical Morality Is Not Bioethics—Medical Ethics in China and the United States." *Perspectives in Biology and Medicine* 27, no. 3: 336–60.

Friedmann, Theodore. 1989. "Progress toward Human Gene Therapy." *Science* 244:1275–81.

———. 1997. "Enhancement of Human Traits." Paper presented at NIH First Gene Therapy Policy Conference, "Human Gene Transfer: Beyond Life-Threatening Disease," Bethesda, Md., 11 September.

Friedmann, Theodore, and Richard Roblin. 1972. "Gene Therapy for Human Genetic Disease?" *Science* 175:949–55.

Garrow, David J. 1994. *Liberty and Sexuality: The Right to Privacy and the Making of* Roe v. Wade. New York: Macmillan/Lisa Drew.

Giddens, Anthony. 1991. *Modernity and Self-Identity.* Stanford, Calif.: Stanford University Press.

Gieryn, Thomas F. 1983. "Boundary-Work and the Demarcation of Science from Non-Science: Strains and Interests in Professional Ideologies of Scientists." *American Sociological Review* 48:781–95.

Gieryn, Thomas F., George M. Bevins, and Stephen C. Zehr. 1985. "Professionalization of American Scientists: Public Science in the Creation/Evolution Trials." *American Sociological Review* 50:392–409.

Gilbert, G. Nigel. 1977. "Referencing as Persuasion." *Social Studies of Science* 7:113–22.

Glass, Bentley. 1965. "The Ethical Basis of Science." *Science* 150:1254–61.

———. 1965. *Science and Ethical Values.* Chapel Hill: University of North Carolina Press.

———. 1970. *The Timely and the Timeless.* New York: Basic Books.

———. 1971. "Science: Endless Horizons or Golden Age?" *Science* 171:23–29.

Glover, Jonathan. 1984. *What Sort of People Should There Be?* New York: Penguin.

Golden, Frederic, and Michael D. Lemonick. 2000. "The Race Is Over." *Time,* 3 July, 18–23.

Gray, Bradford H. 1995. "Bioethics Commissions: What Can We Learn from Past Successes and Failures?" Pp. 261–306 in *Society's Choices: Social and Ethical Decision Making in Biomedicine,* edited by Ruth Ellen Bulger, Elizabeth Meyer Bobby, and Harvey V. Fineberg. Washington, D.C.: National Academy Press.

Grobstein, Clifford. 1979. *A Double Image of the Double Helix: The Recombinant DNA Debate.* San Francisco: W. H. Freeman.

Grobstein, Clifford, and Michael Flower. 1984. "Gene Therapy: Proceed with Caution." *Hastings Center Report* 14, no. 2:13–17.

Gusfield, Joseph. 1981. *The Culture of Public Problems.* Chicago: University of Chicago Press.

Gustafson, James M. 1978. "Theology Confronts Technology and the Life Sciences." *Commonweal,* 16 June, 386–92.

———. 1990. "Moral Discourse about Medicine: A Variety of Forms." *Journal of Medicine and Philosophy* 15:125–42.
Gutmann, Amy, and Dennis Thompson. 1996. *Democracy and Disagreement.* Cambridge, Mass.: Harvard University Press.
———. 1997. "Deliberating about Bioethics." *Hastings Center Report* 27, no. 3: 38–41.
Habermas, Jürgen. 1970. *Toward a Rational Society: Student Protest, Science and Politics.* Boston: Beacon Press.
———. 1987. *The Theory of Communicative Action, Vol. 2: Lifeworld and System: A Critique of Functionalist Reason.* Boston: Beacon Press.
———. 1989. *The Structural Transformation of the Public Sphere.* Cambridge, Mass.: MIT Press.
Hadden, Jeffrey K. 1969. *The Gathering Storm in the Churches.* Garden City, N.Y.: Doubleday.
Haldane, J. B. S. 1963. "Biological Possibilities for the Human Species in the Next Ten Thousand Years." Pp. 337–61 in *Man and His Future,* edited by Gordon Wolstenholme. London: J. and A. Churchill Ltd.
Hamilton, Michael. 1972. *The New Genetics and the Future of Man.* Grand Rapids, Mich.: William B. Eerdmans.
Handler, Philip. 1970. "Science's Continuing Role." *BioScience* 20, no. 20:1101–6.
Hart, Stephen. 1995. "Cultural Sociology and Social Criticism." *Newsletter of the Sociology of Culture Section of the American Sociological Association* 9, no. 3.
Howard, Ted, and Jeremy Rifkin. 1977. *Who Should Play God? The Artificial Creation of Life and What It Means for the Future of the Human Race.* New York: Dell.
Human Gene Therapy Subcommittee, Recombinant DNA Advisory Committee. 1985. "Points to Consider in the Design and Submission of Human Somatic Cell Gene Therapy Protocols." *Recombinant DNA Technical Bulletin* 8, no. 4:181–86.
Hunter, James. 1983. *American Evangelicalism: Conservative Religion and the Quandary of Modernity.* New Brunswick, N.J.: Rutgers University Press.
Huntley, Ola Mae. 1984. "A Mother's Perspective." *Hastings Center Report* (April): 14–15.
Huxley, Aldous. 1932. *Brave New World.* London: Chatto and Windus.
Huxley, Julian. 1961. "The Humanist Frame." Pp. 11–48 in *The Humanist Frame,* edited by Julian Huxley. London: George Allen and Unwin Ltd.
———. 1963. "Eugenics in Evolutionary Perspective." *Perspectives in Biology and Medicine* (Winter): 155–87.
Institute of Society, Ethics and the Life Sciences. 1973. *Bibliography of Society Ethics and the Life Sciences.* Hastings on Hudson, N.Y.: Institute of Society Ethics and the Life Sciences.
———. 1974. *Bibliography of Society Ethics and the Life Sciences 1974.* Hastings on Hudson, N.Y.: Institute of Society Ethics and the Life Sciences.
Jacoby, Russell. 1987. *The Last Intellectuals.* New York: Farrar, Straus and Giroux.
Jennings, Bruce. 1991. "Possibilities of Consensus: Toward Democratic Moral Discourse." *Journal of Medicine and Philosophy* 16, no. 4:447–63.
Jepperson, Ronald L. 1991. "Institutions, Institutional Effects, and Institutionalism." Pp. 143–63 in *The New Institutionalism in Organizational Anal-*

ysis, edited by Walter W. Powell and Paul J. DiMaggio. Chicago: University of Chicago Press.

Jonsen, Albert R. 1991. "American Moralism and the Origin of Bioethics in the United States." *Journal of Medicine and Philosophy* 16:113–30.

———. 1993. "The Birth of Bioethics." *Hastings Center Report* (November–December): S1–S4.

———. 1994. "Foreword." Pp. ix–xvii in *A Matter of Principles? Ferment in U.S. Bioethics,* edited by Edwin R. DuBose, Ronald P. Hamel, and Laurence J. O'Connell. Valley Forge, Pa.: Trinity Press International.

———. 1998. *The Birth of Bioethics.* New York: Oxford University Press.

Jonsen, Albert R., and Andrew Jameton. 1995. "Medical Ethics, History of: The Americas." In *Encyclopedia of Bioethics,* edited by Warren Thomas Reich. New York: Macmillan.

Jonsen, Albert R., and Stephen Toulmin. 1988. *The Abuse of Casuistry: A History of Moral Reasoning.* Berkeley: University of California Press.

Juengst, Eric T. 1991. "Germ-Line Gene Therapy: Back to Basics." *Journal of Medicine and Philosophy* 16:587–92.

Kass, Leon R. 1971. "Babies by Means of In Vitro Fertilization: Unethical Experiments on the Unborn?" *New England Journal of Medicine* 285, no. 21:1174–79.

———. 1971. "The New Biology: What Price Relieving Man's Estate?" *Science* 174:779–88.

———. 1972. "New Beginnings in Life." Pp. 15–63 in *The New Genetics and the Future of Man,* edited by Michael Hamilton. Grand Rapids, Mich.: William B. Eerdmans.

———. 1997. "The Wisdom of Repugnance." *New Republic,* 2 June, 17–26.

Kaye, Howard L. 1997. *The Social Meaning of Modern Biology: From Social Darwinism to Sociobiology.* New Brunswick, N.J.: Transaction Publishers.

Kevles, Daniel. 1985. *In the Name of Eugenics: Genetics and the Uses of Human Heredity.* Berkeley: University of California Press.

Kolata, Gina. 1998. "Scientists Brace for Changes in Path of Human Evolution." *New York Times,* 20 March, A1.

———. 1999. "Scientists Place Jellyfish Genes into Monkeys." *New York Times,* 23 December, A1.

Krimbas, Costas B. 1994. "The Evolutionary Worldview of Theodosius Dobzhansky." Pp. 179–94 in *The Evolution of Theodosius Dobzhansky,* edited by Mark B. Adams. Princeton, N.J.: Princeton University Press.

Krimsky, Sheldon. 1982. *Genetic Alchemy: The Social History of the Recombinant DNA Controversy.* Cambridge, Mass.: MIT Press.

———. 1991. *Biotechnics and Society: The Rise of Industrial Genetics.* New York: Praeger.

Lammers, Stephen E. 1996. "The Marginalization of Religious Voices in Bioethics." Pp. 19–43 in *Religion and Medical Ethics: Looking Back, Looking Forward,* edited by Allen Verhey. Grand Rapids, Mich.: William B. Eerdmans.

Laney, James T. 1970. "The New Morality and the Religious Communities." *Annals of the American Academy of Political and Social Science* 387:14–21.

Lappé, Marc. 1972. "Moral Obligations and the Fallacies of 'Genetic Control.'" *Theological Studies* 33, no. 3:411–27.

Lasch, Christopher. 1979. *The Culture of Narcissism: American Life in an Age of Diminishing Expectations*. New York: Warner Books.

Law, J., et al. 1988. "Policy and the Mapping of Scientific Change: The Co-Word Analysis of Research into Environmental Acidification." *Scientometrics* 14, nos. 3–4:251–64.

Lebacqz, Karen. 1984. "The Ghosts Are on the Wall: A Parable for Manipulating Life." Pp. 22–41 in *The Manipulation of Life,* edited by Robert Esbjornson. San Francisco: Harper and Row.

———. 1999. "Twenty Years Older But Are We Wiser?" Paper presented at the "Belmont Revisited" conference, University of Virginia, Charlottesville, 18 April.

Lederberg, Joshua. 1963. "Biological Future of Man." Pp. 263–73 in *Man and His Future,* edited by Gordon Wolstenholme. London: J. and A. Churchill Ltd.

Levine, Donald N. 1985. *The Flight from Ambiguity.* Chicago: University of Chicago Press.

Levine, Louis. 1995. *Genetics of Natural Populations*. New York: Columbia University Press.

Lindsey, D. 1989. "Using Citations Counts as a Measure of Quality in Science." *Scientometrics* 15:189–203.

Long, D. Stephen. 1993. *Tragedy, Tradition, Transformism: The Ethics of Paul Ramsey.* Boulder, Colo.: Westview Press.

Long, Edward LeRoy. 1982. *A Survey of Recent Christian Ethics*. New York: Oxford University Press.

Long, J. Scott. 1997. *Regression Models for Categorical and Limited Dependent Variables*. Thousand Oaks, Calif.: Sage Publications.

Lorr, Maurice. 1983. *Cluster Analysis for Social Scientists*. San Francisco: Jossey-Bass.

Lukes, Steven. 1974. *Power.* London: Macmillan.

Luria, S. E. 1965. "Directed Genetic Change: Perspectives from Molecular Genetics." Pp. 1–19 in *The Control of Human Heredity and Evolution,* edited by T. M. Sonneborn. New York: Macmillan.

MacNamara, Vincent. 1985. *Faith and Ethics: Recent Roman Catholicism.* Washington, D.C.: Georgetown University Press.

MacRoberts, M. H., and Barbara R. MacRoberts. 1987. "Testing the Ortega Hypothesis: Facts and Artifacts." *Scientometrics* 12, nos. 5–6:293–95.

Madden, Ward. 1970. "Foreword." Pp. v–ix in *The Timely and the Timeless: The Interrelationships of Science, Education and Society.* New York: Basic Books.

Maugh, Thomas H. 2000. "Gene Therapy May Have Cured Three Infants." *Los Angeles Times,* 28 March, A1.

McCain, Katherine W. 1990. "Mapping Authors in Intellectual Space: A Technical Overview." *Journal of the American Society for Information Science* 41, no. 6:433–43.

Medawar, P. B. 1960. *The Future of Man.* New York: Basic Books.

Meilaender, Gilbert. 1990. "Mastering Our Gen(i)es: When Do We Say No?" *Christian Century,* 3 October, 872–75.

———. 1995. *Body, Soul, and Bioethics.* Notre Dame, Ind.: University of Notre Dame Press.

Mickelson, Claudia. 1997. "Meeting Overview and Historical Background: Human Gene Transfer Research in the U.S." Paper presented at NIH First

Gene Therapy Policy Conference, "Human Gene Transfer: Beyond Life-Threatening Disease," Bethesda, Md., 11 September.

Miles, Matthew B., and A. Michael Huberman. 1994. *Qualitative Data Analysis.* 2d ed. Thousand Oaks, Calif.: Sage Publications.

Milewski, Elizabeth A. 1985. "Development of a Points to Consider Document for Human Somatic Cell Gene Therapy." *Recombinant DNA Technical Bulletin* 8, no. 4:176–86.

Miller, A. Dusty, Douglas J. Jolly, and Inder M. Verma. 1984. "Expression of a Retrovirus Encoding Human HPRT in Mice." *Science* 225:630–32.

Miringoff, Marque-Luisa. 1991. *The Social Costs of Genetic Welfare.* New Brunswick, N.J.: Rutgers University Press.

Moore, Kelly. 1996. "Organizing Integrity: American Science and the Creation of Public Interest Organizations, 1955–1975." *American Journal of Sociology* 101, no. 6:1592–1627.

Moreno, Jonathan D. 1995. *Deciding Together: Bioethics and Moral Consensus.* New York: Oxford University Press.

Morris, Aldon, and Carol McClurg Mueller. 1992. *Frontiers of Social Movement Theory.* New Haven, Conn.: Yale University Press.

Muller, Hermann. 1950. "Our Load of Mutations." *American Journal of Human Genetics* 2, no. 2:111–76.

———. 1959. "The Guidance of Human Evolution." *Perspectives in Biology and Medicine* 3, no. 1:1–43.

———. 1963. "Genetic Progress by Voluntarily Conducted Germinal Choice." Pp. 247–362 in *Man and His Future,* edited by Gordon Wolstenholme. London: J. and A. Churchill Ltd.

———. 1965. "Means and Aims in Human Betterment." Pp. 100–122 in *The Control of Human Heredity and Evolution,* edited by T. M. Sonneborn. New York: Macmillan.

Mullins, Nicholas C., et al. 1977. "The Group Structure of Cocitation Clusters: A Comparative Study." *American Sociological Review* 42:552–62.

Murray, Thomas H. 1991. "Ethical Issues in Human Genome Research." *Federation of American Societies of Experimental Biology Journal* 5:55–60.

National Bioethics Advisory Commission. 1997. *Cloning Human Beings: Report and Recommendations of the National Bioethics Advisory Commission.* Rockville, Md.: NBAC.

National Commission for the Protection of Human Subjects of Biomedical and Behavioral Research. 1978. *The Belmont Report: Ethical Principles and Guidelines for the Protection of Human Subjects of Research.* Washington, D.C.: Government Printing Office.

National Council of Churches, Panel on Bioethical Concerns. 1984. *Genetic Engineering: Social and Ethical Consequences.* New York: Pilgrim Press.

National Faculty Directory. 1990. Vol. 20. Detroit: Gale Research.

Norman, Colin. 1983. "Clerics Urge Ban on Altering Germline Cells." *Science* 220:1360–61.

Norusis, Marija J. 1990. *SPSS Base System User's Guide.* Chicago: SPSS, Inc.

Offe, Claus. 1985. "New Social Movements: Challenging the Boundaries of Institutional Politics." *Social Research* 52:663–716.

Orkin, Stuart H., and Arno G. Motulsky. 1995. *Report and Recommendations of the Panel to Assess the NIH Investment in Research on Gene Therapy.* Available on-line at http://www4.od.nih.gov/oba/panelrep.htm.

Palmiter, Richard D., et al. 1982. "Dramatic Growth of Mice That Develop from Eggs Microinjected with Metallothionein-Growth Hormone Fusion Genes." *Nature* 300:611–15.

Parens, Erik. 1997. "Tools from and for Democratic Deliberations." *Hastings Center Report* 27, no. 5:20–22.

Paul, Diane B. 1987. "'Our Load of Mutations' Revisited." *Journal of the History of Biology* 20, no. 3:321–35.

Pellegrino, Edmund D. 1995. "Toward a Virtue-based Normative Ethics for the Health Professions." *Kennedy Institute of Ethics Journal* 5, no. 3:253–77.

Peterson, Richard A. 1976. "The Production of Culture: A Prolegomenon." Pp. 7–22 in *The Production of Culture*, edited by Richard A. Peterson. Beverly Hills, Calif.: Sage Publications.

Pitkin, Hanna Fenichel. 1967. *The Concept of Representation.* Berkeley: University of California Press.

Pontecorvo, G. 1965. "Prospects for Genetic Analysis in Man." Pp. 80–89 in *The Control of Human Heredity and Evolution*, edited by T. M. Sonneborn. New York: Macmillan.

Porter, Theodore M. 1995. *Trust in Numbers: The Pursuit of Objectivity in Science and Public Life.* Princeton, N.J.: Princeton University Press.

Powell, Walter W., and Paul J. DiMaggio, eds. 1991. *The New Institutionalism in Organizational Analysis.* Chicago: University of Chicago Press.

President's Commission for the Study of Ethical Problems in Medicine and Biomedical and Behavioral Research. 1983. *Splicing Life: A Report on the Social and Ethical Issues of Genetic Engineering with Human Beings.* Washington, D.C.: Government Printing Office.

———. 1983. *Summing Up.* Washington, D.C.: Government Printing Office.

Provine, William B. 1992. "Progress in Evolution and Meaning in Life." Pp. 165–80 in *Julian Huxley: Biologist and Statesman of Science*, edited by C. Kenneth Waters and Albert Van Helden. Houston: Rice University Press.

Pusey, Michael. 1991. *Economic Rationalism in Canberra.* New York: Cambridge University Press.

Quinley, Harold. 1974. *The Prophetic Clergy: Social Activism among Protestant Ministers.* New York: John Wiley and Sons.

Ramsey, Paul. 1970. *Fabricated Man: The Ethics of Genetic Control.* New Haven, Conn.: Yale University Press.

———. 1970. *The Patient as Person.* New Haven, Conn.: Yale University Press.

———. 1972. "Genetic Therapy: A Theologian's Response." In *The New Genetics and the Future of Man*, edited by Michael Hamilton. Grand Rapids, Mich.: William B. Eerdmans.

Rawls, John. 1987. "The Idea of Overlapping Consensus." *Oxford Journal of Legal Studies* 7, no. 1:1–25.

Reich, Warren Thomas. 1994. "The Word 'Bioethics': Its Birth and the Legacies of Those Who Shaped It." *Kennedy Institute of Ethics Journal* 4, no. 4: 319–35.

———. 1995. "The Word 'Bioethics': The Struggle over Its Earliest Meanings." *Kennedy Institute of Ethics Journal* 5, no. 1:19–34.

———. 1996. "Revisiting the Launching of the Kennedy Institute: Re-visioning the Origins of Bioethics." *Kennedy Institute of Ethics Journal* 6, no. 4: 323–27.

Rifkin, Jeremy. 1983. *Algeny.* New York: Penguin Books.

———. 1983. *The Theological Letter concerning the Moral Arguments against Genetic Engineering of the Human Germline Cells.* Washington, D.C.: Foundation on Economic Trends.
Robertson, John. 1985. "Genetic Alteration of Embryos: The Ethical Issues." Pp. 115–27 in *Genetics and the Law III,* edited by Aubrey Milunsky and George Annas. New York: Plenum Press.
Rosenberg, Charles E. 1999. "Meanings, Policies, and Medicine: On the Bioethical Enterprise and History." *Daedalus* 128, no. 4:27–46.
Rosenfeld, Albert. 1969. *The Second Genesis: The Coming Control of Life.* Englewood Cliffs, N.J.: Prentice Hall.
Rothman, David J. 1991. *Strangers by the Bedside: A History of How Law and Bioethics Transformed Medical Decision Making.* New York: Basic Books.
Ruse, Michael. 1996. *Monod to Man: The Concept of Progress in Evolutionary Biology.* Cambridge, Mass.: Harvard University Press.
Singer, Ethan. 1974. "Gene Manipulation: Progress and Prospects." Pp. 217–41 in *Macromolecules Regulating Growth and Development,* edited by Elizabeth D. Hay, Thomas J. King, and John Papaconstantinou. New York: Academic Press.
Singer, Peter. 1975. *Animal Liberation.* New York: Random House.
———. 1979. *Practical Ethics.* New York: Cambridge University Press.
Sinsheimer, Robert. 1969. "The Prospect for Designed Genetic Change." *American Scientist* 57, no. 1:134–42.
———. 1970. "Genetic Engineering: The Modification of Man." *Impact of Science on Society* 20, no. 4:279–91.
———. 1975. "Troubled Dawn for Genetic Engineering." *New Scientist* 16: 148–51.
———. 1994. *The Strands of a Life: The Science of DNA and the Art of Education.* Berkeley: University of California Press.
Small, Henry G., and Diana Crane. 1979. "Specialties and Disciplines in Science and Social Science: An Examination of Their Structure Using Citation Indexes." *Scientometrics* 1, nos. 5–6:445–61.
Smith, David H. 1993. "On Paul Ramsey: A Covenant-Centered Ethic for Medicine." Pp. 7–29 in *Theological Voices in Medical Ethics,* edited by Allen Verhey and Stephen E. Lammers. Grand Rapids, Mich.: William B. Eerdmans.
Sonneborn, Tracy M., ed. 1965. *The Control of Human Heredity and Evolution.* New York: Macmillan.
Sorenson, James R. 1973. *Social and Psychological Aspects of Applied Human Genetics: A Bibliography.* Washington, D.C.: Government Printing Office.
Spector, Malcolm, and John I. Kitsuse. 1973. "Social Problems: A Reformulation." *Social Problems* 21:145–59.
Stock, Gregory, and John Campbell. 1998. "Summary Report: Engineering the Human Germline Symposium." University of California, Los Angeles, 20 March. Available on-line at http://www.ess.ucla.edu:80/huge/report.html.
———. 2000. *Engineering the Human Germline.* New York: Oxford University Press.
Stolberg, Sheryl Gay. 2000. "Teenager's Death Is Shaking Up Field of Human Gene-Therapy Experiments." *New York Times,* 27 January, A20.
Stout, Jeffrey. 1988. *Ethics after Babel.* Boston: Beacon Press.

Swetlitz, Marc. 1995. "Julian Huxley and the End of Evolution." *Journal of the History of Biology* 28:181–217.

Taylor, Gordon Rattray. 1968. *The Biological Time Bomb.* New York: World Publishing.

Teilhard de Chardin, Pierre. 1955. *The Phenomenon of Man.* New York: Harper and Brothers.

Tipton, Steven M. 1982. *Getting Saved from the Sixties: Moral Meaning in Conversion and Cultural Change.* Berkeley: University of California Press.

Toulmin, Stephen. 1982. "How Medicine Saved the Life of Ethics." *Perspectives in Biology and Medicine* 25, no. 4:736–50.

Tubbs, James B. 1996. *Christian Theology and Medical Ethics.* Dordrecht: Kluwer Academic Publishers.

Tuchings, Terrence R. 1979. *Rhetoric and Reality: Presidential Commissions and the Making of Public Policy.* Boulder, Colo.: Westview Press.

U.S. Congress, Office of Technology Assessment. 1993. *Biomedical Ethics in U.S. Public Policy—Background Paper.* Washington, D.C.: Government Printing Office.

U.S. Department of Health and Human Services. 1981. "Protection of Human Subjects." *Code of Federal Regulations,* Part 46, Subtitle A.

U.S. Library of Congress. 1973. *Genetic Engineering: Evolution of a Technological Issue.* Washington, D.C.: Government Printing Office.

Veatch, Robert M. 1981. *A Theory of Medical Ethics.* New York: Basic Books.

———. 1995. "Resolving Conflicts among Principles: Ranking, Balancing, and Specifying." *Kennedy Institute of Ethics Journal* 5, no. 3:199–218.

Verhey, Allen. 1995. "'Playing God' and Invoking a Perspective." *Journal of Medicine and Philosophy* 20:347–64.

Wade, Nicholas. 1977. *The Ultimate Experiment: Man-Made Evolution.* New York: Walker.

Wadman, Meredith. 1998. "Cloned Mice Fail to Rekindle Ethics Debate." *Nature* 394:408–9.

———. 1998. "Germline Gene Therapy 'Must Be Spared Excessive Regulation.'" *Nature* 392:317.

Wallace, Bruce. 1970. *Genetic Load: Its Biological and Conceptual Aspects.* Englewood Cliffs, N.J.: Prentice-Hall.

———. 1991. *Fifty Years of Genetic Load: An Odyssey.* Ithaca, N.Y.: Cornell University Press.

Walters, LeRoy. 1975. *Bibliography of Bioethics.* Detroit: Gale Research Co.

———. 1985. "Religion and the Renaissance of Medical Ethics in the United States." Pp. 3–16 in *Theology and Bioethics,* edited by E. E. Shelp. Boston: D. Reidel Publishing.

———. 1986. "The Ethics of Human Gene Therapy." *Nature* 320:225–27.

———. 1991. "Human Gene Therapy: Ethics and Public Policy." *Human Gene Therapy* 2:115–22.

Walters, LeRoy, and Tamar Joy Kahn. 1995. *Bibliography of Bioethics.* Detroit: Gale Research Co.

Walters, LeRoy, and Julie Gage Palmer. 1997. *The Ethics of Human Gene Therapy.* New York: Oxford University Press.

Walzer, Michael. 1994. *Thick and Thin.* Notre Dame, Ind.: University of Notre Dame Press.

Waters, C. Kenneth. 1992. "Introduction: Revising Our Picture of Julian Huxley." Pp. 1–30 in *Julian Huxley: Biologist and Statesman of Science,* edited by C. Kenneth Waters and Albert Van Helden. Houston: Rice University Press.

Watson, James D. 1990. "The Human Genome Project: Past, Present and Future." *Science* 248:44–49.

Weatherall, D. J. 1985. *The New Genetics and Clinical Practice.* New York: Oxford University Press.

Weber, Max. 1946. *From Max Weber: Essays in Sociology.* Translated by H. H. Gerth and C. Wright Mills. New York: Oxford University Press.

———. 1958. *The Protestant Ethic and the Spirit of Capitalism.* New York: Charles Scribner's Sons.

———. 1968. *Economy and Society.* 2 vols. Berkeley: University of California Press.

Weiss, Rick. 1998. "Science on the Ethical Frontier: Engineering the Unborn: The Code of Cross-Generation Cures." *Washington Post,* 22 March, A1.

Wertz, Dorothy C., and John C. Fletcher. 1989. "Fatal Knowledge? Prenatal Diagnosis and Sex Selection." *Hastings Center Report* 19:21–27.

Whittaker, John. 1989. "Creativity and Conformity in Science: Titles, Keywords and Co-Word Analysis." *Social Studies of Science* 19:473–96.

Who's Who in America. 1990. Chicago: Marquis.

Willard, Hunt. 1997. "Panel Discussion." NIH First Gene Therapy Policy Conference, "Human Gene Transfer: Beyond Life-Threatening Disease," Bethesda, Md., 11 September.

Wolfe, Alan. 1989. *Whose Keeper? Social Science and Moral Obligation.* Berkeley: University of California Press.

Wolstenholme, Gordon, ed. 1963. *Man and His Future.* London: J. and A. Churchill Ltd.

Wright, Susan. 1994. *Molecular Politics: Developing American and British Regulatory Policy for Genetic Engineering, 1972–1982.* Chicago: University of Chicago Press.

Wuthnow, Robert. 1988. *The Restructuring of American Religion.* Princeton, N.J.: Princeton University Press.

———. 1989. *Communities of Discourse: Ideology and Social Structure in the Reformation, the Enlightenment and European Socialism.* Cambridge, Mass.: Harvard University Press.

———. 1991. *Acts of Compassion.* Princeton, N.J.: Princeton University Press.

———. 1991. *Between States and Markets: The Voluntary Sector in Comparative Perspective.* Princeton, N.J.: Princeton University Press.

Zimmerman, Burke K. 1991. "Human Germ-Line Therapy: The Case for Its Development and Use." *Journal of Medicine and Philosophy* 16:593–612.

Zuckerman, Harriet. 1987. "Citation Analysis and the Complex Problem of Intellectual Influence." *Scientometrics* 12, nos. 5–6:329–38.

Zussman, Robert. 1992. *Intensive Care: Medical Ethics and the Medical Profession.* Chicago: University of Chicago Press.

Index